Electrolytic Capacitorless Permanent Magnet Synchronous Motor Drives

Dawei Ding · Binxing Li · Qiwei Wang ·
Gaolin Wang · Guoqiang Zhang · Dianguo Xu

Electrolytic Capacitorless Permanent Magnet Synchronous Motor Drives

Springer

Dawei Ding
Harbin Institute of Technology
Harbin, Heilongjiang, China

Binxing Li
Harbin Institute of Technology
Harbin, Heilongjiang, China

Qiwei Wang
Harbin Institute of Technology
Harbin, Heilongjiang, China

Gaolin Wang
Harbin Institute of Technology
Harbin, Heilongjiang, China

Guoqiang Zhang
Harbin Institute of Technology
Harbin, Heilongjiang, China

Dianguo Xu
Harbin Institute of Technology
Harbin, Heilongjiang, China

ISBN 978-981-95-5600-7 ISBN 978-981-95-5601-4 (eBook)
https://doi.org/10.1007/978-981-95-5601-4

This Springer imprint is published by the registered company Springer Nature Singapore Pte Ltd.
The registered company address is: 152 Beach Road, #21-01/04 Gateway East, Singapore 189721, Singapore

Contents

Chapter 1
Basic Knowledge of the Electrolytic Capacitorless Motor drive

1.1 Characteristics of Electrolytic Capacitorless Motor Drives

The motors are important devices for converting mechanical and electrical energy, which play a crucial role in the economic operation. For the continuous reduction of global resources, the methods for reducing the industrial power consumption have received the attention from various countries. The energy loss of the motor system is the main component of the electrical energy loss. In order to reduce the energy consumption and meet the standards of the energy conservation and emission reduction, the research on high-performance motor drive systems has become the main development trend. Permanent magnet synchronous motors (PMSMs) are gradually replacing the asynchronous motors, which have been widely applied in the industrial fields, such as transportation and home appliances. PMSMs have the advantages of simple structure, high power density, and convenient control. Meanwhile, several PMSM drive technologies such as vector control, direct torque control, and other advanced control theories have been widely applied, which greatly improve the performance of drive. PMSM drive has become an essential development direction for modern motor systems, which plays a more important role in the national economy and social development [1].

With the continuous development of the drive technology, the lifetime, reliability, and cost of the system have become the main issues for PMSM drive. Usually, the large electrolytic capacitors are applied in the DC-link to ensure the stable DC-link voltage. However, the electrolytic capacitor occupies the large volume in the drive board, which limits the improvement of the power density. Meanwhile, its lifetime is greatly affected by the environment temperature, which causes the inverter failures. Nowadays, the electrolytic capacitorless PMSM drivers are widely applied in the household appliances and industrial drivers. The film capacitors are applied in the DC-link, which can reduce the DC-link capacitance exactly. Hence, the drive has the

D. Ding et al., *Electrolytic Capacitorless Permanent Magnet Synchronous Motor Drives*, https://doi.org/10.1007/978-981-95-5601-4_1

advantages of the low cost and small size. In addition, the total harmonic distortion (THD) of the input current is lower [2].

Although the electrolytic capacitorless drives have many advantages, they face significant challenges in the motor control. As the capacitance reduces, the harmonics near the resonant frequency are significantly amplified in the drive. Hence, the suppression of current harmonic at the grid and DC sides is the fundamental issue for the stable operation of the electrolytic capacitorless drives. As the grid voltage deviates from the ideal condition, the grid power quality is further deteriorated, which influences the control performance of the conventional strategies. The DC-link voltage ripples introduce additional harmonics into the motor current. The interaction with the fundamental current causes the beat phenomenon, including the current distortion, high-frequency noise, and power loss. The DC-link voltage utilization determines the output capability of the motor voltage [3]. The DC-link voltage fluctuation generates the additional harmonics in the amplitude and duration time of the voltage vector, which results in the decrease of the system maximum performance in the overmodulation region. The reduction of the linear modulation margin further improves the DC-link voltage utilization. Hence, the research on the electrolytic capacitorless drives with the high reliability, high power density, and low cost is of great significance.

1.2 Key Technologies of Electrolytic Capacitorless Motor Drives

1.2.1 Harmonic Suppression for Grid Currents

In order to prevent high-frequency harmonics related to switching frequency from flowing into the power grid and reduce the grid power quality, the electrolytic capacitorless drives usually need to be adapted with DC or AC reactors. The AC motor driven by the frequency converter can be regarded as the constant power load (CPL), which exhibits the negative impedance characteristics. The negative impedance characteristic exacerbates the LC resonance, which leads to the harmonic amplification near the resonance frequency and causes the unstable operation under the weak grid. Therefore, the resonance suppression has been widely studied as a fundamental issue for the electrolytic capacitorless drives.

At present, the resonance suppression methods can be mainly divided into two categories. The first type is realized according to the system characteristic equations, and the negative impedance characteristics can be adjusted by varying the output power based on the DC-link voltage information, which can further improve the system stability. The second type is realized according to the system damping characteristics, and the system damping is increased by constructing the control loops of the DC-link voltage and current to suppress the resonance.

(1) Resonance suppression based on the negative impedance regulation

Through varying the motor current or voltage reference, the disturbance power related to the DC-link voltage fluctuation is injected into the motor, which can regulate the negative impedance characteristics to improve the system stability. The implementation method is shown in Fig. 1.1. K. Pietilainen et al. from ABB added the proportional term of the DC-link voltage harmonics in the *q*-axis current reference and optimized the proportional gain, which aimed to meet the requirements at different operation powers. However, this method may cause severe torque fluctuations at the motor side [4]. Therefore, J. H. Jung et al. from Pusan National University injected the disturbance power into the *d*-axis current reference, which not only achieved the damping control but reduced the negative impact of the disturbance power on the torque [5]. The above methods are based on the current references, which are shown in Fig. 1.1a.

However, the frequency of the DC-link voltage fluctuation is relatively high. Due to the limitation of the current loop bandwidth, the feedback current cannot track the current reference accurately. The resonance suppression method based on the voltage reference shown in Fig. 1.1b is more effective, which can generate the more accurate damping power. R. K. Maheshwari et al. from Aalborg University introduced the

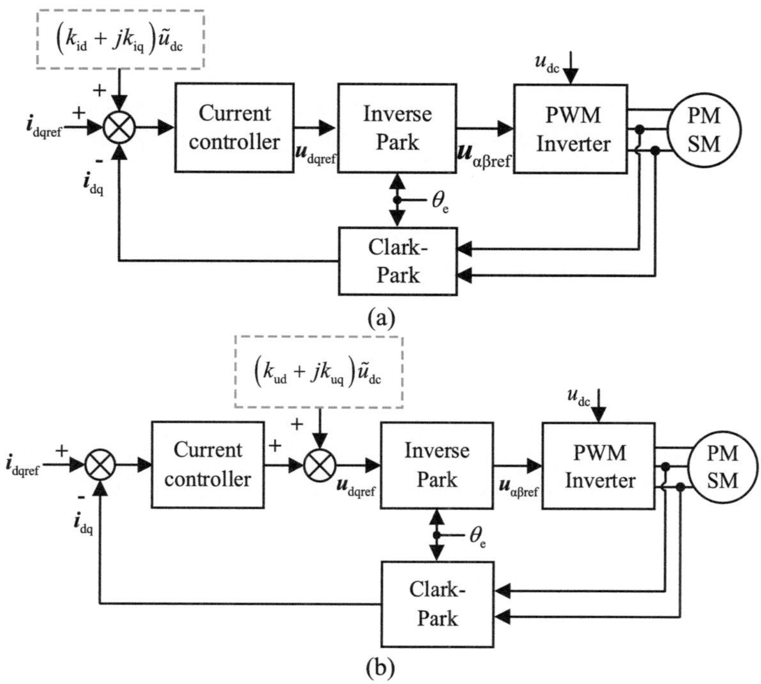

Fig. 1.1 Scheme of changing the negative characteristics by injecting disturbance power. **a** Block diagram of stability control based on current references. **b** Block diagram of stability control based on voltage references

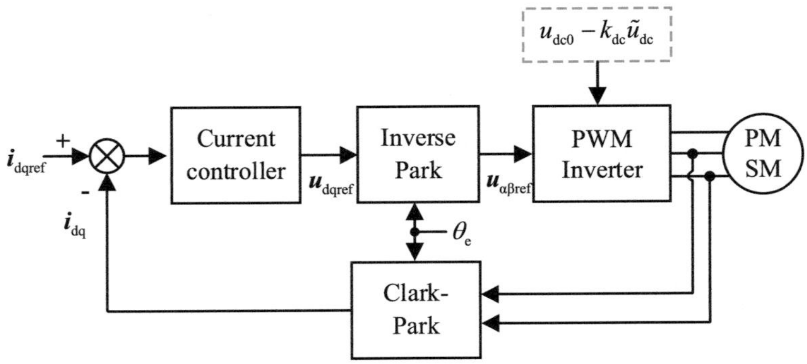

Fig. 1.2 Block diagram of stability control method based on virtual positive impedance

proportional term of the DC-link voltage harmonics into the voltage references at *dq*-axis, and the damper parameters are designed theoretically based on the proportionality coefficient of the current loop [6–8]. M. Hinkkanen et al. from Aalto University regulated the *d*-axis voltage reference with the DC-link voltage harmonics, which is combined with the flux-weakening controller to achieve the comprehensive control effect [9].

Through reversing the DC-link voltage harmonic during the modulation process, the drive can be converted into the positive impedance characteristic. The implementation method is shown in Fig. 1.2. L. Mathe et al. from Danfoss obtained the reconstructed value of the DC-link voltage through delaying the harmonics by half a cycle, which is applied to compensate the sampling delay of the DC-link voltage [10, 11]. D. Wang et al. from Aalborg University utilized the reverse fluctuation of the DC-link voltage in the modulation, which can achieve the system stability control [12]. In order to further improve the damping control effect, Dong Wang et al. used a phase-locked loop (PLL) to reconstruct the rectified voltage. The DC-link voltage harmonics amplified by resonance were reversed [13].

The above scheme is based on the stable conditions of control theory. The negative impedance characteristics can be improved through establishing the feedback control loop of the DC-link voltage, which can suppress the resonance of the drive. However, due to the respective problems in practical applications, the strategy based on the current reference regulation is difficult to achieve the accurate power adjustment, which results in the poor damping control effect. The strategy based on the voltage reference regulation may cause overcurrent problems.

(2) Resonance suppression based on the parallel virtual impedance

The resonance suppression method based on the parallel virtual impedance has clear physical significance and is widely used in the grid connected inverters, which can suppress the peak values of the LCL resonance effectively and improve the drive system stability [14, 15]. Similarly, the active damping method based on the parallel virtual impedance is also applicable for the electrolytic capacitorless drives.

The virtual impedance control schemes are shown in Fig. 1.3. W. J. Lee from Seoul National University proposed a damping control method of parallelling virtual resistors, which are shown in Fig. 1.3a. The damping current was generated with the difference between the rectified voltage and the DC-link voltage, and the Lomberg observer was designed to obtain the rectified voltage. In order to reduce the impact of the damping control on the current control loop, the damping voltage was applied in the direction of the motor current vector [16].

In Fig. 1.3b, paralleling the virtual resistors with the DC-link capacitor introduces the proportional term of the DC-link voltage into the motor. X. Liu et al. from University of Manchester introduced the DC-link voltage harmonics into the motor

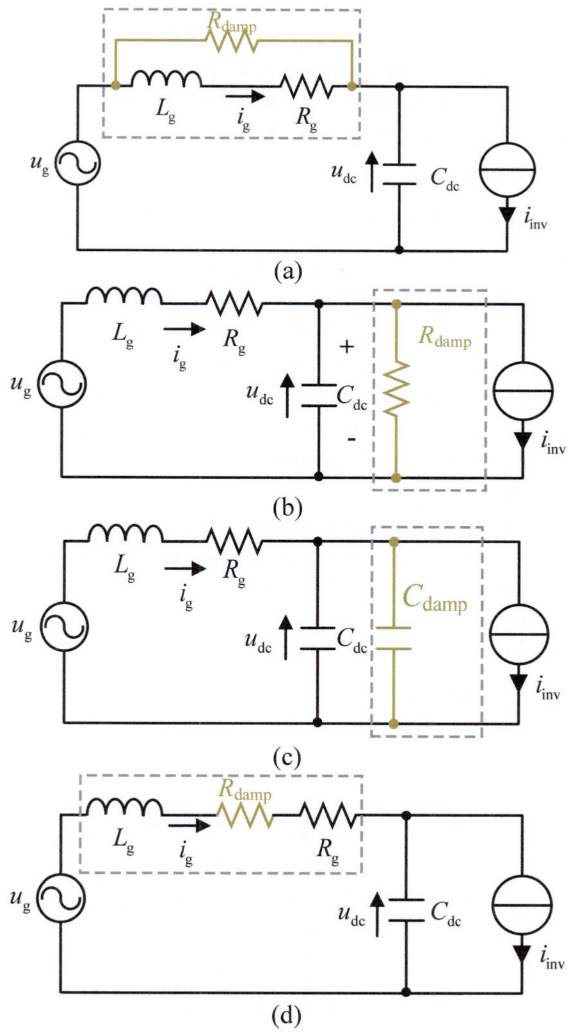

Fig. 1.3 Virtual impedance control scheme. **a** Virtual resistor is paralleled with grid inductor. **b** Virtual resistor is paralleled with DC-link capacitor. **c** Virtual capacitor is paralleled with DC-link capacitor. **d** Virtual resistor is in series with grid inductor

current reference to achieve stable control of the drive system, and analyzed the influence of the damper parameters on the motor torque [17]. P. Magne et al. from GREEN Laboratory proposed the stable controller through paralleling the virtual capacitors, which is shown in Fig. 1.3c. Meanwhile, the large signal modeling was conducted on the system before and after applying the stable controller, and the domain of the attraction under different virtual capacitor capacitances was analyzed [18]. By analyzing the frequency characteristics, the series resistance scheme in Fig. 1.3d can achieve the control effectiveness. On this basis, an active damping stability control method based on inductor current feedback was proposed in [19].

Through simulating the parallel passive components, the resonance of the electrolytic capacitorless drives can be effectively suppressed. However, the selection of damper parameters in such methods depends on the driving system model. As the impedance or motor power varies widely, the control performance cannot be maintained. In addition, the above methods simplified the drive system to CPL in the modeling process. The influence of the motor controller bandwidth and DC-link voltage sampling delay on the damping control performance need to be further researched.

(3) **Resonance suppression based on the auxiliary circuit**

The above schemes mainly adjusted the motor power with control algorithms to achieve the resonance suppression of the drive system. In recent years, the schemes of changing the drive topology have also received attention. H. Shin et al. from Seoul National University paralleled an active compensator at the DC-link to suppress the resonance in the drive, which is shown in Fig. 1.4. The waveform of the grid current could be also reshaped according to the harmonic standard of IEC61000-3-12, which reduced the partial weighted harmonic distortion (PWHD) of the grid current effectively [20, 21].

The unbalanced grid is one of the common faults in the power systems, which are caused by several factors such as the high-power single-phase load, the uneven distribution of the single-phase load, and the randomness of the single-phase power consumption. The grid improvement method based on the balanced grid conditions cannot suppress the harmonics caused by the negative sequence components effectively. To improve the input grid power quality under the unbalanced grid conditions,

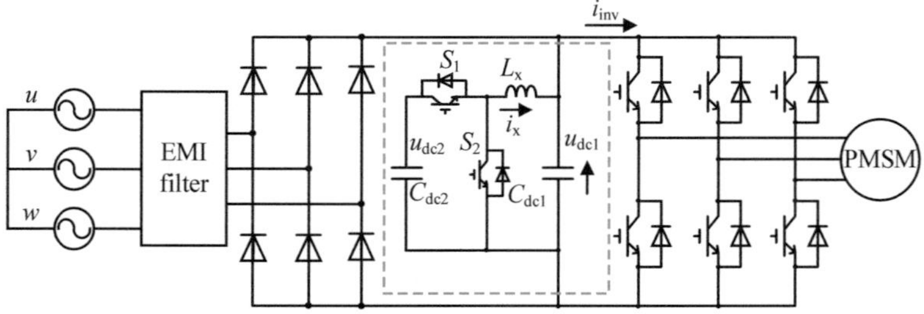

Fig. 1.4 Topology of electrolytic capacitorless motor drives with paralleled compensator

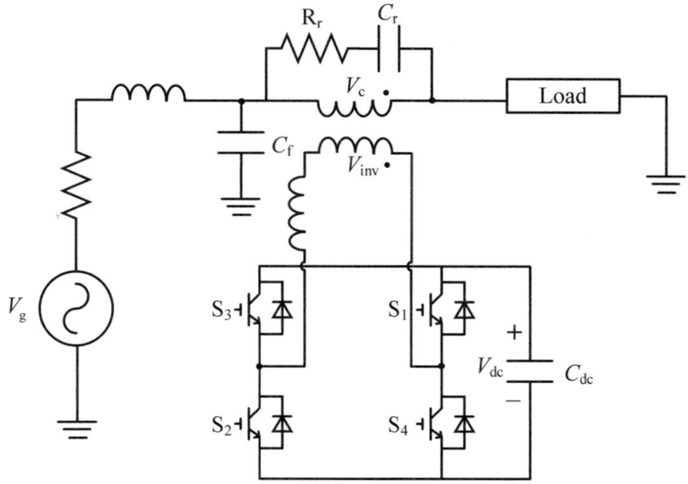

Fig. 1.5 Single-phase equivalent circuit with DVR

the various topology improved methods have been proposed. In [22], a dynamic voltage recovery device (DVR) is proposed to improve the grid voltage when the three-phase input is unbalanced. The single-phase equivalent circuit with DVR is shown in Fig. 1.5. The compensation voltage was generated from the positive sequence component of the two-phase unbalanced line voltage, which was injected into the distribution line through an H-bridge voltage-source inverter. The topology of the electrolytic capacitorless drive system with the dual inverter was proposed in [23], which was shown in Fig. 1.6. The harmonic voltage caused by the grid disturbance was generated through a secondary inverter, whose difference with the output voltage of the main inverter was applied to the motor side. The harmonic current from the grid disturbance could be reduced effectively.

Varying the topology of the drive system can avoid introducing disturbance power into the motor, which can also improve the motor speed expansion and anti-interference performance. However, the system cost and volume are increased. Hence, the suppression strategies based on the control algorithms are paid more attention by the industry.

(1) **Resonance suppression based on the negative sequence regulation**

As the grid voltage is unbalanced, the positive and negative sequence components are existed in the input voltage and current, which result in the double-frequency fluctuations of the active and reactive power in the grid input power. Based on different control objectives, five different expressions of the instantaneous power and reference current were determined in [24]. It is difficult to eliminate the active power fluctuations, the reactive power fluctuations, and the current harmonics at the same time. Regarding the distortion of grid connected current caused by unbalanced grid

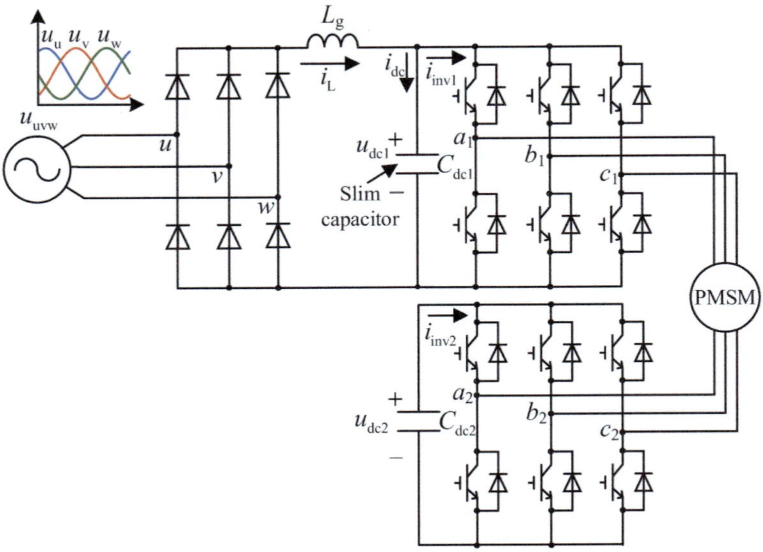

Fig. 1.6 Topology of the electrolytic capacitorless drive system with the dual inverter

side. Aiming at the grid current distortion caused by the unbalanced grid, a negative sequence component compensation method at *dq*-axis was proposed in [25]. Through designing a bandpass filter based on the second-order generalized integral (SOGI), the positive and negative sequence voltages at *αβ*-axis could be obtained, which can be applied for the negative sequence voltage compensation to suppress the current harmonics. However, the method required the decomposition of the positive and negative sequence components, which reduced the dynamic performance of the drive system. In [26], the proportional integral resonant (PIR) controller was proposed, which was equivalent to superimposing an SOGI in the PI controller. The error control of the positive sequence DC components and negative sequence double-frequency harmonics in the rotating coordinate system of the positive sequence. A reduced order generalized integrator (ROGI) was proposed in [27] to replace the SOGI of [26], and the reduced order resonant controller (ROR) was obtained, which achieved the beatless control of the negative sequence components at *αβ*-axis. The switching between different control objectives could be realized through only varying one parameter in the controller, which improved the adaptability for the operation conditions. However, this method required the application of the positive and negative sequence ROR controllers to regulate the current, which improved the complexity of the system design. In [28], the digital implementation of the ROR controller was applied for the rotating coordinate system, which simplified the computational complexity and improved the dynamic performance.

The harmonic suppression methods based on the negative sequence regulation were usually applied under the positive sequence coordinate system or the positive and negative sequence rotating coordinate system based on the dual synchronous reference frame-based phase-locked loop (DSRF-PLL). The input quantity of the controller could be obtained by separating the positive and negative sequence components of the grid current, which increased the computational complexity and control system complexity.

(2) **Resonance suppression based on the inverter admittance reshaping**

As the power grid is unbalanced, the background harmonics of the grid voltage and current are increased. Through reshaping the admittance of the grid-connected inverter, the background harmonics at grid side can be suppressed, which can also improve the grid current quality. Currently, the methods of reshaping the inverter admittance can be divided into three categories. The first type is realized by establishing the sequence impedance or complex vector admittance model of the grid-connected inverter, which aims to eliminate the frequency coupling caused by the asymmetric control structures and reduce the coupling current components brought by the grid disturbances. In [29], a complex vector admittance model of the three-phase grid-connected inverter was built considering the current loop, PLL, and the DC voltage loop at $\alpha\beta$-axis. The compensation method was proposed for the frequency coupling caused by the asymmetry in PLL and the voltage loop, which reduced the frequency coupling components and improved the system stability. In [30, 31], the impact of the asymmetric structure on the inverter frequency coupling was analyzed. Through introducing the decoupling factors into the inner and outer loops of the inverter, the system is transformed from a dual-input-output system to the single-input-output system, which eliminates the frequency coupling and reduces the grid current harmonics. Meanwhile, the admittance reshaping factors are analyzed and selected to reduce the inverter admittance and enhance the suppression ability for the background harmonics at grid side.

The second type involves establishing the resonant controllers for the specific frequency harmonics of the grid current under the unbalanced grid conditions, which increases the output impedance of the grid-connected inverter for harmonic suppression. In [32, 33], the gains of the grid-connected inverter at specific frequencies were increased through applying the resonant controllers and multi-resonant harmonic compensators at $\alpha\beta$-axis, or PI and multi-resonant harmonic compensators at *dq*-axis. This method could reduce the inverter output admittance and eliminate the harmonics at specific frequency. However, the control performance was influenced by the parameter drift. The harmonic current generated by the dead-time effect of switching devices and the grid current distortion caused by the background harmonics of the grid voltage were analyzed in [34]. On this basis, the improved strategies, including paralleling the virtual notch impedance and leading resonant impedance in series, were proposed, which could suppress the harmonic current caused by the aforementioned issues effectively.

The third type is to extract the dominant harmonics or all harmonics from the DC-link voltage or grid current for feedforward, which can increase the inverter output admittance. In [35], a control method based on virtual resistors was proposed. With the fast fourier transform (FFT) analysis of the grid voltage and current, the dominant harmonics were obtained and fed forward. The inverter was equivalent to the virtual resistor, which could absorb the harmonic power from the grid side and suppress the harmonics effectively. However, the computational burden of this method is high. The feedforward strategy of the grid voltage was proposed in [36], whose function was derived through the equivalent transformation of the drive system. The proposed function could eliminate the impact of the grid disturbance on the grid current completely, which can also suppress multiple harmonics in the grid current. The control structure was simple, and the method could maintain high dynamic performance. However, the accurate grid information was difficult to obtain through sampling methods. The point of common coupling (PCC) voltage can reflect the harmonic information of the grid voltage, which is easy to obtain. However, considering PWM updating delay, sampling delay, and weak-grid variation, applying the PCC voltage feedforward strategy causes the output impedance phase of the grid connected inverter to be less than – 90°, which results in the system instability [37]. In [38–40], the feedforward method of PCC voltage was proposed, which increased the phase of the inverter output impedance and enhanced the system stability by varying the weighting coefficient of the feedforward function. However, varying the weighting coefficient reduced the output impedance amplitude of the grid connected inverter, which influenced the suppression ability for the background harmonics at grid side. A dual sampling mode for the grid voltage was designed in [40] on the basis of the variation of the inverter duty cycle, which could obtain the grid voltage accurately. The sampled grid voltage was applied to replace the common coupling point voltage for feedforward, which can suppress the background harmonics and improve the stability under the weak grid conditions.

The LC resonance suppression for electrolytic capacitorless drives has been widely studied as the fundamental issue for stable control and harmonic suppression. Common solutions can be divided into three categories: regulating the motor negative impedance characteristics, paralleling the virtual impedance, and applying the auxiliary circuits. Under the unbalanced grid condition, the software solutions can be divided into two categories: regulating the negative sequence component and reshaping the inverter admittance. Considering the cost and volume of the drive system, the software solutions are preferred. The resonance suppression of the drive system can be achieved through varying the motor current and voltage commands and regulate the motor power with software control algorithms. However, the existing solutions still need to be improved in terms of the accuracy and effectiveness of power regulation. Chapters 3 and 4 in this book will consider the non-ideal characteristics such as the DC-link voltage sampling delay and the limited bandwidth of the current loop during the damper design process, which can lead the LC resonance suppression to be not affected by the grid conditions.

1.2.2 Beat Suppression Method for Motor Side

Beat phenomenon is one of the main issues in the power electronic systems [41–44]. The low-frequency impedance of the power electronic converters is usually

Fig. 1.7 Diagram of time sequence of sampling and update of PWM

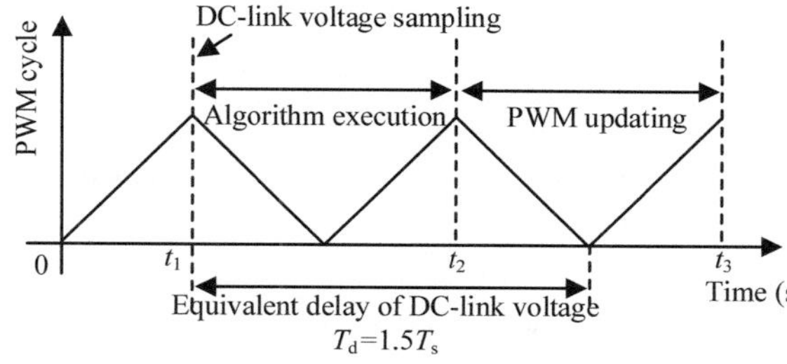

small. Hence, even small low-frequency disturbance can cause severe waveform distortion. The beat problem is common in the track traction systems, which is caused by the interaction of DC-link voltage and inverter output voltage [45–47]. The beat phenomenon also exists in the electrolytic capacitorless system. D. Wang et al. from Aalborg University illustrated the influence of the DC-link fluctuation and sampling delay on the output voltage error of pulse width modulation (PWM). The errors generated the harmonics related to the DC-link ripples in the motor current, which causes the problem of the beat phenomenon [48]. Meanwhile, the periodic fluctuating load can also interact with the additional harmonics of the motor current, which results in the beat phenomenon.

Figure 1.7 shows the diagram of time sequence of sampling and update of PWM. The DC-link voltage is sampled at time t_1, and the PWM update begins at t_2. Hence, the equivalent delay of the DC-link voltage is 1.5 switching cycles. Due to the severe fluctuation of the DC-link voltage, the sampling delay has a significant impact on the drive system. For example, as the switching frequency and sampling frequency are both 6 kHz, the voltage sampling error can reach 9%, and the DC-link voltage sampling error increases with the reduction of the switching frequency.

In order to suppress the beat problem caused by the sampling delay of the DC-link voltage, Y. Kang et al. from Huazhong University of Science and Technology designed a repetitive observer for DC-link voltage, which predicted the average value of the DC-link voltage at the next switching cycle for Space Vector Pulse Width Modulation (SVPWM) update. The beat phenomenon can be suppressed effectively in the track traction system [47]. However, the DC-link voltage in the electrolytic capacitorless drive system is not a strictly periodic signal, and the effectiveness of the repetitive observer is reduced.

The electrolytic capacitorless PMSM drive system is commonly applied in heating ventilation and air conditioning (HVAC) systems. Figure 1.8 shows the load curve of the single rotor compressor. As can be seen, the load torque varies periodically with the mechanical angle, which can also lead to the beat phenomenon in the drive system. Aiming at this, the beatless strategy based on the notch filter was designed in [49], which eliminated the beat phenomenon by constructing a control loop between motor current and voltage. The characteristics of different beat suppression methods are shown in Table 1.1.

The beat phenomenon is caused by the interaction of two or more harmonics, whose frequencies are not proportional to each other. The beat phenomenon is

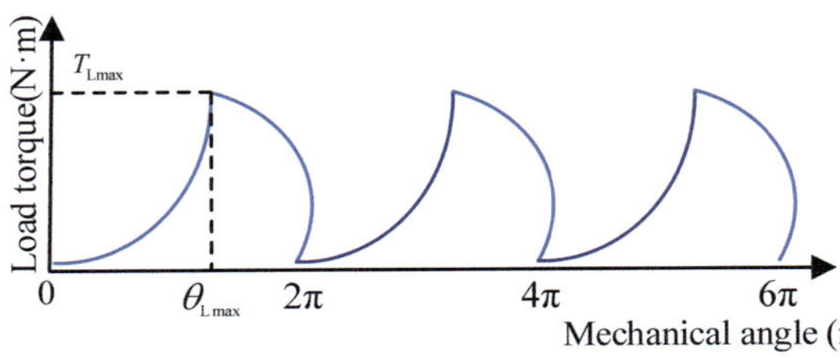

Fig. 1.8 Load curve of compressor

Table 1.1 Character of comparison of different beat suppression methods

Methods	Advantages	Disadvantages
DC-link voltage repetitive observer method	Accurate voltage prediction, high suppression effectiveness	Performance reduction with the low periodicity of DC-link voltage
Method based on the notch filter	Flexible design, closed-loop control, high stability	Low robustness of the controller parameter to the load ripple frequency

usually manifested as the low-frequency fluctuation of the waveform envelope, and the beat frequency is determined by the varying period of the harmonic phase difference. Hence, the beat signal cannot be obtained through Fast Fourier analysis (FFT). In order to suppress the beat phenomenon, the main harmonic components which causes the beat issue should be obtained and suppressed. Nowadays, the applied methods for beat frequency suppression contained the feedforward compensation and the designed feedback control loops. The feedforward method requires high compensation accuracy, which is suitable for systems with high signal periodicity. The feedback control method has the parameter adaptability issues. Chapter 5 in this book will provide the beat suppression strategies, which are suitable for the operation conditions inside and outside the flux-weakening region, The methods also have adaptability to the variation in operating conditions.

1.2.3 Voltage Utilization Improvement Method

The utilization of the DC-link voltage determines the voltage output capacity of the motor, whose control is the key technology in the electrolytic capacitorless drive system [50]. The DC-link voltage ripples introduce additional fluctuations into the voltage vector duration time and motor voltage, which lead the drive system to reach the overmodulation region earlier [51–53]. The reduction of linear modulation margin reduces the voltage utilization [54]. Therefore, several methods have been proposed to improve the utilization of the DC-link voltage.

In order to utilize the DC-link voltage effectively, overmodulation control strategies play an important role [55–57]. In [55], an overmodulation technique was proposed based on FFT analysis, which could realize a seamless transition from linear modulation region to six-step operation. By adjusting the holding angle of each operating point, the voltage output capability can be improved. In [56], an overmodulation strategy based on the voltage vector space division was proposed, which replaces the conventional center projection by parallel projection. On this basis, the DC-link voltage utilization and the output torque capability of the field weakening region can be improved. In [57], a cooperative control scheme for the six-step operation was proposed, which can realize the six-step operation by ensuring the instantaneous current control capability. However, the harmonics of the output voltage and current increase in deep modulation region, which increases the torque ripple.

Different solutions have been proposed to suppress torque ripple in the overmodulation region [58–60]. In [58], an adaptive overmodulation strategy was proposed based on the working quadrant. By designing three types of the reference voltages, the control requirements of the defined working quadrant can be met, and the torque harmonics in the steady state can be reduced. In [59], a method based on the minimum nonlinear voltage was proposed to satisfy torque requirements with minimum overmodulation-voltage nonlinearity. The quantitative relationship between torque ripple and demand was calculated, which can suppress the torque ripple under low torque conditions. In [60], an overmodulation method based on the vector analysis was proposed to suppress torque ripple under low load conditions. By reallocating the stator current, the maximum electromagnetic torque can be output under dynamic processes and high load conditions. The above strategies were realized based on constant DC-link voltage, which makes it difficult to achieve ideal control effect in electrolytic capacitorless motor drives.

Currently, researchers have investigated the limitations of overmodulation control strategies under the large fluctuation of DC-link voltage in electrolytic capacitorless drives [53, 61, 62]. In order to enhance the output voltage capability of electrolytic capacitorless motor drives, a novel strategy for extending the linear region based on voltage vector angle regulation was proposed in [61]. In [62], an overmodulation strategy was proposed through analyzing the harmonics and root mean square values of the motor voltage, which realized a linear transition of the output voltage vector from overmodulation region I to the six-step operation. To further reduce the stator voltage harmonics, the stator voltage distortion of the conventional overmodulation strategy was analyzed in [53]. By setting a fixed voltage boundary, the voltage distortion due to the phase jump back of output voltage vectors is eliminated.

However, the conventional strategies have not paid attention to the periodic fluctuations of the reference voltage vector in the overmodulation region, which are also difficult to be applied to improve the utilization of the DC-link voltage in the linear modulation region. Chapters 6 and 7 in this book will analyze the modulation process under the fluctuated DC-link voltage. On this basis, the proposed voltage utilization improvement methods are applicable in both the linear modulation and the overmodulation regions.

1.3 Summary

This chapter reviews the development of motor drive systems. In order to reduce the energy consumption in the field of electric power transmission, PMSM drive system is studied. With the continuous development of the motor drive technology, the electrolytic capacitorless drive system can improve the system lifetime and reliability, which can also reduce system volume and cost. This chapter provides a detailed analysis of the key technologies of the electrolytic capacitorless drive system, including the harmonic suppression methods for grid current, the beat suppression method for motor side, and the voltage utilization improvement methods.

References

1. S. Kuang, X. Zhang, P. Liu, G. Zhang, Z. Zhang. Sensorless control method for switched reluctance motors based on locations of phase inductance characteristic points [J]. Transactions of China Electrotechnical Society, 2020, 35(20): 4296–4305.
2. H. Lin, W. Song, L. Zhao. Analysis and control of DC-link voltage stabilization in small capacitor motor drive system [J]. Proceedings of the CSEE, 2020, 40(18): 5997–6006.
3. Y. Son and J. I. Ha. Direct power control of a three-phase inverter for grid input current shaping of a single-phase diode rectifier with a small DC-link capacitor [J]. IEEE Transactions on Power Electronics, 2015, 30(7): 3794–3803.
4. K. Pietilainen, L. Harnefors, A. Petersson, and H. P. Nee. DC-link stabilization and voltage sag ride-through of inverter drives [J]. IEEE Transactions on Industrial Electronics, 2006, 53(4): 1261–1268.
5. J. H Jung, H. J Heo, J. M. Kim, and S. U. Choi. DC-link voltage stabilization and source THD improvement using d-axis current injection in reduced DC-link capacitor system [C]. 42nd Annual Conference of the IEEE Industrial Electronics Society, 2016: 2737–2742.
6. R. K. Maheshwari and S. M. Nielsen. Closed loop control of active damped small DC-link capacitor based drive [C]. IEEE Energy Conversion Congress and Exposition, 2010: 4187–4191.
7. R. K. Maheshwari, S. M. Nielsen, B. Henriksen, P. M. Obel and H. Kragh. Active damping technique for small DC-link capacitor based drive system [C]. IEEE International Symposium on Industrial Electronics, 2010: 1205–1209.
8. R. K. Maheshwari, S. M. Nielsen, K. Lu. An active damping technique for small DC-Link capacitor based drive system [J]. IEEE Transactions on Industrial Informatics, 2013, 9(2): 848–858.
9. M. Hinkkanen, L. Harnefors and J Luorni. Control of induction motor drives equipped with small DC-Link capacitance [C]. European Conference on Power Electronics and Applications, 2007: 1–10.
10. L. Mathe, H. R. Andersen, R. Lazar and M. Ciobotaru. DC-link compensation method for slim DC-link drives fed by soft grid [C]. IEEE International Symposium on Industrial Electronics, 2010: 1236–1241.
11. L. Mathe, L. Totok, D. Wang and D. Sera. Resonance reduction for AC drives with small capacitance in the DC link [J]. IEEE Transaction on Industry Application, 2017, 53(4):3814–3820.
12. D. Wang, K. Lu, P. O. Rasmussen, L. Mathe and Y. Feng. Analysis of voltage modulation based active damping techniques for small DC-link drive system [C]. IEEE Energy Conversion Congress and Exposition (ECCE), 2015: 2927–2934.

13. D. Wang, K. Lu, P. O. Rasmussen, L. Mathe, Y. Feng and F. Blaabjerg. Voltage modulation using virtual positive impedance concept for active damping of small DC-Link drive system [J]. IEEE Transactions on Power Electronics, 2018, 33(12): 10611–10621.
14. Y. Wang, W. Ye, R. Song, J. Bi, L. Chen, J. Chen. Design of Additional Active Damping for Three-phase LCL Grid Connected Inverter Based on Impedance Analysis Method [J]. High Voltage Engineering, 2021, 47(08): 2645–2656.
15. W. Yao, Y. Yang, Y. Xu, F. Blaabjerg, S. Liu and G. Wilson. Phase reshaping via all-pass filters for robust LCL-filter active damping [J]. IEEE Transactions on Power Electronics, 2020, 35(3): 3114–3126.
16. W. J. Lee and S. K. Sul. DC-Link voltage stabilization for reduced DC-Link capacitor inverter [J]. IEEE Transaction on Industry Application, 2014, 50(1): 404–414.
17. X. Liu, A. J. Forsyth and A. M. Cross. Negative input-resistance compensator for a constant power load [J]. IEEE Transactions on Industrial Electronics, 2007, 53(6): 3188–3196.
18. P. Magne, D. Marx, B. N. Mobarakeh and S. Pierfederici. Large-signal stabilization of a DC-link supplying a constant power load using a virtual capacitor: impact on the domain of attraction [J]. IEEE Transaction on Industry Application, 2012, 48(3): 878–887.
19. N. Zhao, G. Wang, R. Zhang, B. Li, Y. Bai and D. Xu. Inductor current feedback active damping method for reduced DC-link capacitance IPMSM drives [J]. IEEE Transactions on Power Electronics, 2019, 34(5): 558–4568.
20. H. Shin, H. G. Choi and J. I. Ha. DC-link shunt compensator for three-phase system with small DC-link capacitor [C]. 9th International Conference on Power Electronics and ECCE Asia (ICPE-ECCE Asia), 2015: 33–39.
21. I. Won, Y. Cho and K. B. Lee. Predictive control algorithm for capacitor-less inverters with fast dynamic response [C]. IEEE International Conference on Power and Energy (PECon), 2016: 479–483.
22. M. Pradhan and M. K. Mishra. Dual P-Q theory based energy-optimized dynamic voltage restorer for power quality improvement in a distribution system [J]. IEEE Transactions on Industrial Electronics, 2019, 66(4): 2946–2955.
23. Y. Ohno and H. Haga. Control method of electrolytic capacitorless dual inverter for harmonic compensation under distorted grid voltage [J]. IEEE Transactions on Industrial Electronics, 2022, 58(1): 375–387.
24. Pedro Rodriguez and Adrian V. Timbus. Flexible active power control of distributed power generation systems during grid faults [J]. IEEE Transactions on Industrial Electronics, 2007, 54(5): 2583–2592.
25. L. Zhou, L. Qiang, H. Li, J. Du, Y. Long and B. Liao. Negative component compensation control scheme for photovoltaic grid connected inverter [J]. High Voltage Engineering, 2013, 39(5): 1197–1203.
26. W. Zhang, H. Wang, Y. Ren, J. Hu and Y. He. Investigation on control of three-phase grid-connected inverters under unbalanced grid voltage conditions [J]. Transactions of China Electrotechnical Society, 2010, 25(12): 103–110.
27. C. A. Busada, S. G. Jorge, A. E. Leon. Current controller based on reduced order generalized integrators for distributed generation systems [J]. IEEE Transactions on Industrial Electronics, 2011, 59(7): 2898–2909.
28. X. Zhao, X. Jin and F. Zhou. Unbalanced control of grid-connected inverters based on proportion integraland reduced order resonant controllers [J]. Proceedings of the CSEE, 2013, 33(19): 84–92.
29. X. Zhang, D. Xia, W. Chen, L. Lie and D. Xu. Frequency coupling suppression control method for three-phase grid-connected converter under weak grid [J]. Transactions of China Electrotechnical Society, 2019, 34(21): 4559–4571.
30. X. Tang, Z. Li, Y. Li and Y. He. Frequency coupling oscillation analysis and decoupling control design of grid connected converter [J]. Proceedings of the CSEE, 2023, 43(8): 3127–3137.
31. Z. Li, X. Tang, Y. He, P. Liu and X. Li. Admittance reshaping method of converter for suppressing background harmonic of grid [J]. Proceedings of the CSEE, 2024, 44(2): 704–714.

32. F. Wang, J. L. Duarte, M. A. M. Hendrix and P. F. Ribeiro. Modeling and analysis of grid harmonic distortion impact of aggregated DG inverters [J]. IEEE Transactions on Power Electronics, 2011, 26(3): 786–797.
33. Y. Jia, J. Zhao and X. Fu. Direct grid current control of LCL-filtered grid-connected inverter mitigating grid voltage disturbance [J]. IEEE Transactions on Power Electronics, 2014, 29(3): 1532–1541.
34. J. Chen, X. Zhang, Z. Yan, Z. Wei and X. Chen. Dead-time effect and background grid-voltage harmonic suppression methods for inverters with virtual impedance control [J]. Transactions of China Electrotechnical Society, 2021, 36(8): 1671–1680.
35. Y. Ling, J. Zeng, Y. Liu and L. Yang. Virtua-resistance-type photovoltaic inverter with low-order harmonies suppression [J]. Automation of Electric Power Systems, 2018, 42(9): 114–119.
36. X. Wang, X. Ruan, S. Liu, and C. K. Tse. Full feedforward of grid voltage for grid-connected inverter with LCL filter to suppress current distortion due to grid voltage harmonics [J]. IEEE Transactions on Power Electronics, 2010, 25(12): 3119–3127.
37. D. Yang, X. Ruan, and H. Wu. Impedance shaping of the grid-connected inverter with LCL filter to improve its adaptability to the weak grid condition [J]. IEEE Transactions on Power Electronics, 2014, 29(11): 5795–5805.
38. Q. Qian, S. Xie, L. Huang, J. Xu, Z. Zhang, and B. Zhang. Harmonic suppression and stability enhancement for parallel multiple grid-connected inverters based on passive inverter output impedance [J]. IEEE Transactions on Industrial Electronics, 2017, 64(9): 7587–7598.
39. M. Li, X. Zhang, H. Zhang, H. Zhang, and W. Guan. An improved grid voltage feedforward compensation scheme of grid-connected inverter in weak grid [C]. IEEE International Power Electronics and Application Conference and Exposition, Shenzhen, China, 2018: 1–6.
40. X. Ruan, X. Wang, D. Pan, D. Yang, W. Lei, C. Bao. Weighted-feedforward scheme of grid voltages for the three-phase LCL-type grid-connected inverters under weak grid condition [J]. Control techniques for LCL-type grid-connected inverters, 2018: 249–270.
41. X. Ge, X. Feng, K. Han, W. Song. Hardware-in-the-Loop Application of Beat-less Control Scheme for Three-Level Traction Inverter System [J]. Transactions of China Electrotechnical Society, 2014, 29(8): 204–210.
42. X. Ge, X. Feng. Research on Beat-less Control Technology in Three Level Inverter [J]. Journal of the China Railway Society, 2010, 32(3): 125–130.
43. W. Kawamura, Y. Chiba, M. Hagiwara and H. Akagi. Experimental verification of an electrical drive fed by a modular multilevel TSBC converter when the motor frequency gets closer or equal to the supply frequency [J]. IEEE Transactions on Industry Application, 2017, 53(3): 2297-2306.
44. J. I. Itoh, G. T. Chiang and K. Maki. Beatless synchronous PWM control for high-frequency single-pulse operation in a matrix converter [J]. IEEE Transactions on Power Electronics, 2013,28(3): 1338–1347.
45. Y. Liu, W. Mei, L. Hu, W. Gan, Y. Lyu. Beat-less Control Method for High-speed EMU [J]. Electric Drive for Locomotives, 2020(2): 48–52+57.
46. Y. Song, W. Song, B. Xu. Digitized Implementation of Beat-frequency Suppression Algorithm Based on One-cycle Control for Traction Motors [J]. Proceedings of the CSEE, 2019, 39(10): 3007–3016.
47. H. Ouyang, K. Zhang, P. Zhang and Y. Kang. Repetitive compensation of fluctuating DC link voltage for railway traction drives [J]. IEEE Transactions on Power Electronics, 2011, 26(8): 2160–2171.
48. D. Wang and K. Lu. Analysis of system interharmonics of VSI-fed small DC-link drive with varying power load [C]. IEEE Energy Conversion Congress and Exposition (ECCE), 2018: 3347–3354.
49. N. Zhao, G. Wang, B. Li, R. Zhang and D. Xu. Beat phenomenon suppression for reduced DC-link capacitance IPMSM drives with fluctuated load torque [J]. IEEE Transactions on Industrial Electronics, 2019, 66(11): 8334–8344.
50. Y. Bai, B. Li, Q. Wang, D. Ding, G. Zhang, G. Wang, and D. Xu. An adaptive-frequency harmonic suppression strategy based on vector reconstruction for current measurement error of PMSM drives [J]. IEEE Trans. Power Electron., 2023, 38(1): 34–40.

51. L. Zheng, R. P. Kandula, and D. Divan. Robust predictive control for modular solid-state transformer with reduced DC-link and parameter mismatch [J]. IEEE Transactions on Power Electronics, 2021, 36(12): 14295–14311.
52. D. Ding, G. Wang, N. Zhao, G. Zhang, and D. Xu. Enhanced flux-weakening control method for reduced dc-link capacitance IPMSM drives [J]. IEEE Transactions on Power Electronics, 2019, 34(8): 7788–7799.
53. G. Wang, H. Hu, D. Ding, N. Zhao, Y. Zou, and D. Xu. Overmodulation strategy for electrolytic capacitorless PMSM drives: voltage distortion analysis and boundary optimization [J]. IEEE Transactions on Power Electronics, 2020, 35(9): 9574–9585.
54. C. Yin, W. Ding, L. Ming, and P. C. Loh. Single-stage active split-source inverter with high DC-link voltage utilization [J]. IEEE Transactions on Power Electronics, 2021, 36(6): 6699–6711.
55. H. Lee, S. Hong, J. Choi, K. Nam, and J. Kim. Sector-based analytic overmodulation method [J]. IEEE Transactions on Industrial Electronics, 2019, 66(10): 7624–7632.
56. R. Jing, G. Zhang, G. Wang, G. Bi, D. Ding, and D. Xu. An overmodulation strategy based on voltage vector space division for high-speed surface-mounted PMSM drives [J]. IEEE Transactions on Power Electronics, 2022, 37(12): 15370–15381.
57. Y. -C. Kwon, S. Kim, and S. -K. Sul. Six-step operation of PMSM with instantaneous current control [J]. IEEE Transactions on Industry Application, 2014, 50(4): 2614–2625.
58. B. Wang, L. Wang, Y. Yu, and D. Xu. Adaptive overmodulation strategy for PMSM field-weakening control based on working quadrant [J]. IEEE Transactions on Industrial Electronics, 2023, 70(10): 9959–9970.
59. X. Zhang, B. Wang, Y. Yu, J. Zhang, and D. Xu. Minimum-nonlinear-voltage method for torque ripple suppression in induction motor overmodulation and field-weakening control [J]. IEEE Transactions on Industrial Electronics, 2022, 69(5): 4495–4509.
60. J. Zhang, B. Wang, Y. Yu, X. Zhang, and D. Xu. Overmodulation harmonic modeling and suppression for induction motor field-weakening control with extended voltage tracking method [J]. IEEE Transactions on Industrial Electronics, 2023, 70(6): 5637–5650.
61. R. Gao, D. Ding, G. Wang, Q. Wang, G. Zhang, and D. Xu. Linear region expansion strategy based on voltage vector angle regulation for PMSM drives with small DC-link capacitor [J]. IEEE Transactions on Power Electronics, 2023, 38(7): 8327–8339.
62. H. Saren, O. Pyrhonen, K. Rauma, and O. Laakkonen. Overmodulation in voltage source inverter with small DC-link capacitor [C]. IEEE 36th Power Electronics Specialists Conference, 2005: 892–898.

Chapter 2
Impedance Model of Reduced DC-Link Capacitance IPMSM Drives

2.1 Introduction

The use of small capacitance film capacitors significantly alters the impedance characteristics of drive, thereby affecting its stability and dynamic performance. In this chapter, the mathematical model of the electrolytic capacitorless PMSM drive is respectively derived based on the constant-power load characteristics and the non-ideal characteristics of motor side. The accuracy of the mathematical model is verified through sweep frequency response analysis, which provides a theoretical foundation for the stability analysis and control optimization discussed in the subsequent chapters.

2.2 Model Based on Constant-Power Load Characteristics

The structure of PMSM drive is shown in Fig. 2.1, which mainly includes DC-link inductor L_g, uncontrolled rectifier, DC-link capacitor C_{dc}, three phase bridge inverter and PMSM [1]. u_{uvw} is the three-phase power supply, *abc* is the three-phase motor winding, L_s is the grid side inductor, i_g is the DC-link current, and u_{dc} is the DC-link voltage. The topology of traditional three-phase input diode rectifier PMSM drives with large DC-link electrolytic capacitor is shown in Fig. 2.1a, in which the large electrolytic capacitor is used in the DC-link to stabilize the DC-link voltage. In the fields of heating, ventilation and air conditioning (HVAC), the reactor needs to be connected in series on the grid side or DC side to improve the power quality of grid current.

The structure of electrolytic capacitorless drive is similar to that of the traditional drive. Figure 2.1b, c show two kinds of electrolytic capacitorless PMSM drive topology, respectively. Different from the traditional drive, the main function of the reactor in the electrolytic capacitorless drive is to filter out the high-frequency

D. Ding et al., *Electrolytic Capacitorless Permanent Magnet Synchronous Motor Drives*, https://doi.org/10.1007/978-981-95-5601-4_2

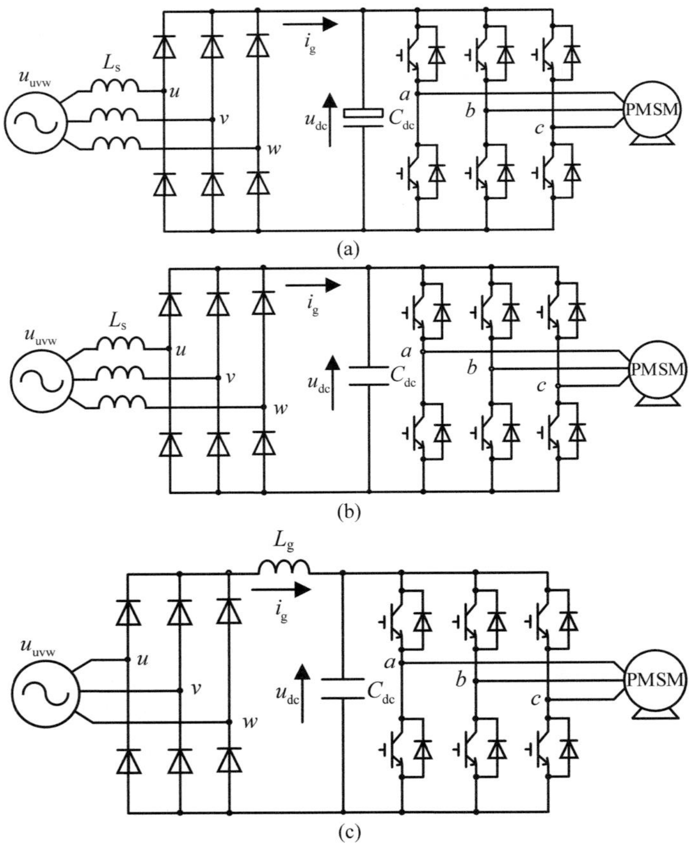

Fig. 2.1 Topology of three-phase input diode rectifier PMSM drives. **a** Topology of traditional drive with large DC-link electrolytic capacitor. **b** Topology I of electrolytic capacitorless drive. **c** Topology IIof electrolytic capacitorless drive

harmonics related to switching frequency, and prevent them from polluting the power grid. Therefore, the inductance is obviously smaller than that of the reactor used in the traditional drive. In fact, both three-phase reactor and DC-link reactor can filter out high-frequency harmonics, thus the two topologies are commonly used.

In the modeling process of electrolytic capacitorless PMSM drive, the motor driven by inverter is typically regarded as a constant power load. This modeling approach can effectively simplify the drive model while reflecting the resonance characteristics of drive. The state equation for drive is given as

$$\begin{cases} L_g \frac{di_g}{dt} = u_g - R_g i_g - u_{dc} \\ C_{dc} \frac{du_{dc}}{dt} = i_g - i_{inv} \end{cases}. \tag{2.1}$$

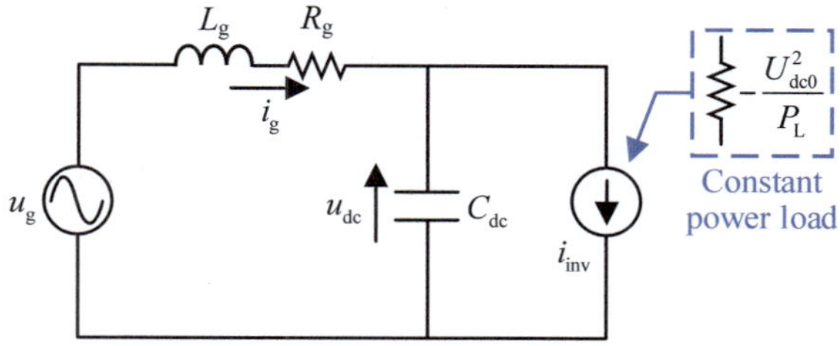

Fig. 2.2 Equivalent model of the drive

where u_g is the rectified voltage, i_{inv} is the inverter input current, and R_g is the DC-link resistance.

Considering the constant power characteristics of the motor, the input current of the inverter can be expressed as

$$i_{inv} = \frac{P_L}{u_{dc}} = \frac{P_L}{U_{dc0} + \Delta u_{dc}}. \tag{2.2}$$

where P_L is the load power, U_{dc0} is the mean value of the DC-link voltage, and Δu_{dc} is the small signal of the DC-link voltage.

Since the fluctuation of the DC-link voltage is relatively small, the input current of the inverter can be linearized as [2]

$$i_{inv} \doteq \frac{P_L}{U_{dc0}} - \frac{P_L}{U_{dc0}^2}\Delta u_{dc}. \tag{2.3}$$

By simplifying the motor as a constant power load and cascading it with an LC filter, the equivalent model of the drive can be obtained as shown in Fig. 2.2 [3]

The transfer function can be expressed as [4]

$$G(s) = \frac{u_{dc}}{u_g} = \frac{U_{dc0}^2}{U_{dc0}^2 L_g C_{dc} s^2 + (U_{dc0}^2 R_g C_{dc} - P_L L_g)s + U_{dc0}^2 - P_L R_g}. \tag{2.4}$$

According to (2.4), the resonance frequency ω_0 of the drive can be derived as

$$\omega_0 = \sqrt{\frac{1}{L_g C_{dc}}\left(1 - \frac{R_g P_L}{U_{dc0}^2}\right)} \doteq \sqrt{\frac{1}{L_g C_{dc}}}. \tag{2.5}$$

According to (2.4), the damping ratio ξ can be expressed as

$$\xi \doteq \frac{1}{2\omega_0}\left(\frac{R_g}{L_g} - \frac{P_L}{C_{dc} U_{dc0}^2}\right). \tag{2.6}$$

Figure 2.3 shows the Bode diagram of (2.4). Due to the relatively low damping of the system, a distinct resonance peak exists at the resonance frequency. The resonance

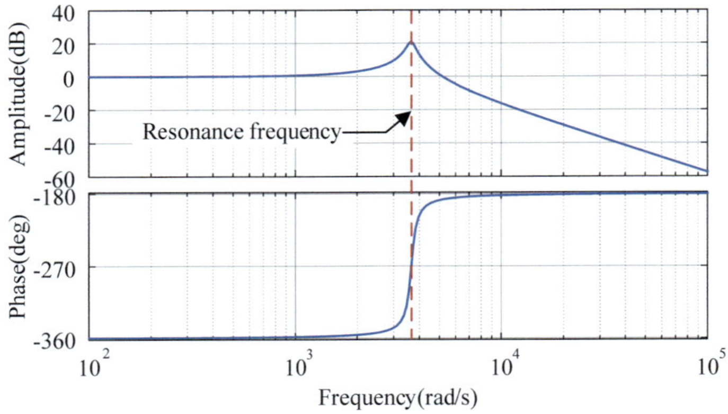

Fig. 2.3 Bode diagram of drive

peak amplifies the harmonics near the resonance frequency, thereby affecting the power quality on the grid side and the stability of the drive.

The drive model based on constant power load can effectively reflect drive characteristics, such as the resonance frequency and the variation tendency of damping, providing theoretical guidance for the subsequent design of virtual damping control method. However, virtual damping control methods typically achieve increased system damping by simulating changes in motor power. Consequently, the system model based on the ideal constant power characteristics cannot achieve accurate damping matching, which limits the effectiveness of damping control. In practical electrolytic capacitorless PMSM drive, the sampling delay of DC-link voltage with fluctuation has a significant impact on the system characteristics. For instance, with the sampling frequency of 6 kHz, the voltage sampling error can reach up to 9% [5]. Therefore, considering the influence of DC-link voltage delay on the motor side characteristics during the modeling process can achieve more effective damping control.

2.3 Model Considering Non-ideal Characteristics of Motor Side

Vector control is a commonly used method for PMSM drive, whose block diagram is shown in Fig. 2.4. The outer speed loop is used to control the motor speed, and the inner current loop is responsible for controlling the torque current i_q and excitation current i_d. The double-closed-loop vector control adopts PI controller, whose parameters are designed based on motor parameters. ω_e is the electrical angular velocity, θ_e is the electrical angle of motor, $i_{a,b,c}$ are the three-phase motor current, $i_{\alpha,\beta}$ are the $\alpha\beta$-axis currents, ω_{eref}, $u_{dref,qref}$, $u_{\alpha ref,\beta ref}$ and $i_{dref,qref}$ are the references of electrical angular velocity, dq-axis voltages, $\alpha\beta$-axis voltages and dq-axis currents, respectively.

Considering the non-ideal characteristics of the motor side, the equivalent model of the electrolytic capacitorless PMSM drive is shown in Fig. 2.5. $\boldsymbol{u}_s$ is the synthesized voltage vector, $\boldsymbol{i}_s$ is the synthesized current vector, R_s is the stator resistance, $\boldsymbol{e}_s$ is the synthesized back electromotive force, d is the duty cycle, $\boldsymbol{u}_{sref}$ and $\boldsymbol{i}_{sref}$ are the vector references of synthesized voltage and synthesized current, respectively. During the modeling process, two primary non-ideal factors of sampling and update delays, and current loop bandwidth are taken into account to enhance the accuracy of the model.

The timing diagram of DC-link voltage sampling and PWM output is shown in Fig. 2.6, where T_s is the switching period (synchronized with the sampling period), u_{dc} is the actual DC-link voltage, u_{dc_s} is the sampled DC-link voltage, and u_{dc_e} is the equivalent sampled DC-link voltage, u_{dc_min} is the minimum value of DC-link voltage [6]. Sampling is performed at kT_s, and the sampled DC-link voltage is maintained for one switching period and then participates in algorithm calculation.

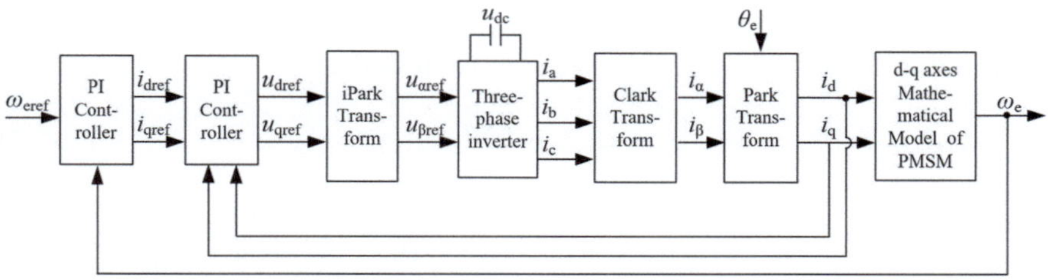

Fig. 2.4 Block diagram of vector control

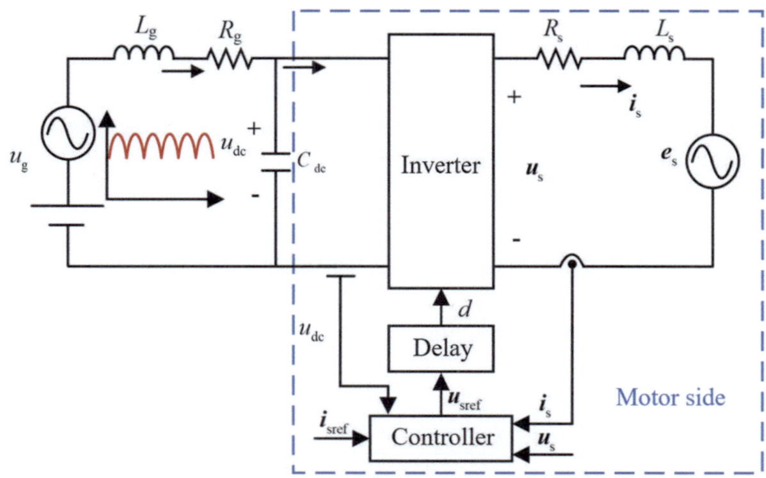

Fig. 2.5 Equivalent model of the electrolytic capacitorless PMSM drive considering motor side characteristics

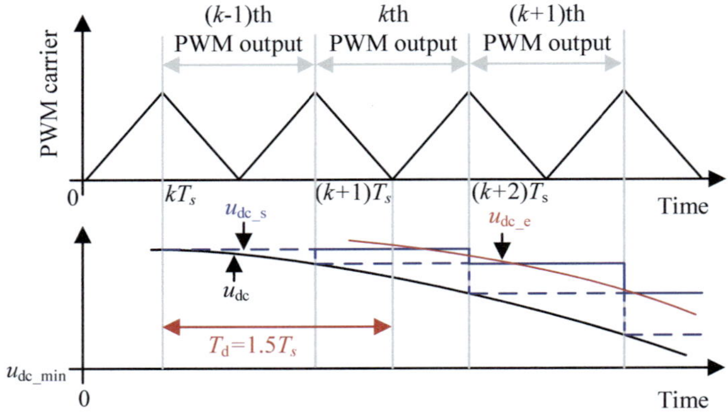

Fig. 2.6 Timing diagram of DC-link voltage sampling and PWM output

The PWM duty cycle signal is output at $(k + 1)T_s$. During the algorithm execution, the equivalent delay of the DC-link voltage is $T_d = 1.5T_s$.

The sampled DC-link voltage u_{dc_s} can be expressed as

$$u_{dc\_s}(s) \approx e^{-0.5sT_s} u_{dc}(s). \tag{2.7}$$

The duty cycle components at $(k + 1)T_s$ can be generated by $\alpha\beta$-axis voltage references and sampled DC-link voltage at kT_s. Thus, the actual $\alpha\beta$-axis voltages $u_{\alpha,\beta}$ at $(k + 1)T_s$ can be expressed as [7]

$$\begin{bmatrix} u_\alpha(k+1) \\ u_\beta(k+1) \end{bmatrix} = u_{dc}(k+1)\begin{bmatrix} D_\alpha(k) \\ D_\beta(k) \end{bmatrix} = \frac{u_{dc}(k+1)}{u_{dc\_s}(k)}\begin{bmatrix} u_{\alpha ref}(k) \\ u_{\beta ref}(k) \end{bmatrix}. \tag{2.8}$$

where $D_{\alpha,\beta}$ is the duty cycle at $\alpha\beta$-axis.

According to (2.8), the relationship between the $\alpha\beta$-axis voltages and the small-signal of DC-link voltage Δu_{dc} can be expressed as

$$\begin{bmatrix} u_\alpha \\ u_\beta \end{bmatrix} = \frac{U_{dc0} + \Delta u_{dc}}{U_{dc0} + \Delta u_{dc} e^{-1.5sT_s}}\begin{bmatrix} e^{-sT_s} u_{\alpha ref} \\ e^{-sT_s} u_{\beta ref} \end{bmatrix}. \tag{2.9}$$

Since the coordinate transformation angle has an equivalent delay of $1.5T_s$, (2.9) can be expressed as

$$\begin{bmatrix} \cos\theta & -\sin\theta_e \\ \sin\theta_e & \cos\theta_e \end{bmatrix}\begin{bmatrix} u_d \\ u_q \end{bmatrix} = \frac{U_{dc0} + \Delta u_{dc}}{U_{dc0} + \Delta u_{dc} e^{-1.5sT_s}}$$

$$\begin{bmatrix} \cos(\theta_e - 1.5\omega_e T_s) & -\sin(\theta_e - 1.5\omega_e T_s) \\ \sin(\theta_e - 1.5\omega_e T_s) & \cos(\theta_e - 1.5\omega_e T_s) \end{bmatrix}\begin{bmatrix} U_{dref0} + e^{-sT_s}\Delta u_{dref} \\ U_{qref0} + e^{-sT_s}\Delta u_{qref} \end{bmatrix}. \tag{2.10}$$

From (2.10), the actual *dq*-axis voltages $u_{d,q}$ can be expressed as

$$\begin{bmatrix} u_d \\ u_q \end{bmatrix} = \frac{U_{dc0} + \Delta u_{dc}}{U_{dc0} + \Delta u_{dc} e^{-1.5sT_s}} C_0 \begin{bmatrix} U_{dref0} + e^{-sT_s} \Delta u_{dref} \\ U_{qref0} + e^{-sT_s} \Delta u_{qref} \end{bmatrix} \tag{2.11}$$

where,

$$C_0 = \begin{bmatrix} \cos 1.5\omega_e T_s & \sin 1.5\omega_e T_s \\ -\sin 1.5\omega_e T_s & \cos 1.5\omega_e T_s \end{bmatrix}. \tag{2.12}$$

where $u_{d,q}$ are the *dq*-axis voltages, $U_{dref0,qref0}$ are the mean values of the *dq*-axis voltages, and $\Delta u_{dref,qref}$ are the small signals of the *dq*-axis voltage references.

Therefore, the actual *dq*-axis voltages at $(k + 1)T_s$ can be obtained as

$$\begin{bmatrix} u_d[k+1] \\ u_q[k+1] \end{bmatrix} = \frac{u_{dc}[k+1]}{u_{dc}[k-0.5]} C_0 \begin{bmatrix} u_{dref}[k] \\ u_{qref}[k] \end{bmatrix}. \tag{2.13}$$

Combining (2.11), the small signal of *dq*-axis voltages $\Delta u_{d,q}$ can be expressed as

$$\begin{bmatrix} \Delta u_d(s) \\ \Delta u_q(s) \end{bmatrix} = C_0 \left\{ e^{-sT_s} \begin{bmatrix} \Delta u_{dref}(s) \\ \Delta u_{qref}(s) \end{bmatrix} + \frac{(1 - e^{-1.5sT_s}) \Delta u_{dc}(s)}{U_{dc0}} \begin{bmatrix} U_{dref0} \\ U_{qref0} \end{bmatrix} \right\}. \tag{2.14}$$

The mathematical model of the motor can be expressed as

$$\begin{bmatrix} u_d \\ u_q \end{bmatrix} = \begin{bmatrix} R_s + L_d s & -\omega_e L_q \\ \omega_e L_d & R_s + L_q s \end{bmatrix} \begin{bmatrix} i_d \\ i_q \end{bmatrix} + \begin{bmatrix} 0 \\ \omega_e \psi_f \end{bmatrix}. \tag{2.15}$$

where $L_{d,q}$ are the *dq*-axis motor inductances.

Therefore, the small signal of the motor voltage can be expressed as

$$\begin{bmatrix} \Delta u_d(s) \\ \Delta u_q(s) \end{bmatrix} = \begin{bmatrix} R_s + L_d s & -\omega_e L_q \\ \omega_e L_d & R_s + L_q s \end{bmatrix} \begin{bmatrix} \Delta i_d(s) \\ \Delta i_q(s) \end{bmatrix}. \tag{2.16}$$

where $\Delta i_{d,q}$ are the *dq*-axis currents.

Considering the factor of current sampling, the motor voltage references can be expressed as

$$\begin{bmatrix} u_{dref} \\ u_{qref} \end{bmatrix} = \begin{bmatrix} -\boldsymbol{G}_d & 0 \\ 0 & -\boldsymbol{G}_q \end{bmatrix} \begin{bmatrix} i_d e^{-0.5sT_s} \\ i_q e^{-0.5sT_s} \end{bmatrix} + \begin{bmatrix} \boldsymbol{G}_d i_{dref} \\ \boldsymbol{G}_q i_{qref} \end{bmatrix} \tag{2.17}$$

$$\boldsymbol{G}_d = \omega_{cb} L_d + \frac{\omega_{cb} R_s}{s}, \boldsymbol{G}_q = \omega_{cb} L_q + \frac{\omega_{cb} R_s}{s}. \tag{2.18}$$

where ω_{cb} is the bandwidth of current loop.

According to (2.17), the small signals of *dq*-axis voltages can be derived as

$$\begin{bmatrix} \Delta u_{dref}(s) \\ \Delta u_{qref}(s) \end{bmatrix} = \begin{bmatrix} -\boldsymbol{G}_d & 0 \\ 0 & -\boldsymbol{G}_q \end{bmatrix} \begin{bmatrix} \Delta i_d(s) \\ \Delta i_q(s) \end{bmatrix} e^{-0.5sT_s}. \tag{2.19}$$

Combining (2.14), (2.16) and (2.19), the relationship between motor currents and DC-link voltage can be derived as

$$\left\{ C_0^{-1} \begin{bmatrix} R_s + L_d s & -\omega_e L_q \\ \omega_e L_d & R_s + L_q s \end{bmatrix} - e^{-1.5sT_s} \begin{bmatrix} -\boldsymbol{G}_d & 0 \\ 0 & -\boldsymbol{G}_q \end{bmatrix} \right\} \begin{bmatrix} \Delta i_d(s) \\ \Delta i_q(s) \end{bmatrix} = \frac{\left(1 - e^{-1.5sT_s}\right)}{U_{dc0}} \begin{bmatrix} U_{dref0} \\ U_{qref0} \end{bmatrix} \Delta u_{dc}(s). \tag{2.20}$$

(2.20) can be simplified as

$$\begin{bmatrix} A_{1d}(s) & A_{1q}(s) \\ A_{2d}(s) & A_{2q}(s) \end{bmatrix} \begin{bmatrix} \Delta i_d(s) \\ \Delta i_q(s) \end{bmatrix} = \frac{\left(1 - e^{-1.5sT_s}\right)}{U_{dc0}} \begin{bmatrix} U_{dref0} \\ U_{qref0} \end{bmatrix} \Delta u_{dc}(s). \tag{2.21}$$

where,

$$\begin{bmatrix} A_{1d}(s) & A_{1q}(s) \\ A_{2d}(s) & A_{2q}(s) \end{bmatrix} = C_0^{-1} \begin{bmatrix} R_s + L_d s & -\omega_e L_q \\ \omega_e L_d & R_s + L_q s \end{bmatrix} - e^{-1.5sT_s} \begin{bmatrix} -\boldsymbol{G}_d & 0 \\ 0 & -\boldsymbol{G}_q \end{bmatrix}. \tag{2.22}$$

Further, the transfer function can be simplified as

$$\begin{aligned} \begin{bmatrix} W_d(s) \\ W_q(s) \end{bmatrix} &= \frac{1}{\Delta u_{dc}(s)} \begin{bmatrix} \Delta i_d(s) \\ \Delta i_q(s) \end{bmatrix} \\ &= \frac{1 - e^{-1.5sT_s}}{U_{dc0}\left[A_{2q}(s)A_{1d}(s) - A_{2d}(s)A_{1q}(s)\right]} \begin{bmatrix} U_{dref0}A_{2q}(s) - U_{qref0}A_{1q}(s) \\ U_{qref0}A_{1d}(s) - U_{dref0}A_{2d}(s) \end{bmatrix}. \end{aligned} \tag{2.23}$$

By substituting (2.16) into (2.23), the relationship between motor voltages and DC-link voltage can be obtained as

$$\frac{1}{\Delta u_{dc}(s)} \begin{bmatrix} \Delta u_d(s) \\ \Delta u_q(s) \end{bmatrix} = \begin{bmatrix} R_s + L_d s & -\omega_e L_q \\ \omega_e L_d & R_s + L_q s \end{bmatrix} \begin{bmatrix} W_d(s) \\ W_q(s) \end{bmatrix}. \tag{2.24}$$

Neglecting inverter losses, the energy relationship between the DC-link side and the motor side can be expressed as

$$u_{\mathrm{dc}} i_{\mathrm{inv}} = \frac{3}{2}\left(u_{\mathrm{d}} i_{\mathrm{d}} + u_{\mathrm{q}} i_{\mathrm{q}}\right). \tag{2.25}$$

The small signal model of (2.25) can be expressed as

$$\Delta u_{\mathrm{dc}} I_{\mathrm{inv0}} + U_{\mathrm{dc0}} \Delta i_{\mathrm{inv}} = \frac{3}{2}\left(\Delta u_{\mathrm{d}} I_{\mathrm{d0}} + U_{\mathrm{d0}} \Delta i_{\mathrm{d}} + \Delta u_{\mathrm{q}} I_{\mathrm{q0}} + U_{\mathrm{q0}} \Delta i_{\mathrm{q}}\right). \tag{2.26}$$

where Δi_{inv} is the small signal of inverter input current, I_{inv0}, $I_{\mathrm{d0,q0}}$ and $U_{\mathrm{d0,q0}}$ are the mean values of inverter input current, *dq*-axis currents and *dq*-axis voltages.

The motor side admittance Y_{m} can be expressed as the ratio of inverter input current to DC-link voltage

$$\begin{aligned} Y_{\mathrm{m}}(s) = &-\frac{I_{\mathrm{inv0}}}{U_{\mathrm{dc0}}} + \frac{3}{2}\left(\frac{\Delta u_{\mathrm{d}}(s)}{\Delta u_{\mathrm{dc}}(s)} \frac{I_{\mathrm{d0}}}{U_{\mathrm{dc0}}} + \frac{U_{\mathrm{d0}}}{U_{\mathrm{dc0}}} \frac{\Delta i_{\mathrm{d}}(s)}{\Delta u_{\mathrm{dc}}(s)}\right. \\ &\left. + \frac{\Delta u_{\mathrm{q}}(s)}{\Delta u_{\mathrm{dc}}(s)} \frac{I_{\mathrm{q0}}}{U_{\mathrm{dc0}}} + \frac{U_{\mathrm{q0}}}{U_{\mathrm{dc0}}} \frac{\Delta i_{\mathrm{q}}(s)}{\Delta u_{\mathrm{dc}}(s)}\right). \end{aligned} \tag{2.27}$$

Substituting (2.23) and (2.24) into (2.27), the motor admittance can be obtained as

$$\begin{aligned} Y_{\mathrm{m}}(s) = \frac{\Delta i_{\mathrm{inv}}(s)}{\Delta u_{\mathrm{dc}}(s)} = &-\frac{I_{\mathrm{inv0}}}{U_{\mathrm{dc0}}} + \frac{3}{2}\frac{W_{\mathrm{d}}(s)(U_{\mathrm{d0}} + \omega_{\mathrm{e}} L_{\mathrm{d}} I_{\mathrm{q0}} + L_{\mathrm{d}} I_{\mathrm{d0}} s + I_{\mathrm{d0}} R_{\mathrm{s}})}{U_{\mathrm{dc0}}} \\ &+ \frac{3}{2}\frac{W_{\mathrm{q}}(s)(U_{\mathrm{q0}} - \omega_{\mathrm{e}} L_{\mathrm{q}} I_{\mathrm{d0}} + L_{\mathrm{q}} I_{\mathrm{q0}} s + I_{\mathrm{q0}} R_{\mathrm{s}})}{U_{\mathrm{dc0}}}. \end{aligned} \tag{2.28}$$

Considering the non-ideal characteristics of the motor side, the transfer function of the drive can be further derived as

$$G(s) = \frac{1}{\left(L_{\mathrm{g}} s + R_{\mathrm{g}}\right)(C_{\mathrm{dc}} s + Y_{\mathrm{m}}) + 1}. \tag{2.29}$$

Taking the parameters shown in Table 2.1 as an example, with the motor speed set to 55 Hz and the load torque set to 15 N m, the Bode diagrams of $W_{\mathrm{d}}(s)$ and $W_{\mathrm{q}}(s)$ are shown in Fig. 2.7. It can be seen that the theoretical results of the impedance model are approximately consistent with the simulation results.

Figure 2.8 shows the Bode diagram of the motor admittance $Y_{\mathrm{m}}(s)$. The impact of the DC-link voltage sampling delay on $Y_{\mathrm{m}}(s)$ is illustrated in Fig. 2.8a, where the current loop bandwidth is set to 400 Hz. It can be seen that when the equivalent DC-link voltage sampling delay is zero, the phase of the $Y_{\mathrm{m}}(s)$ remains at 180°, indicating an ideal negative impedance characteristic on the motor side. However, as the equivalent DC-link sampling delay increases, the phase of the $Y_{\mathrm{m}}(s)$ deviates from 180°. Figure 2.8b shows the impact of the current loop bandwidth on $Y_{\mathrm{m}}(s)$. It can be seen that the higher the current loop width, the closer the motor side approaches constant power characteristics in the low and mid frequency domain. In

Table 2.1 Main parameters of simulation

Parameters	Value
Switching frequency	8 kHz
DC-link capacitance	30 μF
DC-link inductance	2.5 mH
Rated speed Rated current	471 rad/s 26 A
Stator resistance	0.265 Ω
D-axis inductance	7.5 mH
Q-axis inductance	17.5 mH
Flux linkage of rotor	0.5 Wb
Number of pole pairs	3

actual systems, due to the limitations of current loop bandwidth and the presence of DC-link voltage sampling delay, the motor side deviates from the ideal constant power characteristics.

2.4 Model Considering Flux-Weakning Control

Voltage closed-loop flux-weakening control is widely used. When the motor voltage vector is higher than the DC-link voltage, the speed can be increased by reducing the *d*-axis current, as shown in Fig. 2.9. i_{dref} can be expressed as

$$i_{\text{dref}} = \left(k_{\text{p}} + \frac{k_{\text{i}}}{s}\right)(K_{\text{dc}} u_{\text{dc\_s}} - |\mathbf{u}_{\text{sref}}|). \tag{2.30}$$

where K_{dc} is the voltage coefficient in flux-weakening control.

Therefore, Δi_{dref} can be expressed as

$$\Delta i_{\text{dref}} = \left(k_{\text{p}} + \frac{k_{\text{i}}}{s}\right)[K_{\text{dc}} e^{-0.5sT_{\text{s}}} \Delta u_{\text{dc}} - (U_{\text{dref0}} \Delta u_{\text{dref}} + U_{\text{qref0}} \Delta u_{\text{qref}})/U_{\text{sref0}}]. \tag{2.31}$$

where U_{sref0} is the mean value of synthesized voltage, k_{p} and k_{i} are the proportional coefficient and integral coefficient of voltage controller, respectively.

According to (2.31), the flux-weakening control introduces extra harmonics in the *d*-axis current reference, which can not be ignored in vector control. Therefore, the small signal of *dq*-axis voltages can be expressed as

$$\begin{bmatrix} \Delta u_{\text{dref}} \\ \Delta u_{\text{qref}} \end{bmatrix} = e^{-0.5sT_{\text{s}}} \begin{bmatrix} -\boldsymbol{G}_{\text{d}} & 0 \\ 0 & -\boldsymbol{G}_{\text{q}} \end{bmatrix} \begin{bmatrix} \Delta i_{\text{d}} \\ \Delta i_{\text{q}} \end{bmatrix} + \begin{bmatrix} \boldsymbol{G}_{\text{d}} \Delta i_{\text{dref}} \\ 0 \end{bmatrix}. \tag{2.32}$$

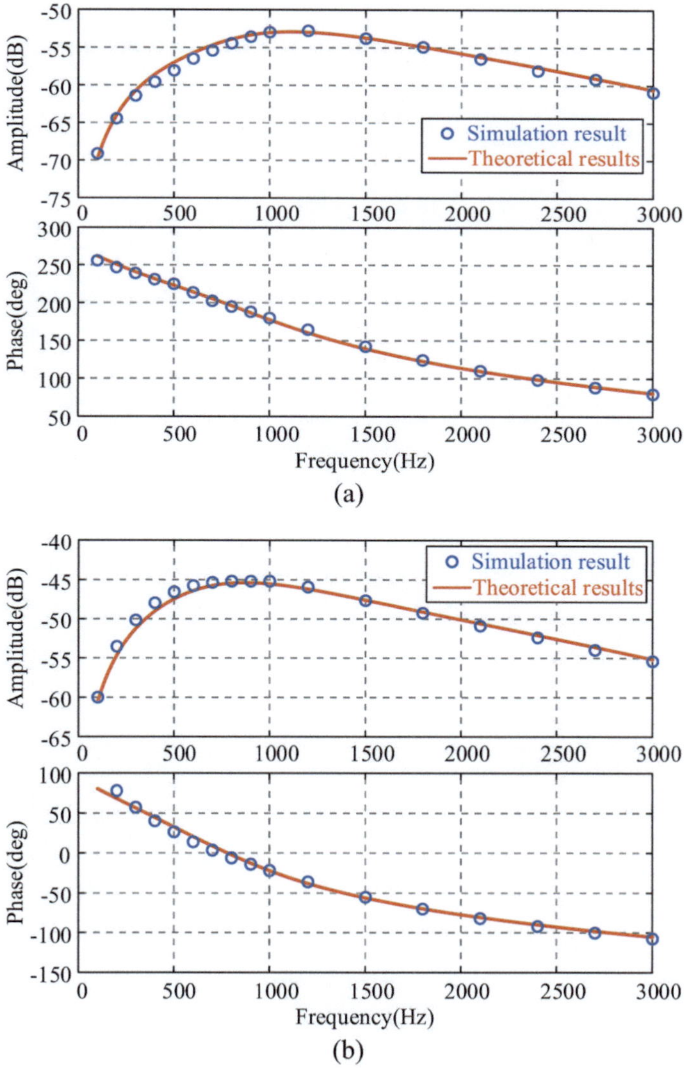

Fig. 2.7 Bode diagrams of $W_{\mathrm{d}}(s)$ and $W_{\mathrm{q}}(s)$. **a** $W_{\mathrm{d}}(s)$. **b** $W_{\mathrm{q}}(s)$

According to (2.31) and (2.32), the small signal of the d-axis current reference can be expressed as

$$\begin{cases} \boldsymbol{G}_{\mathrm{d}}\Delta i_{\mathrm{dref}} = \boldsymbol{G}_{\mathrm{dc}}e^{-0.5sT_{\mathrm{s}}}\left(K_{\mathrm{dc}}U_{\mathrm{sref0}}\Delta u_{\mathrm{dc}} + U_{\mathrm{dref0}}\boldsymbol{G}_{\mathrm{d}}\Delta i_{\mathrm{d}} + U_{\mathrm{qref0}}\boldsymbol{G}_{\mathrm{q}}\Delta i_{\mathrm{q}}\right) \\ \boldsymbol{G}_{\mathrm{dc}} = \boldsymbol{G}_{\mathrm{d}}\left(k_{\mathrm{p}} + \frac{k_{\mathrm{i}}}{s}\right)/(U_{\mathrm{sref0}} + U_{\mathrm{dref0}}\boldsymbol{G}_{\mathrm{d}}\boldsymbol{G}_{\mathrm{f}}) \end{cases}. \quad (2.33)$$

where G_{f} is the voltage controller.

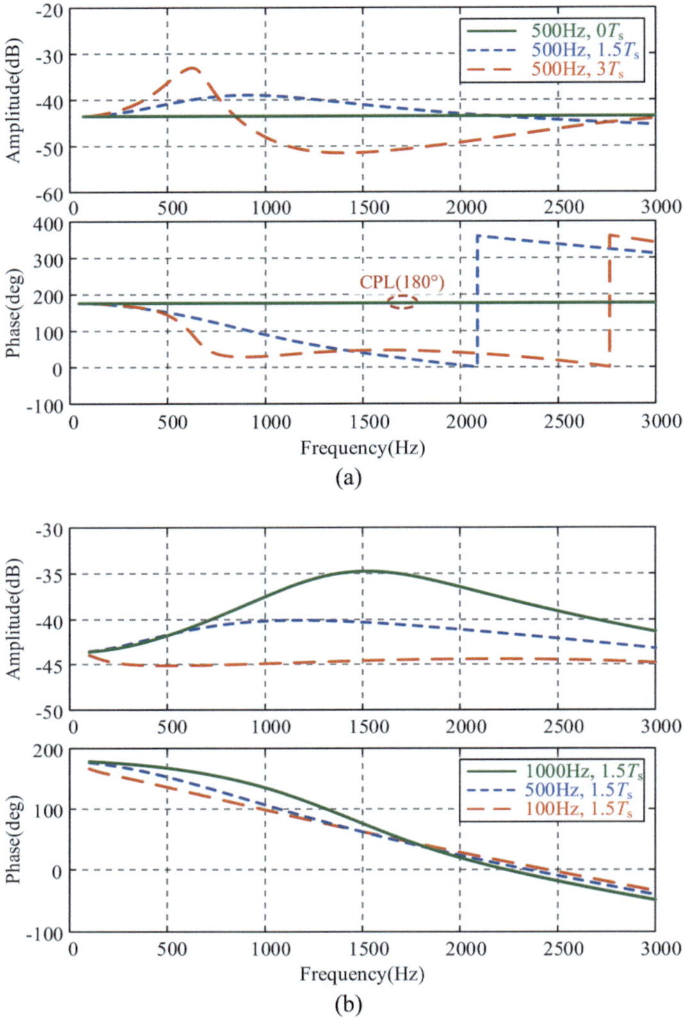

Fig. 2.8 Bode diagram of motor admittance $Y_{\mathrm{m}}(s)$. **a** Impact of the DC-link voltage sampling delay on $Y_{\mathrm{m}}(s)$. **b** Impact of the current loop bandwidth on $Y_{\mathrm{m}}(s)$

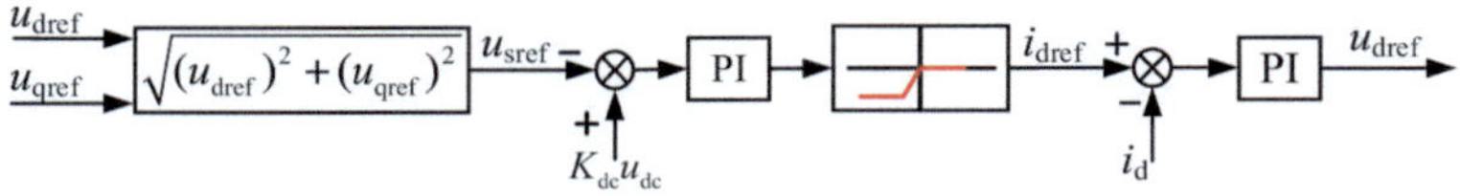

Fig. 2.9 Bode diagram of voltage closed-loop flux-weakening control

According to (2.14), (2.16), (2.32) and (2.33), the relationship between DC-link voltage and motor current can be expressed as

$$\begin{bmatrix} \mathbf{A}_{1d} & \mathbf{A}_{1q} \\ \mathbf{A}_{2d} & \mathbf{A}_{2q} \end{bmatrix} \begin{bmatrix} \Delta i_d \\ \Delta i_q \end{bmatrix} = \left\{ \frac{1 - e^{-1.5sT_s}}{U_{dc,0}} \begin{bmatrix} U_{dref0} \\ U_{qref0} \end{bmatrix} + \begin{bmatrix} K_{dc} U_{sref0} e^{-1.5sT_s} \boldsymbol{G}_{dc} \\ 0 \end{bmatrix} \right\} \Delta u_{dc} \tag{2.34}$$

where

$$\begin{bmatrix} \mathbf{A}_{1d} & \mathbf{A}_{1q} \\ \mathbf{A}_{2d} & \mathbf{A}_{2q} \end{bmatrix} = \mathbf{C}_0^{-1} \begin{bmatrix} R_s + L_d s & -\omega_e L_q \\ \omega_e L_d & R_s + L_q s \end{bmatrix} + e^{-1.5sT_s} \left\{ \begin{matrix} \begin{bmatrix} \boldsymbol{G}_d & 0 \\ 0 & \boldsymbol{G}_q \end{bmatrix} \\ -\boldsymbol{G}_{dc} \begin{bmatrix} U_{dref0} \boldsymbol{G}_d & U_{qref0} \boldsymbol{G}_q \\ 0 & 0 \end{bmatrix} \end{matrix} \right\}. \tag{2.35}$$

Therefore, the transfer function from DC-link voltage to motor current can be expressed as

$$\begin{aligned} \begin{bmatrix} W_d(s) \\ W_q(s) \end{bmatrix} = \frac{1}{\Delta u_{dc}(s)} \begin{bmatrix} \Delta i_d(s) \\ \Delta i_q(s) \end{bmatrix} &= \frac{1}{A_{2q}(s) A_{1d}(s) - A_{2d}(s) A_{1q}(s)} \left\{ K_{dc} U_{sref0} e^{-1.5sT_s} \right. \\ &\cdot \left. \begin{bmatrix} G_{dc}(s) A_{2q}(s) \\ -G_{dc}(s) A_{2d}(s) \end{bmatrix} + \frac{1 - e^{-1.5sT_s}}{U_{dc,0}} \begin{bmatrix} U_{dref0} A_{2q}(s) - U_{qref0} A_{1q}(s) \\ U_{qref0} A_{1d}(s) - U_{dref0} A_{2d}(s) \end{bmatrix} \right\}. \end{aligned} \tag{2.36}$$

Accordingly, if (2.16) is substituted into (2.36), the relationship between motor voltage and DC-link voltage can be obtained as

$$\frac{1}{\Delta u_{dc}(s)} \begin{bmatrix} \Delta u_d(s) \\ \Delta u_q(s) \end{bmatrix} = \begin{bmatrix} R_s + L_d s & -\omega_e L_q \\ \omega_e L_d & R_s + L_q s \end{bmatrix} \begin{bmatrix} W_{id}(s) \\ W_{iq}(s) \end{bmatrix}. \tag{2.37}$$

Substituting (2.36) and (2.37) into (2.27), the motor admittance can be derived as

$$\begin{aligned} Y_m(s) = \frac{\Delta i_{inv}}{\Delta u_{dc}} &= -\frac{I_{inv0}}{U_{dc0}} + \frac{3}{2} \frac{W_d(s)(U_{d0} + \omega_e L_d I_{q0} + L_d I_{d0} s + I_{d0} R_s)}{U_{dc0}} \\ &+ \frac{3}{2} \frac{W_q(s)(U_{q0} - \omega_e L_q I_{d0} + L_q I_{q0} s + I_{q0} R_s)}{U_{dc0}}. \end{aligned} \tag{2.38}$$

Figure 2.10 shows the Bode diagram of $Y_m(s)$ considering the weak magnetic condition. It can be seen that the theoretical results of the impedance model are close to the simulation results.

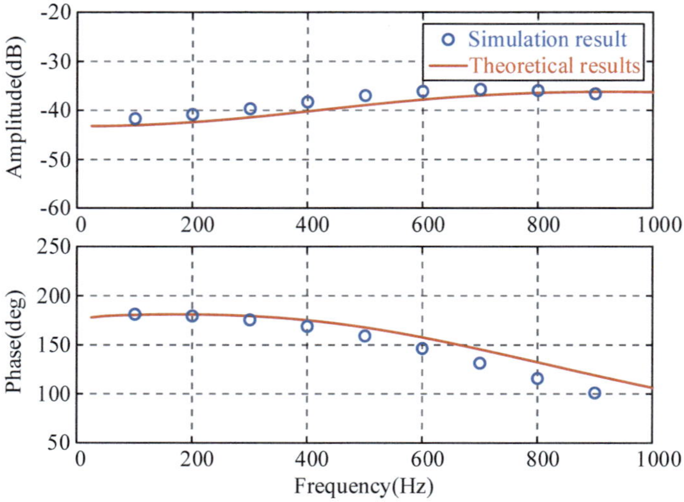

Fig. 2.10 Bode diagram of $Y_{\mathrm{m}}(s)$ considering the weak magnetic condition

2.5 Summary

Accurate drive model is beneficial for characterizing the resonance characteristics and the energy flow characteristics of the drive, and providing theoretical guidance for the design of control algorithm. The main work and conclusions are as follows:

(1) The model of drive based on the ideal constant power characteristics has been established, which can effectively simplify the drive model while reflecting the resonance characteristics of drive. However, the accuracy of the model is limited, which affects the control performance.

(2) The impedance model of drive during whole speed region has been established, which reveals the influence of DC-link voltage sampling delay and current loop width on motor side characteristics.

References

1. D. Ding, G. Zhang, G. Wang, Z. Ren, N. Mijatovic, T. Dragičević, D. Xu. Impedance Reshaping for Inherent Harmonics in PMSM Drives With Small DC-Link Capacitor [J]. IEEE Transactions on Power Electronics, 2022, 37(12): 14265–14279.
2. N. Zhao, G. Wang, D. Ding, G. Zhang, D. Xu. Impedance Based Stabilization Control Method for Reduced DC-Link Capacitance IPMSM Drives [J]. IEEE Transactions on Power Electronics, 2019, 34(10): 9879–9890.
3. W. J. Lee and S. K. Sul. DC-Link voltage stabilization for reduced DC-Link capacitor inverter [J]. IEEE Transaction on Industry Application, 2014, 50(1): 404–414.

4. K. Pietilainen, L. Harnefors, A. Petersson, H. P. Nee. DC-Link Stabilization and Voltage Sag Ride-Through of Inverter Drives [J]. IEEE Transactions on Industrial Electronics, 2006, 53(4): 1261–1268.
5. L. Mathe, H. Rosendal, R. Lazar, and M. Ciobotaru. DC-link compensation method for slim DC link drives fed by soft grid [C]. in IEEE International Symposium on Industrial Electronics, 2010: 1236–1241.
6. B. H. Bae and S. K. Sul. A compensation method for time delay of full-digital synchronous frame current regulator of PWM AC drives [J]. IEEE Transaction on Industry Application, 2003, 39(3): 802–810.
7. L. Mathe, L. Torok, D. Wang and D. Sera [J]. Resonance reduction for AC drives with small capacitance in the DC link. IEEE Transaction on Industry Application, 2017, 53(4): 3814–3820.

Chapter 3
Power Quality Improvement Methods on the Grid Side

3.1 Introduction

Due to the reduced DC-link capacitance, the drive faces a severe LC resonance issue, resulting in the deterioration of grid-side power quality, which fails to meet the IEC61000-3-12 standard. This presents both a research hotspot and a major challenge in the development of electrolytic capacitorless PMSM drive. Based on the impedance model of the electrolytic capacitorless PMSM drive established in the previous chapter, this chapter improves grid-side current quality from two perspectives: virtual admittance and inductor current control. In addition, a preliminary investigation into resonance suppression methods for electrolytic capacitorless PMSM drive with low DC-link inductance is conducted.

3.2 Virtual Admittance Reshaping-Based LC Resonance Suppression Strategy

3.2.1 Stability Control Mechanism of Parallel Virtual Resistors and Capacitors

Multiple ways of parallelling virtual resistors R_{damp} [1, 2] or capacitors C_{damp} [3, 4] can effectively suppress the LC resonance of the system. As shown in Fig. 3.1, this study takes DC-link capacitors in parallel with resistors and capacitors as examples for analysis, respectively. Figure 3.2a shows the frequency characteristics of the system when the DC-link capacitor is connected in parallel with virtual resistors with different values. It can be seen that the resonance peak can be effectively suppressed after the parallel connection with virtual resistor. And the suppression effect will be better with the increase of the parallel resistor value. Figure 3.2b shows the system

D. Ding et al., *Electrolytic Capacitorless Permanent Magnet Synchronous Motor Drives*, https://doi.org/10.1007/978-981-95-5601-4_3

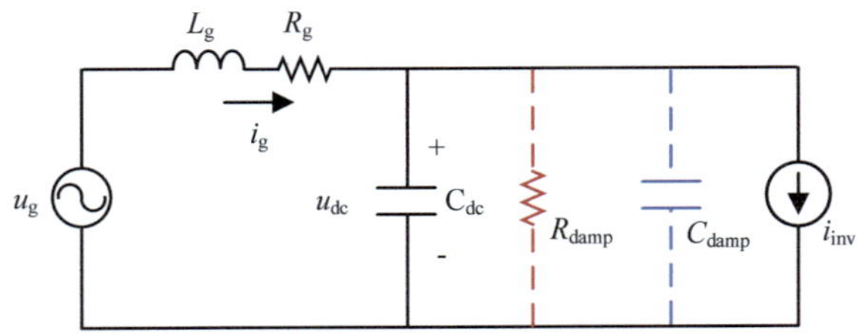

Fig. 3.1 Equivalent model of active damping control strategy, reprinted from Ref. [5], copyright 2022, with permission from Nature

frequency characteristics when the DC-link capacitor is connected in parallel with capacitors with different values. With the increase of the DC-link capacitor, the resonant frequency shifts forward, and the high-frequency harmonics of the system can be effectively suppressed.

The damping current i_{damp} in paralleling virtual resistor and capacitor can be expressed as

$$i_{\text{damp}} = \begin{cases} \frac{1}{R_{\text{damp}}} u_{\text{dc}}, \text{ parallel } R_{\text{damp}} \\ C_{\text{damp}} \frac{du_{\text{dc}}}{dt}, \text{ parallel } C_{\text{damp}} \end{cases}. \tag{3.1}$$

The corresponding block diagram is shown in Fig. 3.3. The damping current is generated by the DC-link voltage without adding additional current and voltage sensors. The active damping control strategy can increase the system damping and effectively suppress the LC resonance peak, but it will also produce a large power disturbance on the motor side.

The output voltage of the rectifier is composed of the dc component and the harmonic components with the integer multiples of six times the grid voltage frequency. These harmonic components are called inherent harmonics. The inherent harmonics near the LC resonant frequency will be amplified. For example, in the experimental platform, the DC inductance is 2.5 mH, the capacitance is 30 μF, the grid frequency is 50 Hz, and thus the resonance frequency is 580 Hz. Figure 3.4 shows the simulation results of DC-link voltage response using parameters in Table 2.1. It can be seen that the inherent harmonics of the system near the resonant frequency are 300 and 600 Hz. By comparing Fig. 3.4a, c, it can be seen that compared with the rectifier voltage, the inherent harmonic amplitude near the resonant frequency in the DC-link voltage is significantly increased.

3.2.2 Impedance Optimization Design Theory of Drive System

In practical application, for the three-phase power supply electrolytic capacitorless motor drive, only the inherent harmonics amplified at the LC resonant frequency need to be suppressed. Different from the active damping method shown in Fig. 3.3,

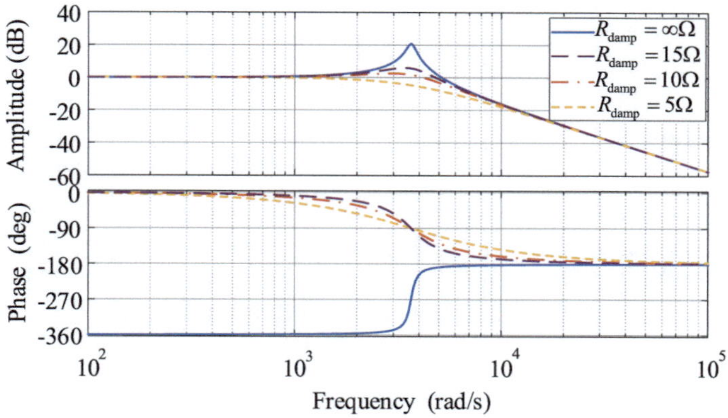

(a)

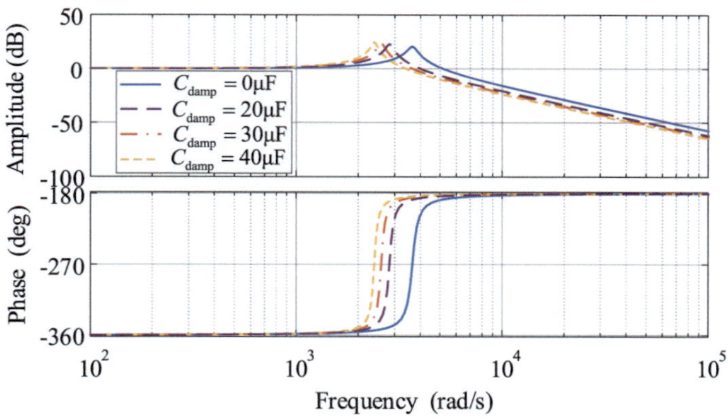

(b)

Fig. 3.2 Frequency characteristics when different virtual resistors and capacitors are paralleled. **a** Paralleling virtual resistor. **b** Paralleling virtual capacitor

the impedance shaping is only carried out for the drive at the inherent harmonic frequency to reduce the power disturbance on the motor side. The equivalent model of the impedance reshaping scheme is shown in Fig. 3.5a. The transfer function at the inherent harmonic frequency can be expressed as

$$G_{\text{damp\_n}}(s) = \frac{u_{\text{dc\_n}}}{u_{\text{g\_n}}} = \frac{1}{(L_g s + R_g)(C_{\text{dc}} s + Y_{\text{m}} + Y_{\text{d\_n}}) + 1}. \tag{3.2}$$

where $Y_{\text{d\_n}}$ is the virtual admittance at the inherent harmonic frequency, $u_{\text{dc\_n}}$ is the inherent DC-link voltage harmonic, $u_{\text{g\_n}}$ is the inherent rectified voltage harmonic.

Thus, the damping current at inherent harmonic frequency $i_{\text{damp\_n}}$ can be expressed as

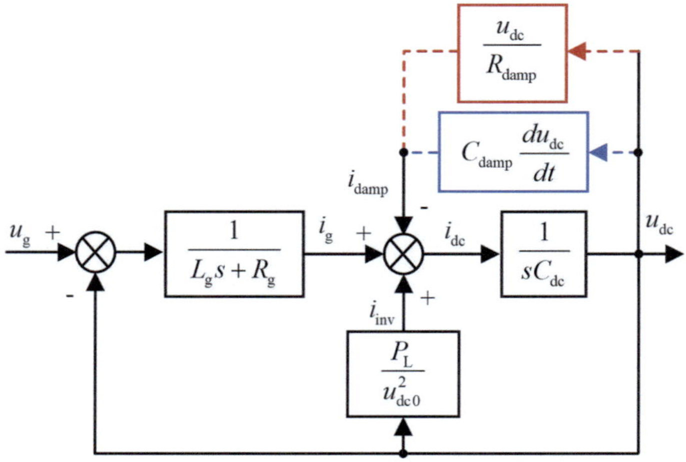

Fig. 3.3 Block diagram of active damping control strategy, reprinted from Ref. [5], copyright 2022, with permission from Nature

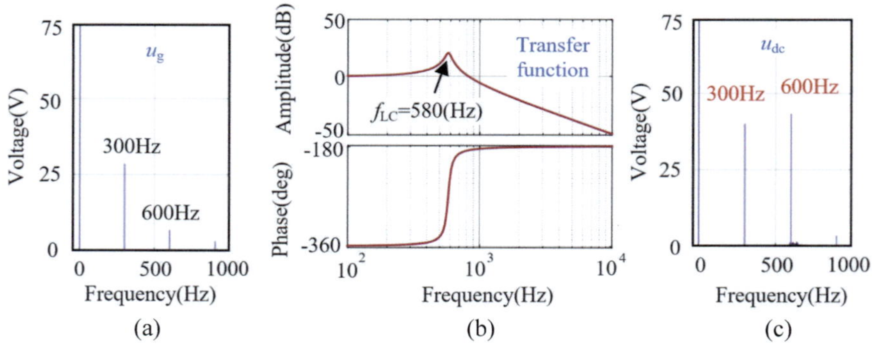

Fig. 3.4 Simulation results of the DC-link voltage. **a** Inherent harmonics in rectified voltage. **b** Bode diagram of the system. **c** Inherent harmonics in DC-link voltage

$$i_{\mathrm{damp\_n}} = u_{\mathrm{dc\_n}} Y_{\mathrm{d\_n}}. \tag{3.3}$$

Virtual admittance can be expressed as

$$Y_{\mathrm{d\_n}} = \left|Y_{\mathrm{d\_n}}\right| \cos\theta_{\mathrm{d\_n}} + j\left|Y_{\mathrm{d\_n}}\right| \sin\theta_{\mathrm{d\_n}}. \tag{3.4}$$

where $\theta_{\mathrm{d\_n}}$ is the virtual admittance angle at inherent harmonic frequency.

Figure 3.5a shows the equivalent model of electrolytic capacitorless PMSM drive based on impedance reshaping, and the virtual admittance is parallel with the DC-link capacitor. The control block diagram is shown in Fig. 3.5b. The damping current is generated by extracting harmonics at the inherent harmonic frequency of the DC-link

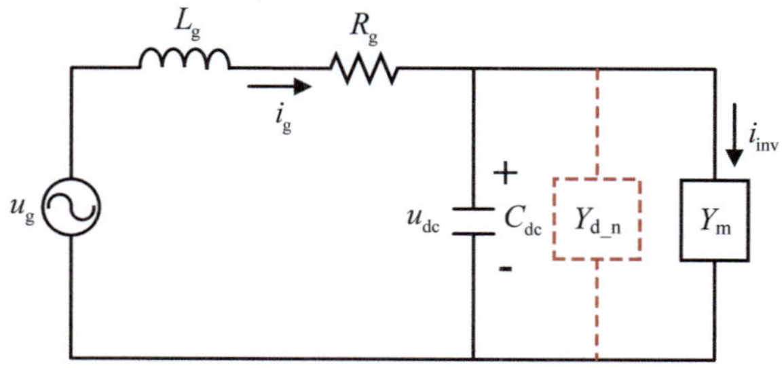

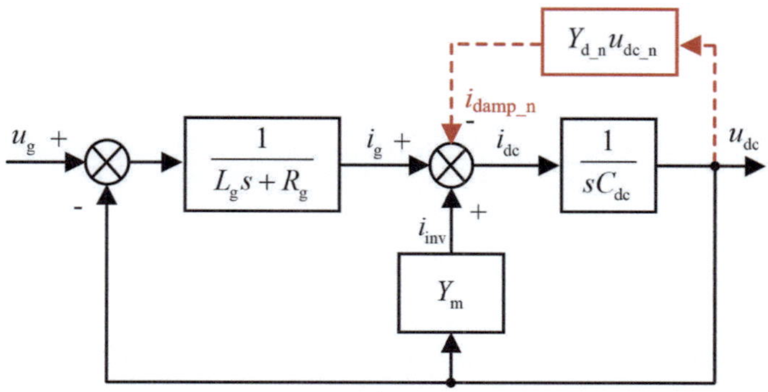

Fig. 3.5 The proposed impedance remodeling strategy. **a** Equivalent model. **b** Control block diagram

voltage near LC resonance frequency, in which the virtual admittance design at each natural harmonic is very critical.

The control block diagram of the electrolytic capacitorless PMSM based on virtual admittance reshaping is shown in Fig. 3.6. $\hat{\omega}_e$ is the estimated electrical angular velocity, $\hat{\theta}_e$ is the estimated electrical angle of motor, $u_{\alpha\_damp},_{\beta\_damp}$ are the damping voltages at $\alpha\beta$-axis, u_{damp_n} is the damping voltage at inherent harmonic frequency, and P_{damp_n} is the damping power at inherent harmonic frequency. The impedance reshaping scheme is integrated into the control architecture of the sensorless PMSM drive [6]. The inherent harmonics of DC-link voltage at LC resonant frequency are obtained by BPF to generate damping current and damping power. Then, according to the power balance principle between the inverter and the motor, the damping voltage is generated on the motor side, where the damping voltage is in the same direction as the motor current vector, so as to reduce the influence of the damping voltage on the motor current control loop. The next section will focus on the construction of virtual admittance.

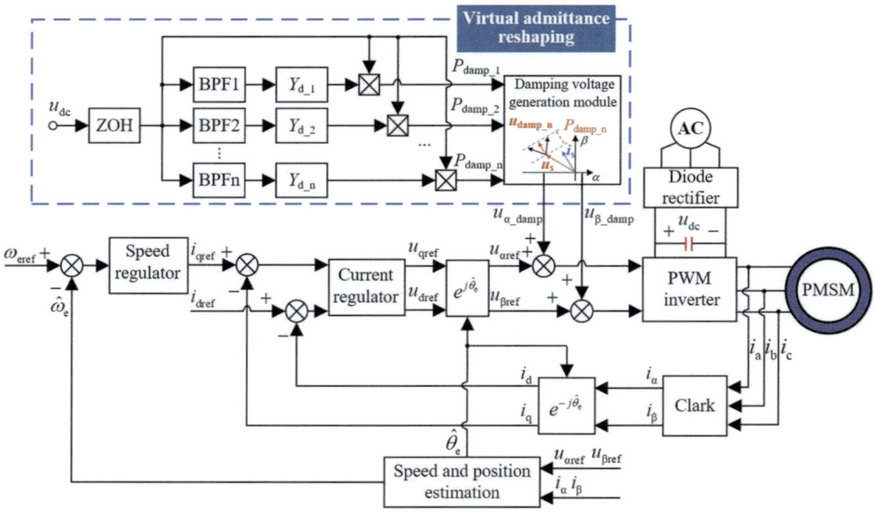

Fig. 3.6 Control block diagram of d electrolytic capacitorless PMSM drive based on virtual admittance reshaping, reprinted from Ref. [5], copyright 2022, with permission from nature

3.2.3 Virtual Admittance Construction Method with Low Disturbance Power

In order to reduce the additional disturbance on the motor side, it is necessary to reduce the damping power as much as possible. The damping power is the product of the damping current and the DC-link voltage which can be expressed as [5]

$$P_{\mathrm{damp\_n}} = i_{\mathrm{damp\_n}} u_{\mathrm{dc}}. \tag{3.5}$$

Ignoring the product of high-order terms, the damping power can be approximated as

$$P_{\mathrm{damp\_n}} \doteq Y_{\mathrm{d\_n}} u_{\mathrm{dc\_n}} U_{\mathrm{dc0}}. \tag{3.6}$$

In (3.6), the DC component of the DC-link voltage can be regarded as a constant value, and the amplitude of damping power is determined by the product of the virtual admittance amplitude and the inherent harmonic amplitude of the DC-link voltage. Therefore, the design of virtual admittance is very important. It is required to keep the virtual admittance amplitude as small as possible to reduce the damping power under the premise of effectively suppressing the inherent harmonics of the DC-link voltage. In order to reconstruct the impedance of each inherent harmonic, the virtual admittance can be expressed as

$$Y_{\mathrm{d\_n}}=\left|Y_{\mathrm{d\_n}}\right|\cos\theta_{\mathrm{d\_n}}+\frac{\left|Y_{\mathrm{d\_n}}\right|}{\omega_{\mathrm{d\_n}}}\sin\theta_{\mathrm{d\_n}}s. \tag{3.7}$$

where $\omega_{\mathrm{d\_n}}$ is the angular velocity of the inherent harmonic.

By substituting (3.7) into (3.2), the transfer function at inherent harmonic frequency can be further expressed as

$$G_{\mathrm{damp\_n}}(s)=\frac{1}{\left(L_g s+R_g\right)\left(C_{\mathrm{dc}}s+Y_{\mathrm{m\_n}}+\left|Y_{\mathrm{d\_n}}\right|\cos\theta_{\mathrm{d\_n}}+\frac{\left|Y_{\mathrm{d\_n}}\right|}{\omega_{\mathrm{d\_n}}}\sin\theta_{\mathrm{d\_n}}s\right)+1}. \tag{3.8}$$

where $Y_{\mathrm{m\_n}}$ is the motor side admittance at inherent harmonic frequency.

ω_g is the grid side angular velocity, which is 100π rad/s in the book. The LC resonance frequency of the experimental platform of the electrolytic capacitorless PMSM drive is 580 Hz, so the impedance reshaping is mainly carried out at the frequencies of $6\omega_g$ and $12\omega_g$. The parameters of the experimental platform are shown in Table 2.1. The following analysis and experimental results are described in the form of per unit value. The voltage base value is 310 V, the current base value is 21 A, the power base value is 11 kW, the frequency base value is 314 rad/s, the angle base value is π, and the impedance base value is 15 Ω.

Figure 3.7 shows the relationship between the amplitude of virtual admittance and the inherent harmonic suppression ability at the frequency $6\omega_g$. The curved surface is the amplitude $|G_{\mathrm{damp\_1}}|$ of transfer function $G_{\mathrm{damp\_1}}$ when the amplitude $|Y_{\mathrm{d\_1}}|$ of virtual admittance and the angle $\theta_{\mathrm{d\_1}}$ take different values. The lower $|G_{\mathrm{damp\_1}}|$ is, the stronger the suppression ability of inherent harmonics will be. Under the same amplitude of $|G_{\mathrm{damp\_1}}|$, the smaller $|Y_{\mathrm{d\_1}}|$ is, the smaller the damping power will be. As can be seen from the Fig. 3.7, when $\theta_{\mathrm{d\_1}}$ is set to -0.51 p.u., $|G_{\mathrm{damp\_1}}|$ is relatively low. And the larger $|Y_{\mathrm{d\_1}}|$ is, the stronger the suppression ability of inherent harmonics will be. Considering the inherent harmonic suppression ability and damping power, $|Y_{\mathrm{d\_1}}|$ is set to 0.60 p.u., in which case $|G_{\mathrm{damp\_1}}| = 0.63$ dB. When the dashed lines are connected in parallel with the 1.67 p.u. virtual resistor, the corresponding admittance value is also 0.60 p.u., and in this case $|G_{\mathrm{damp\_1}}|$ is only 2.55 dB. It can be concluded that the impedance reshaping method using parallel virtual admittance can reduce the damping power at $6\omega_g$ to 83.3% and increase the harmonic suppression effect to 120%.

Figure 3.8 shows the relationship between the amplitude of virtual admittance at the frequency $12\omega_g$ and the inherent harmonic suppression ability. In accordance with the above principle of virtual admittance parameter selection, $|Y_{\mathrm{d\_2}}|$ and $\theta_{\mathrm{d\_2}}$ are set as 0.45 p.u. and − 1.08 p.u., respectively. In this case, the amplitude $|G_{\mathrm{damp\_2}}|$ of the transfer function is 8.20 dB. The dashed line plane in the Fig. 3.8 is the case when the 2.20 p.u. virtual resistor is connected in parallel, and the transmission amplitude is 13.80 dB.

The theoretical value of virtual admittance at the inherent harmonics near the LC resonant frequency can be obtained through Figs. 3.7 and 3.8. However, the influence

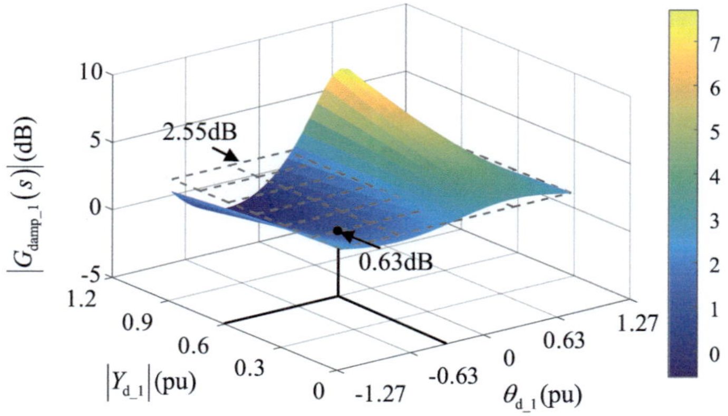

Fig. 3.7 Amplitude of Bode diagram at the frequency of $6\omega_g$ with different virtual admittances, reprinted from Ref. [5], copyright 2022, with permission from Nature

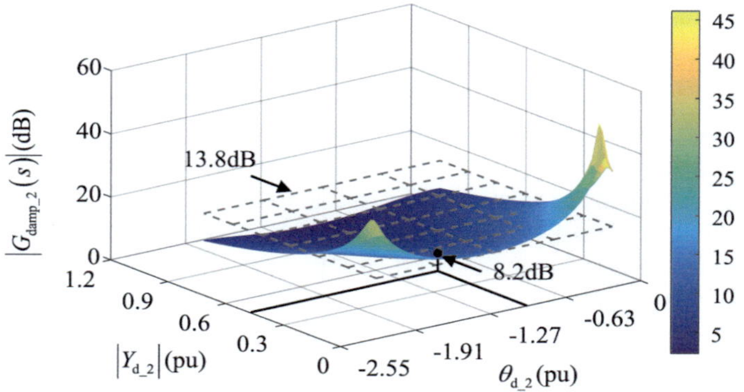

Fig. 3.8 Amplitude of Bode diagram at the frequency of $12\omega_g$ with different virtual admittances, reprinted from Ref. [5], copyright 2022, with permission from Nature

of the DC-link voltage sampling update delay still needs to be considered in the actual system. As shown in (3.6), if the influence of DC-link voltage delay is not considered when damping current is generated, this part of delay will cause the angle of virtual admittance to deviate from the theoretical value, resulting in poor damping effect.

Figure 3.9a shows the influence of DC-link voltage delay at $6\omega_g$, u_{dc_1} is the DC-link voltage harmonic at $6\omega_g$, u_{dc_e1} is the DC-link voltage equivalent sampling value at $6\omega_g$ whose angle lag is φ_{e_1}. Therefore, it is necessary to subtract the DC-link voltage delay angle from the theoretical value of virtual admittance θ_{d_1} to accurately reshape the impedance at the frequency of $6\omega_g$, where the damping current can be expressed as

$$i_{damp\_n} = u_{dc\_en}\left|Y_{d\_n}\right|\angle\left(\theta_{d\_n} - \varphi_{e\_n}\right). \tag{3.9}$$

Figure 3.9b, c show the amplitude shift of the transfer function caused by DC-link voltage delay at the frequency $6\omega_g$ and $12\omega_g$, respectively. It can be seen that the DC-link voltage sampling delay will increase $|G_{damp\_1}(s)|$ and $|G_{damp\_2}(s)|$, thus weakening the harmonic suppression ability at the corresponding frequency.

3.2.4 Experimental Verification

Figure 3.10 shows the harmonic test results at the frequency of $6\omega_g$ when virtual admittance is not used, and the output power of the motor is 0.46 p.u. The effect of the virtual admittance on the inherent harmonic suppression ability and the motor disturbance is observed by changing the amplitude of virtual admittance. In Figs. 3.11, 3.12 and 3.13, the angle of virtual admittance increases from − 0.76 p.u. to − 0.51 p.u. and − 0.22 p.u. And it can be seen that the decrease of inherent harmonic u_{dc_1} at the DC-link voltage frequency $6\omega_g$ is 17%, 35% and 15%, respectively, which is consistent with the theoretical analysis in Fig. 3.7. The standard deviation of current harmonics on the motor side changes from 16 to 27% and 145%. Therefore, setting the amplitude of virtual admittance to $|Y_{d\_1}| = 0.60$ p.u. and $\theta_{d\_1} = -0.51$ p.u. can effectively suppress the inherent harmonics at frequency $6\omega_g$, and disturbance at the motor side will not be too large.

Similarly, Fig. 3.14 shows the harmonic test results at the frequency of $12\omega_g$ under different virtual admittance. In Figs. 3.15, 3.16 and 3.17, the virtual admittance angle decreases from − 0.83 p.u. to − 1.08 p.u. and − 1.27 p.u.. And it can be seen that the harmonic u_{dc_2} at the frequency $12\omega_g$ of the DC-link voltage decreases by 6%, 32% and 34%, respectively, which is consistent with the theoretical analysis results in Fig. 3.8. The current harmonics of the motor side varied from 39 to 48% and 84%. Therefore, setting the virtual admittance value to $|Y_{d\_1}| = 0.45$ p.u. and $\theta_{d\_1} = -1.08$ p.u. can effectively suppress the inherent harmonics at the frequency of $12\omega_g$.

The above experimental results show that the virtual admittance reshaping method can effectively suppress the inherent harmonics near the LC resonant frequency in the electrolytic capacitorless drive. The optimal virtual admittance amplitude and angle are consistent with the theoretical analysis. So, the admittance reshaping method can effectively reduce the introduction of motor side disturbance power and suppress the natural frequency harmonics of the system at the same time.

Figure 3.18 shows the waveforms of DC-link voltage and grid current before and after the study method is adopted, and the output power of the motor is 0.5 p.u. It can be seen from Fig. 3.18a that after virtual admittance is enabled, the fluctuation of DC-link grid voltage and current will both decrease. Figure 3.18c shows the amplified waveforms of DC-link voltage and grid current after enabling Y_{d_1}, the grid current THD can be reduced from 51 to 39%, and the voltage fluctuation of the DC-link can be reduced from 0.42 p.u. to 0.34 p.u.. When Y_{d_2} is further enabled, the grid current THD drops to 35%, and the voltage fluctuation of the DC-link is 0.29 p.u., as shown in Fig. 3.18d.

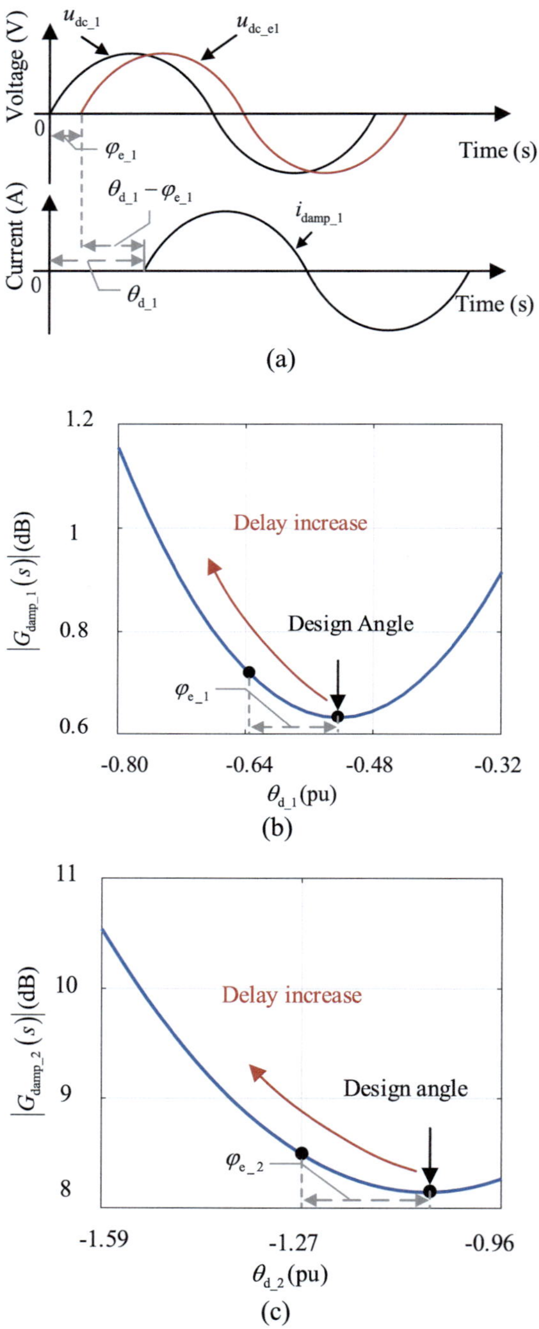

Fig. 3.9 Analysis results of influence of DC-link voltage delay. **a** Relationship between angle of virtual admittance and voltage delay. **b** Amplitude shift due to voltage delay at the frequency of $6\omega_g$. **c** Amplitude divergence due to voltage delay at the frequency of $12\omega_g$

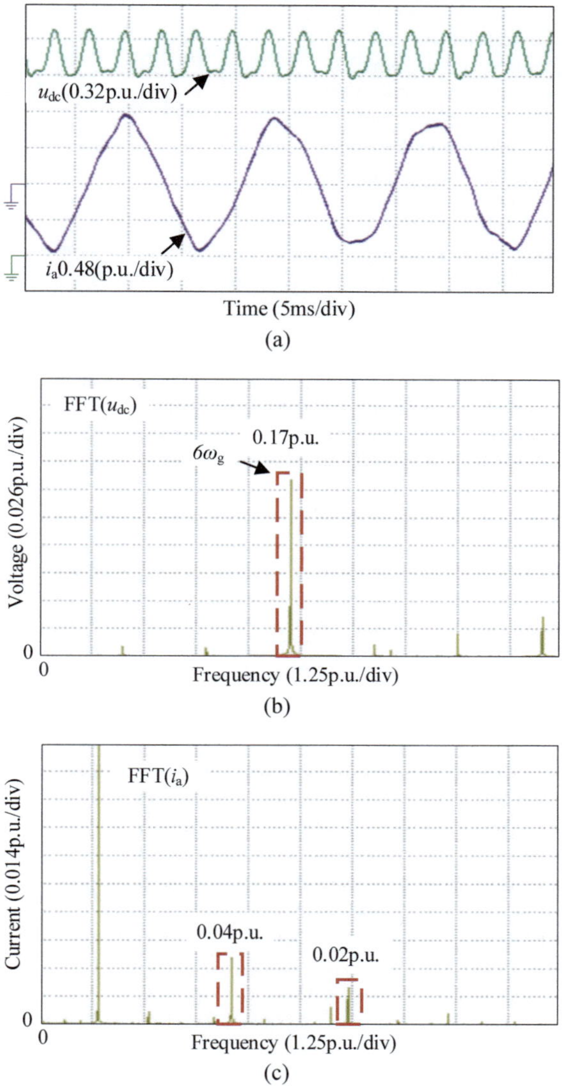

Fig. 3.10 Voltage and current harmonics at $6\omega_g$ frequency without applying virtual admittance, reprinted from Ref. [5], copyright 2022, with permission from Nature. **a** DC-link voltage and motor current. **b** Amplitude of DC-link voltage harmonic at frequency of $6\omega_g$. **c** Amplitude of motor current harmonics related to frequency of $6\omega_g$

Figure 3.19 shows the statistical results of grid current THD before and after the algorithm is enabled under different output power conditions. At present, the current harmonics regulation algorithm is only applicable to the middle and low power segment. In the case of insufficient DC-link voltage margin in high power segment, the boundary voltage of weak magnetic controller needs to be redesigned. It can be seen that the proposed impedance reshaping method can effectively reduce the grid current THD under different power conditions. When the output power is low, the decrease of grid THD is larger.

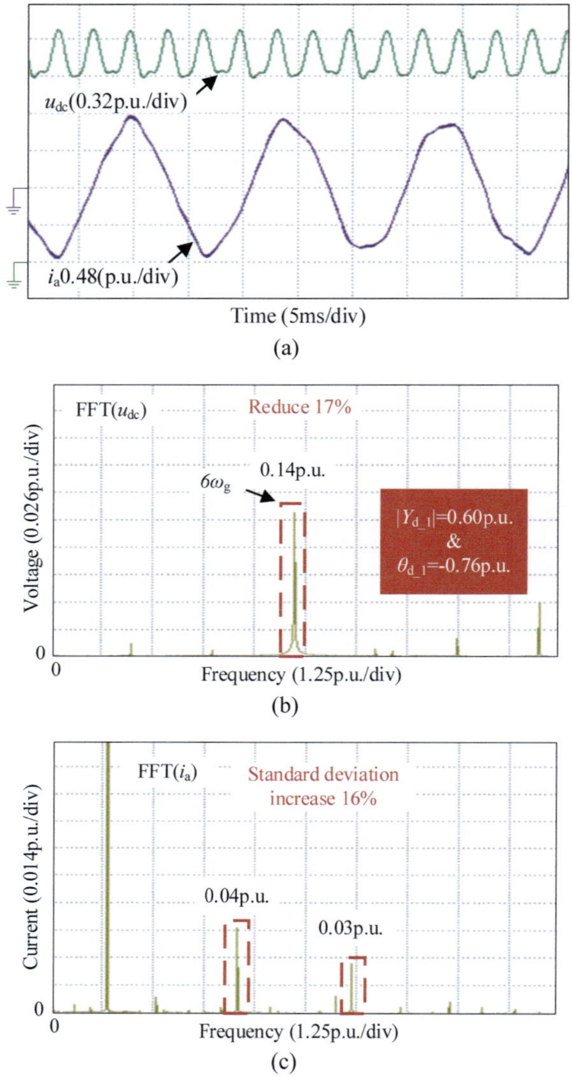

Fig. 3.11 Voltage and current harmonics at frequency of $6\omega_g$ when applying virtual admittance. ($|Y_{d\_2}| = 0.60$ p.u. and $\theta_{d\_2} = -0.76$ p.u.) **a** DC-link voltage and motor current. **b** Amplitude of DC-link voltage harmonic at frequency of $6\omega_g$. **c** Amplitude of motor current harmonics related to frequency of $6\omega_g$

3.3 Rectified Current Harmonic Suppression Method Based on DC-Link Voltage Decoupling

3.3.1 Rectified Current Harmonic Suppression Method

The traditional damping control increases the impedance of the drive by the form of DC-link voltage harmonic feedforward, and indirectly achieves the harmonic suppression effect of the grid side current. Since the harmonics of rectified current

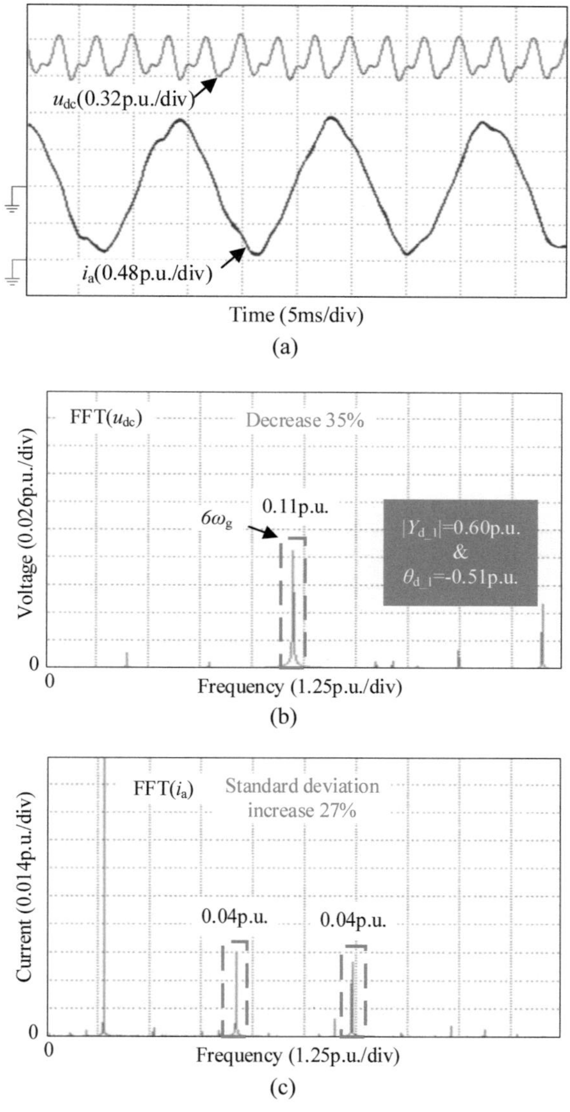

Fig. 3.12 Voltage and current harmonics at frequency of $6\omega_g$ when applying virtual admittance ($|Y_{d\_2}| = 0.60$ p.u. and $\theta_{d\_2} = -0.51$ p.u.), reprinted from Ref. [5], copyright 2022, with permission from Nature. **a** DC-link voltage and motor current. **b** Amplitude of DC-link voltage harmonic at frequency of $6\omega_g$. **c** Amplitude of motor current harmonics related to frequency of $6\omega_g$

are closely related to the grid current harmonics, the quality of the grid-side input power can be directly improved by controlling the harmonics of rectified current.

The maximum diode on-angle of the three-phase uncontrolled rectifier bridge is 120°. So, in half a period, the grid current non-zero region is 120°. At this time, the grid side input reference current i_{grid_ref} can be expressed as [7]

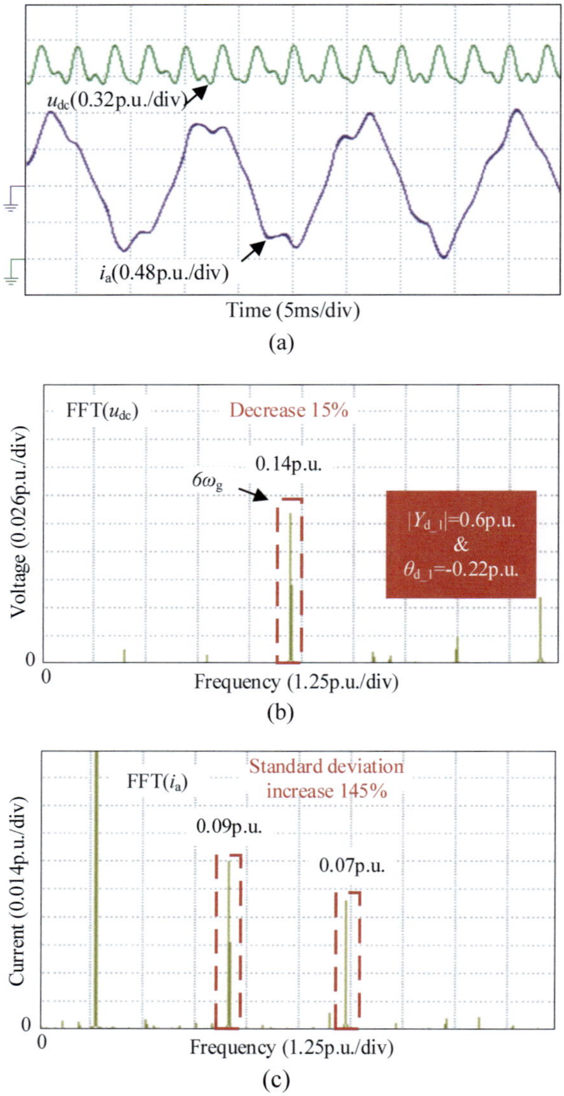

Fig. 3.13 Voltage and current harmonics at frequency of $6\omega_g$ when applying virtual admittance ($|Y_{d\_2}| = 0.60$ p.u. and $\theta_{d\_2} = -0.22$ p.u.), reprinted from Ref. [5], copyright 2022, with permission from Nature. **a** DC-link voltage and motor current. **b** Amplitude of DC-link voltage harmonic at frequency of $6\omega_g$. **c** Amplitude of motor current harmonics related to frequency of $6\omega_g$

$$i_{grid\_ref} = \begin{cases} 0 \ \ \theta_g \in [0, \pi/6) \cup [5\pi/6, 7\pi/6) \cup [11\pi/6, 2\pi) \\ 1 + \alpha\left(\left|\cos(3\theta_g)\right| - \frac{2}{\pi}\right) \ \theta_g \in [\pi/6, 5\pi/6) \\ -1 - \alpha\left(\left|\cos(3\theta_g)\right| - \frac{2}{\pi}\right) \ \theta_g \in [7\pi/6, 11\pi/6) \end{cases} \quad (3.10)$$

where θ_g is the grid current phase and α is the weight coefficient of the ratio between the peak-to-peak value and the effective value of the grid-side input reference current of the 120° conduction area. (3.11) is obtained by Fourier series expansion of (3.10) as

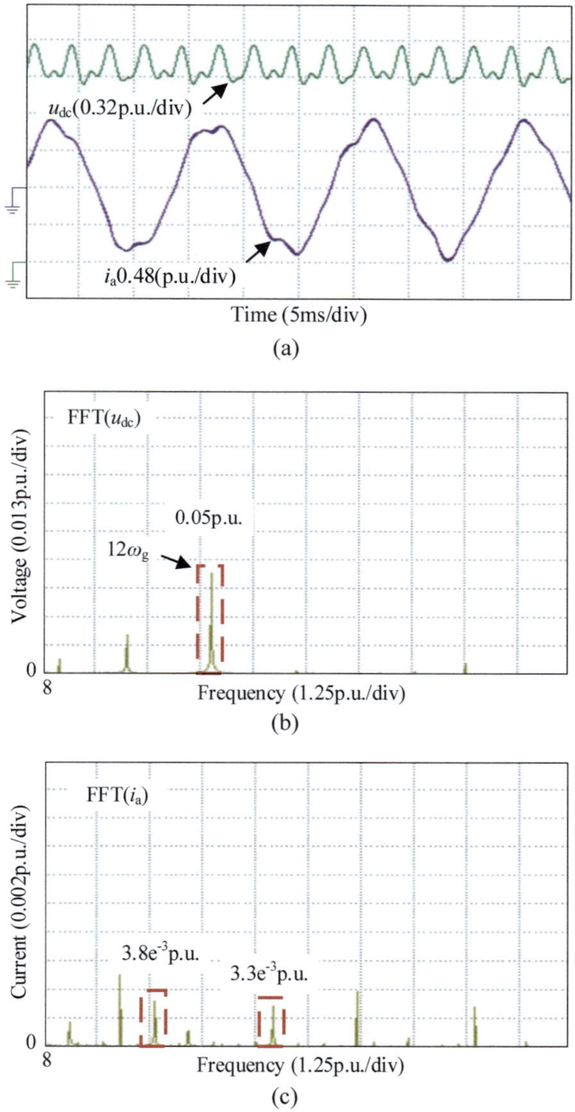

Fig. 3.14 Voltage and current harmonics at frequency of 12ωg without applying virtual admittance, reprinted from Ref. [5], copyright 2022, with permission from Nature. **a** DC-link voltage and motor current. **b** Amplitude of DC-link voltage harmonic at frequency of $12\omega_g$. **c** Amplitude of motor current harmonics related to frequency of $12\omega_g$

$$i_{\text{grid\_ref}} = \sum_{n=1}^{\infty} (k_n \alpha + b_n) \sin(n\theta_g) \tag{3.11}$$

where

$$k_n = \begin{cases} \frac{4\left[\cos\left(\frac{5n\pi}{6}\right) - \cos\left(\frac{n\pi}{6}\right)\right]}{n\pi^2} + \frac{6\left[\sin\left(\frac{n\pi}{2}\right) + \sin\left(\frac{n\pi}{6}\right) + \sin\left(\frac{5n\pi}{6}\right)\right]}{\pi(9-n^2)} & n \neq 3 \\ 0 & n \neq 3 \end{cases} \tag{3.12}$$

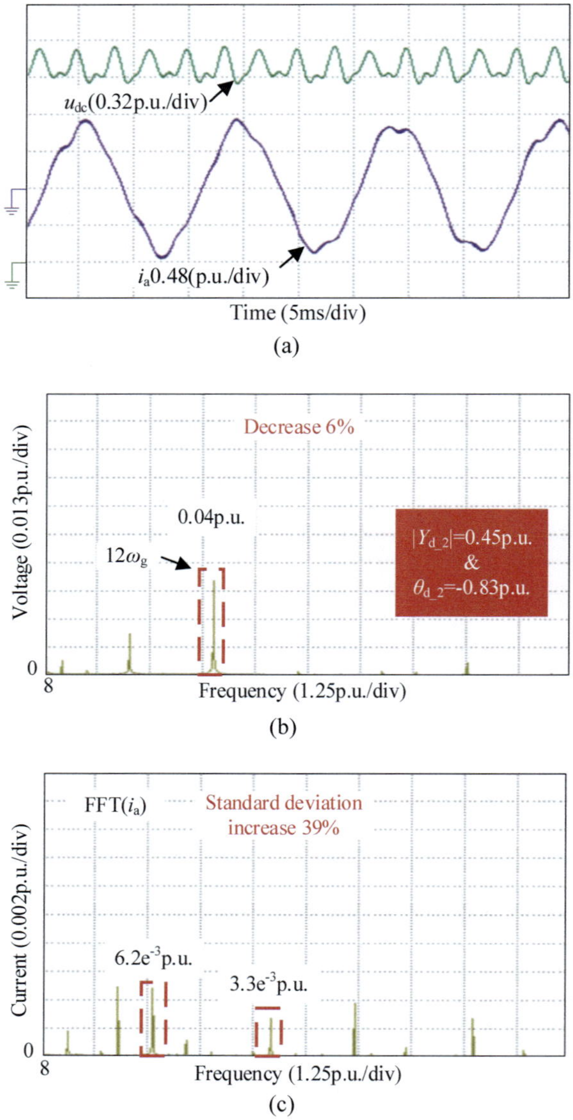

Fig. 3.15 Voltage and current harmonics at frequency of $12\omega_g$ when applying virtual admittance ($|Y_{d\_2}| = 0.45$ p.u. and $\theta_{d\_2} = -0.83$ p.u.), reprinted from Ref. [5], copyright 2022, with permission from Nature. **a** DC-link voltage and motor current. **b** Amplitude of DC-link voltage harmonic at frequency of $12\omega_g$. **c** Amplitude of motor current harmonics related to frequency of $12\omega_g$

$$b_n = \frac{2\left[\cos\left(\frac{n\pi}{6}\right) - \cos\left(\frac{5n\pi}{6}\right)\right]}{n\pi} \tag{3.13}$$

Therefore, the THD of grid current can be expressed as

$$\mathrm{THD} = \frac{\sqrt{\sum_{n=2}^{n=40}(k_n\alpha + b_n)^2}}{k_1\alpha + b_1} = \frac{\sqrt{0.12186\alpha^2 - 0.01822\alpha + 0.1071}}{0.01422\alpha + 1.1027} \tag{3.14}$$

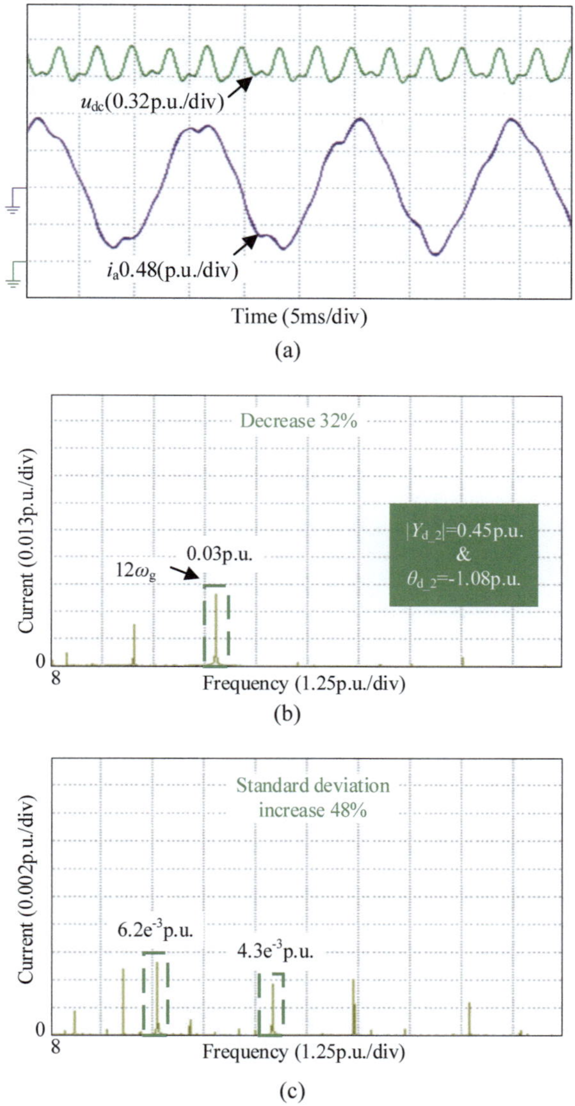

Fig. 3.16 Voltage and current harmonics at frequency of $12\omega_g$ when applying virtual admittance ($|Y_{d\_2}| = 0.45$ p.u. and $\theta_{d\_2} = -1.08$ p.u.), reprinted from Ref. [5], copyright 2022, with permission from Nature. **a** DC-link voltage and motor current. **b** Amplitude of DC-link voltage harmonic at frequency of $12\omega_g$. **c** Amplitude of motor current harmonics related to frequency of $12\omega_g$

Under different weight coefficients α, the THD of grid input reference current and waveforms are shown in Fig. 3.20a, b, respectively. It can be seen from Fig. 3.20a that when the weight coefficient is $0 \leq \alpha \leq 0.2$, the grid-side current has a lower THD. When the weight coefficient $\alpha = 0$, the reference current waveform of grid side is a rectangular wave with a conduction angle of 120°, and the rectified currennt waveform is a DC component with the same amplitude as the reference current of grid side, as shown in Fig. 3.21. The THD of grid current is 29.7% [8]. Therefore, the rectified current with only DC component is selected as the reference value, and

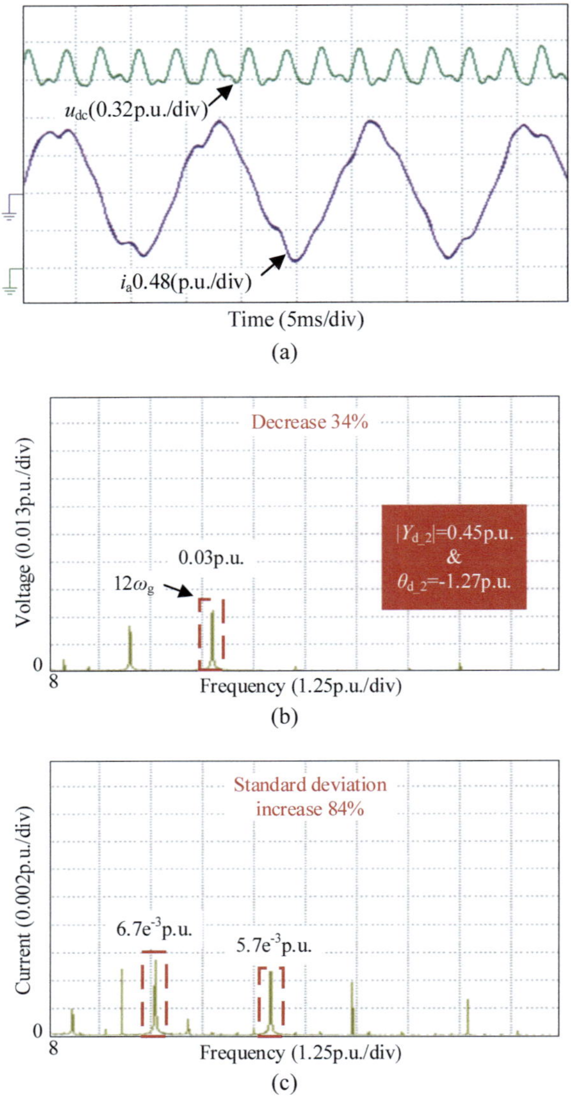

Fig. 3.17 Voltage and current harmonics at frequency of $12\omega_g$ when applying virtual admittance ($|Y_{d\_2}|$=0.45 p.u. and θ_{d_2} = -1.27 p.u.), reprinted from Ref. [5], copyright 2022, with permission from Nature. **a** DC-link voltage and motor current. **b** Amplitude of DC-link voltage harmonic at frequency of $12\omega_g$. **c** Amplitude of motor current harmonics related to frequency of $12\omega_g$

the power quality of the grid current can be improved by controlling the 6th and 12th harmonic components of the rectified current.

Figure 3.22 is the closed-loop control block diagram of the traditional rectified current harmonic suppression method, which extracts the 6th and 12th harmonics of the rectified current through the bandpass filter. S_{abc} is working state of power switching device, i_{g_r} is reconstructed rectified current, and Δi_{gref} is the small signal of reference rectified current. The difference between the current harmonic of rectified current and the reference value is taken as the input of the current resonant controller.

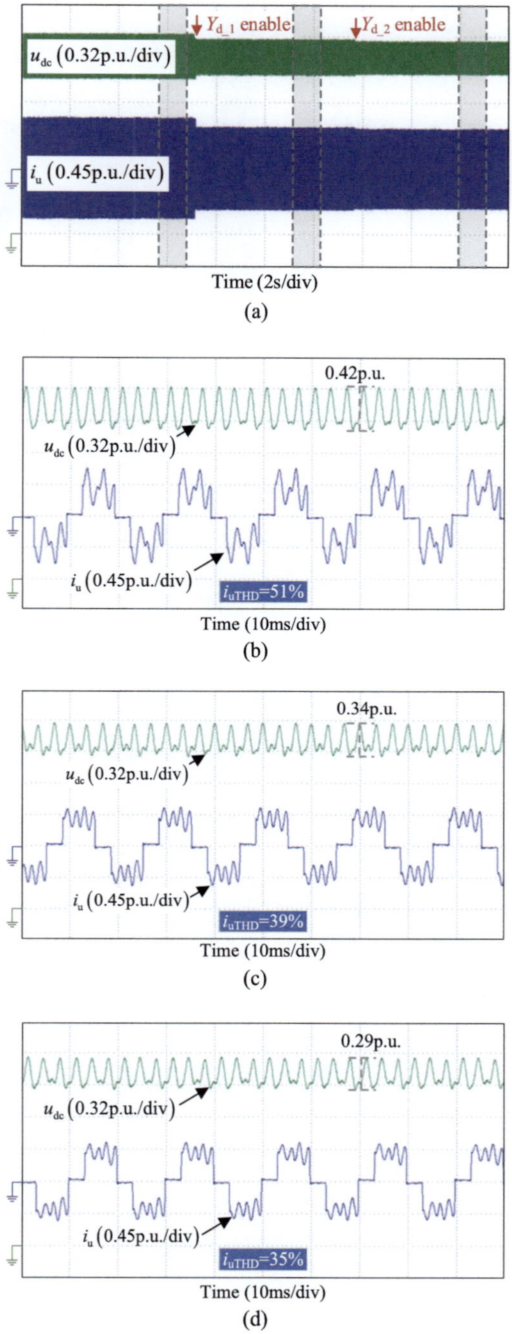

Fig. 3.18 Experimental results of DC-link voltage and grid current, reprinted from Ref. [5], copyright 2022, with permission from Nature. **a** The whole waveform. **b** Enlarged waveform before applying the impedance reshaping method. **c** Enlarged waveform after enabling first inherent-harmonic virtual admittance. **d** Enlarged waveform after enabling second inherent-harmonic virtual admittance

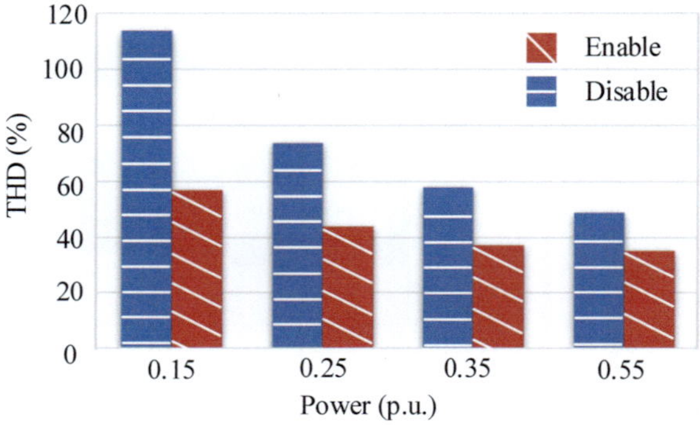

Fig. 3.19 Experimental results of the grid current THD under different output powers

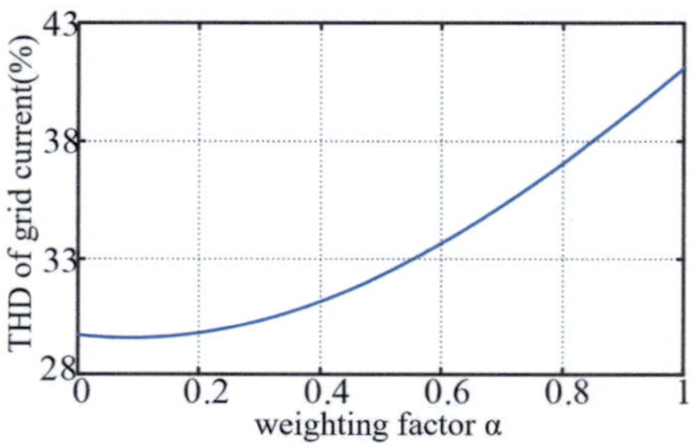

(a) THD of grid current under different α

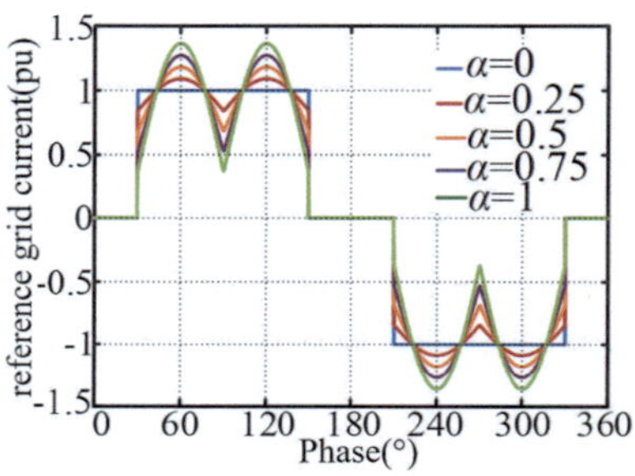

(b) reference grid current under different α

Fig. 3.20 Input reference current waveform and THD of grid current with different weight coefficients

Fig. 3.21 Reference grid current(i_{grid_ref}) and reference rectified current(i_{g_ref})

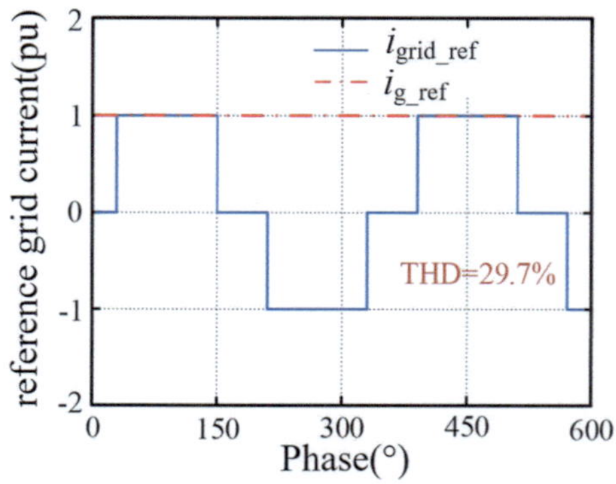

Then, the output of the current resonant controller is superimposed to the reference voltage of the q axis to achieve the effect of suppressing the current harmonic of rectified current and improving the power quality of the grid side [9].

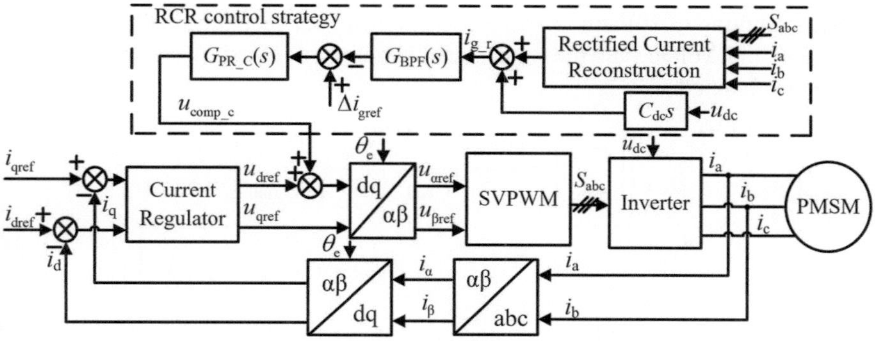

Fig. 3.22 Closed-loop control block diagram of the traditional rectified current harmonic suppression method, reprinted from Ref. [10], copyright 2022, with permission from Nature

It can be seen from Fig. 3.22 that with the harmonic suppression method of rectified current, the compensation voltage $u_{comp\_c}(s)$ superimposed to the q-axis reference voltage is expressed as

$$u_{comp\_c}(s) = -G_{BPF}(s)G_{PR\_C}(s)G_{dealy}(s)i_{g\_r}(s) \tag{3.15}$$

where $G_{BPF}(s)$ is the band-pass filter, $G_{PR\_C}(s)$ is current proportional resonant controller, and $G_{delay}(s)$ is delay transfer function of the control system. $G_{PR\_C}(s)$ and $G_{BPF}(s)$ can be expressed as

$$G_{BPF}(s) = \frac{2\xi\omega_0 s}{s^2 + 2\xi\omega_0 s + \omega_0^2} \tag{3.16}$$

$$G_{PR\_C}(s) = K_{RP\_C} + K_{RR\_C}\frac{2\omega_c s}{s^2 + 2\omega_c s + \omega_0^2} \tag{3.17}$$

where K_{RP_C}, K_{RR_C} and ω_c are proportional gain, resonant gain and resonant bandwidth of current proportional resonant controller, respectively.

$$i_{g\_r} = C_{dc}\frac{du_{dc}}{dt} + D_a i_a + D_b i_b + D_c i_c \tag{3.18}$$

where D_a, D_b and D_c are the duty cycle of phase *a*, *b* and *c* of the inverter, respectively.

With the traditional current control strategy of rectified current, the reference small signal of *dq*-axis voltage can be expressed as

$$\begin{bmatrix} \Delta u_{dref}(s) \\ \Delta u_{qref}(s) \end{bmatrix} = \begin{bmatrix} -\mathbf{G}_d & 0 \\ 0 & -\mathbf{G}_q \end{bmatrix}\begin{bmatrix} \Delta i_d(s) \\ \Delta i_q(s) \end{bmatrix} + e^{-sT_s}\begin{bmatrix} 0 \\ -G_{BPF}(s)G_{PR\_C}(s)i_{g\_r}(s) \end{bmatrix} \tag{3.19}$$

By combining (3.19) and (2.16), the relationship between the small signal variation of DC-link voltage and the small signal variation of dq-axis current with control method is obtained as

$$\begin{bmatrix} A_{1d}(s) & A_{1q}(s) \\ A_{2d}(s) & A_{2q}(s) \end{bmatrix}\begin{bmatrix} \Delta i_d \\ \Delta i_q \end{bmatrix} = \frac{\left(1-e^{-1.5sT_s}\right)\Delta u_{dc}}{U_{dc0}}\begin{bmatrix} U_{dref0} \\ U_{qref0} \end{bmatrix} + e^{-sT_s}\begin{bmatrix} 0 \\ -G_{BPF}(s)G_{PR\_C}(s)i_{g\_r}(s) \end{bmatrix} \tag{3.20}$$

After adding the control method, the transfer function $W_{d,qmad}$ between the small signal of the dq-axis currents and the small signal of DC-link voltage can be expressed as

$$\begin{bmatrix} W_{dmad}(s) \\ W_{qmad}(s) \end{bmatrix} = \frac{1}{\Delta u_{dc}}\begin{bmatrix} \Delta i_d \\ \Delta i_q \end{bmatrix} = \begin{bmatrix} W_d(s) \\ W_q(s) \end{bmatrix} + G_{mad}(s)\frac{\Delta i_{g\_r}}{\Delta u_{dc}}\begin{bmatrix} A_{1q} \\ -A_{1d} \end{bmatrix} \tag{3.21}$$

where

$$G_{mad}(s) = \frac{e^{-sT_s}G_{PR\_C}(s)G_{BPF}(s)}{A_{2q}(s)A_{1d}(s) - A_{2d}(s)A_{1q}(s)} \tag{3.22}$$

The small signal of rectified current can be expressed as

$$\Delta i_g = C_{dc}\Delta u_{dc}(s)s + \Delta i_{inv}(s) \tag{3.23}$$

where Δi_g is the small signal of DC-link current.

By substituting (3.20) and (3.23) into (2.26), the input impedance of the motor drive Y_{mad} after applying the traditional rectified current control strategy can be expressed as

$$Y_{mad}(s) = \frac{Y_m + \frac{3}{2}\frac{G_{mad}A_{1q}(s)C_{dc}s\left(U_{d0} + \omega_e L_d I_{q0} + L_d I_{d0}s + I_{d0}R_s\right)}{U_{dc0}} - \frac{3}{2}\frac{G_{mad}A_{1d}(s)C_{dc}s\left(U_{q0} - \omega_e L_q I_{d0} + L_q I_{q0}s + I_{q0}R_s\right)}{U_{dc0}}}{1 - \frac{3}{2}\frac{G_{mad}A_{1q}(s)\left(U_{d0} + \omega_e L_d I_{q0} + L_d I_{d0}s + I_{d0}R_s\right)}{U_{dc0}} + \frac{3}{2}\frac{G_{mad}A_{1d}(s)\left(U_{q0} - \omega_e L_q I_{d0} + L_q I_{q0}s + I_{q0}R_s\right)}{U_{dc0}}} \tag{3.24}$$

According to (3.21) and (3.24), the transfer function between the small signal of q-axis current and the small signal of inverter current can be obtained as

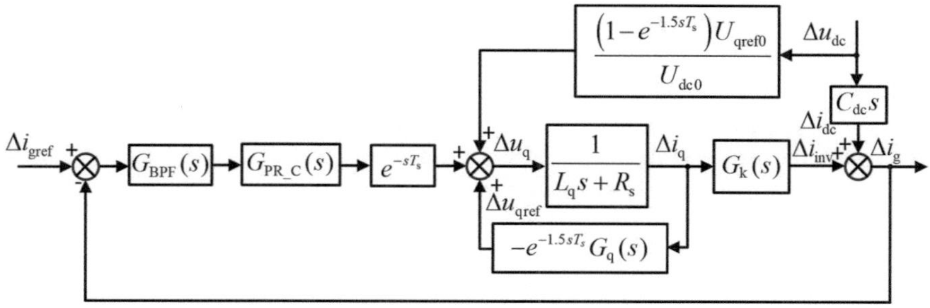

Fig. 3.23 Small signal block diagram of rectified current regulation

$$G_{\text{K}}(s)=\frac{\Delta i_{\text{inv}}}{\Delta i_{\text{q}}}=\frac{\Delta i_{\text{inv}}}{\Delta u_{\text{dc}}}\cdot\frac{\Delta u_{\text{dc}}}{\Delta i_{\text{q}}}=\frac{Y_{\text{mad}}(s)}{W_{\text{qmad}}(s)} \tag{3.25}$$

The diagram of small signal closed-loop control of the rectified current harmonic suppression method can be obtained as shown in Fig. 3.23.

The input of the small signal closed-loop control loop is the reference small signal of the rectified current and the small signal of the DC-link voltage, and the open-loop transfer function G_{L}(s) is

$$G_{\text{L}}(s)=\frac{e^{-sT_s}G_{\text{BPF}}(s)G_{\text{PR\_C}}(s)Y_{\text{mad}}(s)}{W_{\text{qmad}}(s)\left[(L_{\text{q}}s+R_{\text{s}})+e^{-sTs}G_{\text{q}}(s)\right]} \tag{3.26}$$

It can be seen that the actual small signal variations of rectified current is

$$\begin{aligned}\Delta i_{\text{g}}(s)=&\frac{G_{\text{L}}(s)}{1+G_{\text{L}}(s)}\Delta i_{\text{Lref}}+\frac{C_{\text{dc}}s}{1+G_{\text{L}}(s)}\Delta u_{\text{dc}}\\&+\frac{(1-e^{-1.5sT_s})U_{\text{qref0}}Y_{\text{mad}}(s)}{U_{\text{dc0}}W_{\text{qmad}}(s)(1+G_{\text{L}}(s))\left[(L_{\text{q}}s+R_{\text{s}})+e^{-sT_s}G_{\text{q}}(s)\right]}\Delta u_{\text{dc}}\end{aligned} \tag{3.27}$$

As can be seen from Fig. 3.23 and (3.27), the DC-link voltage sampling delay and small signal fluctuation will introduce disturbance term into the rectified current closed-loop control loop, which makes it difficult for the small signal of rectified current to track the reference. According to (3.27), the transfer function between the small signal of rectified current and the small signal of DC-link voltage is

$$G_{\text{D}}(s)=\frac{C_{\text{dc}}s}{1+G_{\text{L}}(s)}+\frac{(1-e^{-1.5sT_s})U_{\text{qref0}}Y_{\text{mad}}(s)}{U_{\text{dc0}}W_{\text{qmad}}(s)(1+G_{\text{L}}(s))\left[(L_{\text{q}}s+R_{\text{s}})+e^{-sT_s}G_{\text{q}}(s)\right]} \tag{3.28}$$

According to the steady-state operating point data of the test motor under rated power measured off-line, the Bode diagram of G_{D}(s) under different current proportional resonant controller parameters is obtained, as shown in Fig. 3.24. Increasing the resonant parameter K_{RR} of the resonant controller can improve the gain of the proportional resonant controller at the resonant frequency. It can be seen from the amplitude-frequency characteristics of $G_D(s)$ as shown in Fig. 3.24 that with the increase of resonance parameters of the current proportional resonance controller, the amplitude of $G_D(s)$ decreases at $6\omega_g$ and $12\omega_g$, and the suppression effect of rectified current harmonic suppression method on harmonic disturbances at $6\omega_g$ and $12\omega_g$ in the DC-link voltage is enhanced.

The permanent magnet synchronous motor controlled by the inverter is regarded as a cascade system with the DC-link capacitor and filter inductor. By analyzing the Nyquist diagram of the admittance ratio of output Y_{LC} of the DC-link capacitor and the admittance of the motor side Y_{mad}, the stability changes of electrolytic capacitor-less drive can be obtained under different proportional gains of current proportional resonant controller. Y_{LC} can be expressed as

$$Y_{LC}(s) = \frac{L_g C_{dc} s^2 + R_g C_{dc} s + 1}{L_g s + R_g} \tag{3.29}$$

Under different resonant gain parameters K_{RR_C} of the current proportional resonant controller, the Nyquist diagram of the admittance ratio Y_{mad}/Y_{LC} is shown in Fig. 3.25. Without the control method, the system is unstable due to the reduction of the DC-link capacitance. According to the stability criterion of the cascade system, when the resonant gain parameter K_{RR_C} of the current proportional resonance controller is gradually increased within a certain range, the intersection point of the Nyquist diagram and the negative real axis is closer to the point $(-1, j0)$ from the left side of $(-1, j0)$, and the stability of the system is increased. When the parameters of the current proportional resonant controller are $K_{RP\_C} = 8$ and $K_{RR\_C} = 50$, the intersection point of the Nyquist diagram of Y_{mad}/Y_{LC} and the negative real axis is located on the right of the point $(-1, j0)$, and the system is stable. However, when

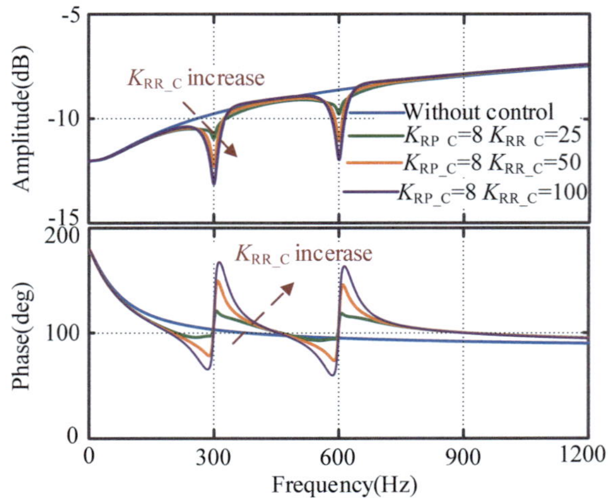

Fig. 3.24 Bode diagrams of the transfer function between Δi_g and Δu_d

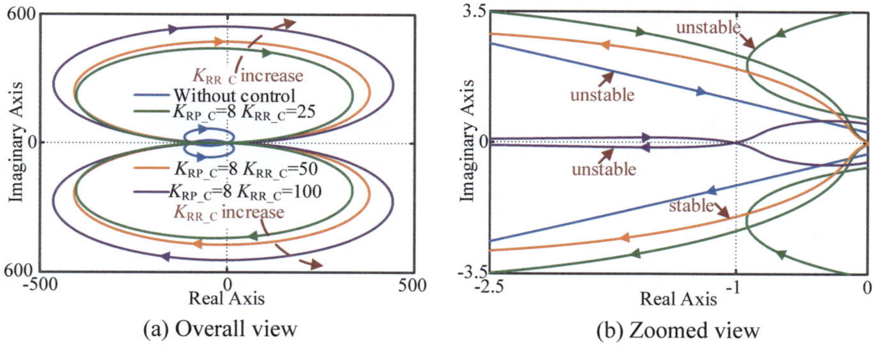

Fig. 3.25 Nyquist diagrams of $Y_{\mathrm{mad}}(s)/Y_{\mathrm{LC}}(s)$ under different resonant parameters, reprinted from Ref. [10], copyright 2022, with permission from Nature

the resonant gain parameter of the current proportional resonant controller continues to increase to $K_{\mathrm{RP\_C}} = 8$ and $K_{\mathrm{RR\_C}} = 100$, the intersection point of the Nyquist diagram of $Y_{\mathrm{mad}}/Y_{\mathrm{LC}}$ and the negative real axis gradually migrates to the left side of $(-1, \mathrm{j}0)$, and the system tends to be unstable.

3.3.2 Harmonic Suppression for Grid Currents Based on DC-Link Voltage Decoupling

In view of the disturbance term in small signal control loop caused by DC-Link voltage sampling, PWM update delay and DC-Link voltage fluctuation, a rectified current harmonic suppression method based on DC-Link voltage decoupling is studied. By predicting the main frequency harmonics in DC-Link voltage, and constructing the DC-Link voltage harmonics closed-loop control loop, the DC-Link voltage disturbance term in the control loop of small signal variations rectified current harmonic suppression method is decoupled to further suppress the rectified current harmonics and improve the power quality of the grid current. The block diagram of rectified current harmonic suppression method based on DC-link voltage decoupling is shown in Fig. 3.26. The control method consists of DC-link voltage harmonic prediction and the closed-loop control of DC-link voltage harmonics.

(1) DC-link voltage harmonics closed loop control

The block diagram of closed-loop control of rectified current harmonics suppression method based on DC-link voltage decoupling is shown in Fig. 3.27. The small signal variations of DC-link voltage which is compensated through harmonics prediction is taken as feedback value, and the q-axis voltage compensation signal is obtained through the voltage proportional resonant controller, and the q-axis voltage compensation signal is superimposed to the reference voltage of q-axis. In this way, the

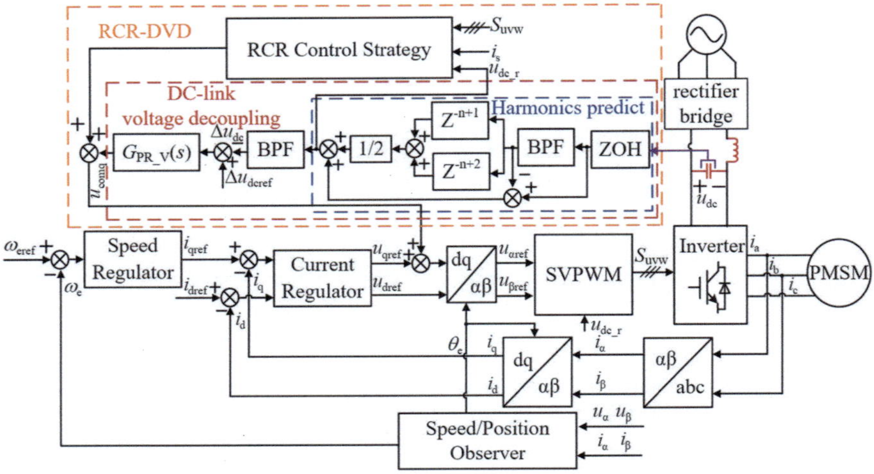

Fig. 3.26 Control block diagram of the proposed rectified current regulation with DC-link voltage decoupling, reprinted from Ref. [10], copyright 2022, with permission from Nature

closed-loop control of main harmonics of DC-link voltage is realized, and the influence of DC-link voltage harmonics on rectified current control is suppressed.

With the harmonics suppression method for rectified current control based on DC-link voltage decoupling, the small signal variations of reference voltage in dq axis is

$$\begin{aligned}\begin{bmatrix}\Delta u_{\mathrm{dref}}(s)\\ \Delta u_{\mathrm{qref}}(s)\end{bmatrix}=&\begin{bmatrix}-G_{\mathrm{d}}(s) & 0\\ 0 & -G_{\mathrm{q}}(s)\end{bmatrix}\begin{bmatrix}\Delta i_{\mathrm{d}}(s)\\ \Delta i_{\mathrm{q}}(s)\end{bmatrix}\\ &+\begin{bmatrix}0\\ -G_{\mathrm{BPF}}(s)G_{\mathrm{PR\_C}}(s)\Delta i_{\mathrm{L\_r}}(s)\end{bmatrix}\\ &+\begin{bmatrix}0\\ -G_{\mathrm{BPF}}(s)G_{\mathrm{PR\_V}}(s)\Delta u_{\mathrm{dc\_r}}(s)\end{bmatrix}\end{aligned} \tag{3.30}$$

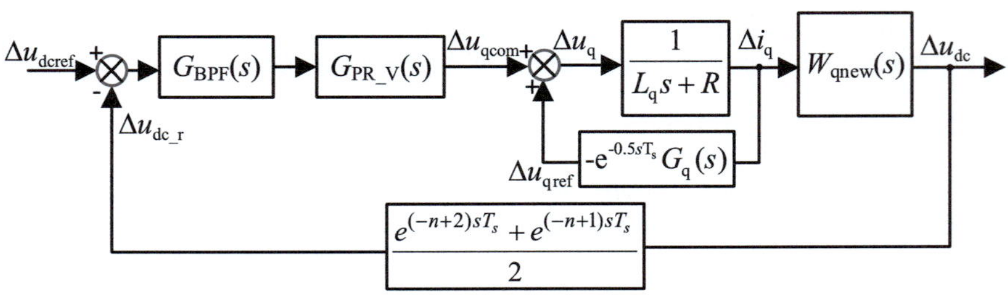

Fig. 3.27 Block diagram of DC-link voltage decoupling

where $G_{PR\_v}(s)$ is the transfer function of the voltage resonant controller. And the transfer function between the small signal variations of dq axis currents and the small signal variations of the DC-link voltage is

$$\begin{aligned}\begin{bmatrix} W_{qnew}(s) \\ W_{qnew}(s) \end{bmatrix} = \frac{1}{\Delta u_{dc}} \begin{bmatrix} \Delta i_d(s) \\ \Delta i_q(s) \end{bmatrix} = \begin{bmatrix} W_d(s) \\ W_q(s) \end{bmatrix} \\ + G_{mad}(s)\frac{\Delta i_g(s)}{\Delta u_{dc}(s)}\begin{bmatrix} A_{1q}(s) \\ -A_{1d}(s) \end{bmatrix} \\ + G_{mad2}(s)\begin{bmatrix} A_{1q}(s) \\ -A_{1d}(s) \end{bmatrix}\end{aligned} \tag{3.31}$$

where

$$G_{mad2}(s) = -G_{BPF}(s)G_{PR\_v}(s) \tag{3.32}$$

At this time, the input impedance of the motor drive Y_{new} can be expressed as

$$Y_{new}(s) = \frac{Y_m + \frac{3}{2}\frac{(G_{mad}A_{1q}(s)C_{dc}s + G_{mad2}A_{1q}(s))(U_{d0} + \omega_e L_d I_{q0} + L_d I_{d0} s + I_{d0} R_s)}{U_{dc0}} - \frac{3}{2}\frac{(G_{mad}A_{1d}(s)C_{dc}s + G_{mad2}A_{1d}(s))(U_{q0} - \omega_e L_q I_{d0} + L_q I_{q0} s + I_{q0} R_s)}{U_{dc0}}}{1 - \frac{3}{2}\frac{G_{mad}A_{1q}(s)(U_{d0} + \omega_e L_d I_{q0} + L_d I_{d0} s + I_{d0} R_s)}{U_{dc0}} + \frac{3}{2}\frac{G_{mad}A_{1d}(s)(U_{q0} - \omega_e L_q I_{d0} + L_q I_{q0} s + I_{q0} R_s)}{U_{dc0}}} \tag{3.33}$$

3.3.3 *Analysis of Harmonic Suppression Effect*

With the harmonic suppression method for rectified current regulation based on DC-link voltage decoupling, the Nyquist diagram of Y_{new}/Y_{LC} is shown in Fig. 3.28. From the above analysis, it can be seen that with the harmonic suppression method for rectified current, when $K_{RP\_C} = 8$ and $K_{RR\_C} = 100$, although the rectified current control can improve the harmonics suppression effect of the DC-link voltage, the system tends to be unstable at this time. As can be seen from Fig. 3.28, under the same current proportional resonant controller parameters, after applying the rectified current regulation with DC-link voltage decoupling, the intersection point of Nyquist diagram of Y_{new}/Y_{LC} and virtual axis moves from the left side of $(-1, j0)$ to the right side, and the control system tends to be stable.

Figure 3.29 shows G_D(s) Bode diagram after applying the rectified current regulation with DC-link voltage decoupling. It can be seen from the amplitude-frequency characteristic curve of G_D(s) that with the increase of resonant parameter K_{RR_V} of the

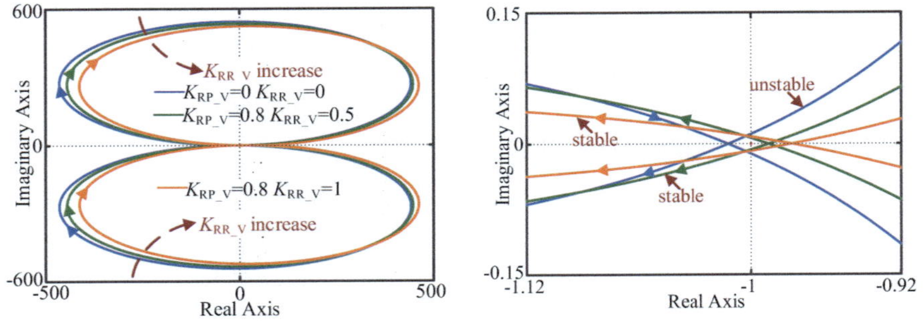

(a) Overall view (b) Zoomed view

Fig. 3.28 Nyquist diagrams of $Y_{\text{new}}(s)/Y_{\text{LC}}(s)$ under different resonant parameters, reprinted from Ref. [10], copyright 2022, with permission from Nature

DC-link voltage proportional resonant controller, the amplitude of $G_{\text{D}}(\text{s})$ decreases at $6\omega_{\text{g}}$ and $12\omega_{\text{g}}$, and the closed-loop control of rectified current enhances its ability to suppress the harmonic disturbance of the DC-link voltage at $6\omega_{\text{g}}$ and $12\omega_{\text{g}}$.

After applying the rectified current regulation with DC-link voltage decoupling, the transfer function $G_{\text{y}}(\text{s})$ between the small signal variations of DC-link voltage Δu_{dc} and the small signal variations of rectifier voltage Δu_{grec} can be expressed as

$$G_{\text{y}}(s) = \frac{1}{\left(L_{\text{g}}s + R_{\text{g}}\right)[sC_{\text{dc}} + Y_{\text{new}}(s)] + 1} \tag{3.34}$$

The input impedance $Z_{\text{gmad}}(\text{s})$ of grid side can be expressed as

Fig. 3.29 Bode diagrams of transfer function between Δi_{g} and Δu_{d}

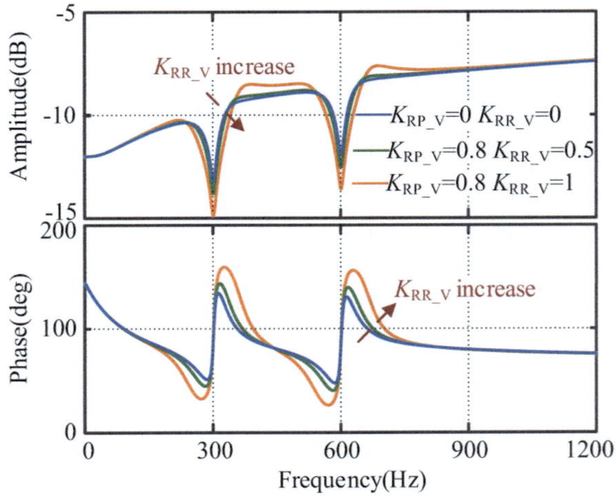

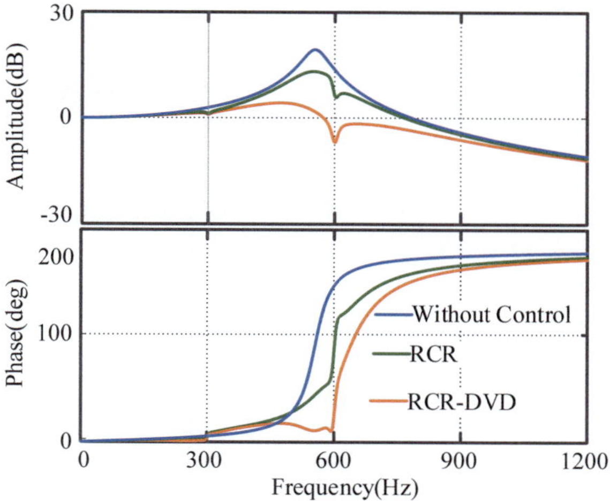

Fig. 3.30 Bode diagram of the transfer function between the DC-link voltage and the rectified voltage

$$Z_{\mathrm{gmad}}(s) = R_{\mathrm{g}} + sL_{\mathrm{g}} + \frac{1}{C_{\mathrm{dc}}s + Y_{\mathrm{new}}(s)} \tag{3.35}$$

The Bode diagram of $G_{\mathrm{y}}(\mathrm{s})$ before and after applying the rectified current regulation with DC-link voltage decoupling is shown in Fig. 3.30. Without control method, the ratio of the small-signal variations of rectified voltage to the small signal variations of DC-link voltage appears voltage peak near the resonant frequency, resulting in the rectified voltage harmonics near the resonant frequency in the DC-link voltage is amplified. The voltage peak is reduced when the rectified current regulation is applied. Moreover, when rectified current regulation with DC-link voltage decoupling is applied, the peak is further reduced and the DC-link voltage harmonic is further suppressed.

The Bode diagram of the input impedance before and after adding the rectified current regulation with DC-link voltage decoupling is shown in Fig. 3.31. Without the control strategy, the input impedance of the grid side is small near the resonant frequency, so even if the output voltage harmonics of the rectifier are small, there will be larger harmonics in the rectified current. After the rectified current regulation is applied, the input impedance at the resonant frequency is enlarged. After the rectified current regulation with DC-link voltage decoupling is applied, the impedance near harmonic frequency of $6\omega_{\mathrm{g}}$ and $12\omega_{\mathrm{g}}$ increases, and the harmonics of grid currents at $6\omega_{\mathrm{g}}$ and $12\omega_{\mathrm{g}}$ can be effectively suppressed.

3.3.4 Experimental Validation

Figure 3.32a shows the experiment results of the harmonics at $6\omega_{\mathrm{g}}$ and $12\omega_{\mathrm{g}}$ in the DC-link voltage before and after applying the control strategies. As shown

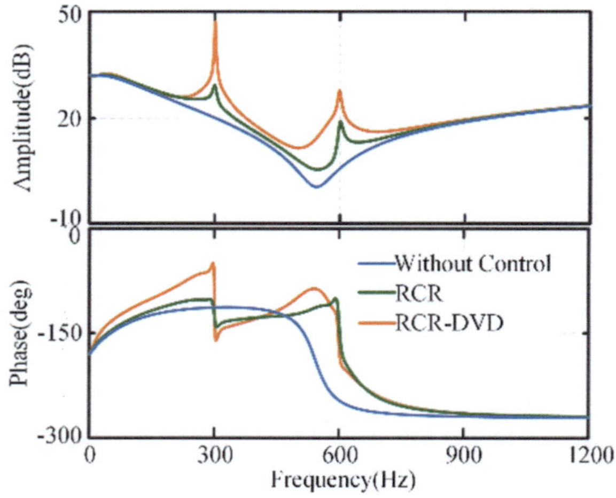

Fig. 3.31 Bode diagram of the grid-side input impedance before and after applying the control strategies

in Fig. 3.32b, The amplitudes of the harmonics voltages at $6\omega_g$ and $12\omega_g$ before applying the control are 49.9 V and 15.3 V, respectively. After applying rectified current regulation, the amplitudes of the corresponding harmonics are decreased to 39.8 V and 11.2 V, respectively, which is shown in Fig. 3.32c. With the harmonics prediction, the amplitudes of the corresponding harmonics decrease to 37.9 V and 6.13 V as shown in Fig. 3.32d. Figure 3.32e shows that after applying rectified current regulation with DC-link voltage decoupling, the amplitudes of the corresponding harmonic are further decreased to 32.6 V and 4.6 V. The results indicate that applying rectified current regulation with DC-link voltage decoupling can significantly reduce the amplitudes of harmonics in DC-link voltage compared to rectified current regulation.

Figure 3.33a shows harmonics at $6\omega_g$ and $12\omega_g$ in the rectified current before and after applying the harmonic suppression based on rectified current regulation with DC-link voltage decoupling control strategy. As shown in the Fig. 3.33b, the amplitudes of harmonics at $6\omega_g$ and $12\omega_g$ in the rectified current before applying the control are 2.6 A and 1.7 A, respectively. After applying rectified current regulation, the amplitudes of the corresponding harmonics decrease to 0.4 A and 1.4 A, which is shown in Fig. 3.33c. With the harmonics prediction, the amplitudes of the corresponding harmonics decrease to 0.5 A and 1 A as shown in Fig. 3.33d. Figure 3.33e shows that after applying rectified current regulation with DC-link voltage decoupling, the amplitudes of corresponding harmonics can be further decrease to 0.05 A and 0.5 A. After applying the harmonic suppression based on rectified current regulation with DC-link voltage decoupling control strategies, the amplitudes of the harmonics at $6\omega_g$ and $12\omega_g$ are significantly reduced, and the power quality on the grid side is further improved.

Figure 3.34 shows the waveforms of the u-phase grid current and DC-link voltage before and after applying the control strategies. The waveforms of the DC-link voltage and grid current before and after applying rectified current regulation are shown in

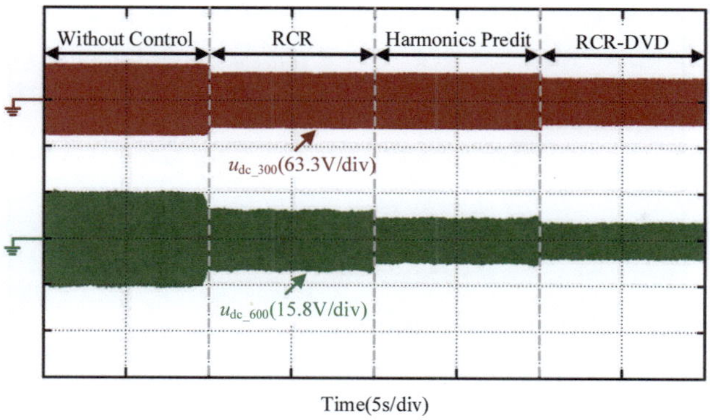

(a) Enable control strategies

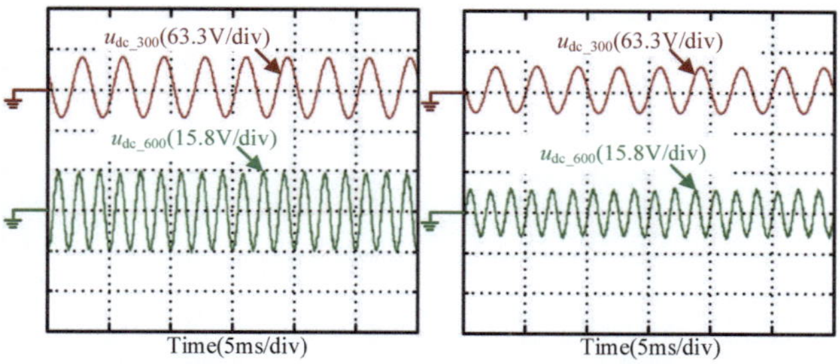

(b) Zoomed view without control (c) Zoomed view with RCR

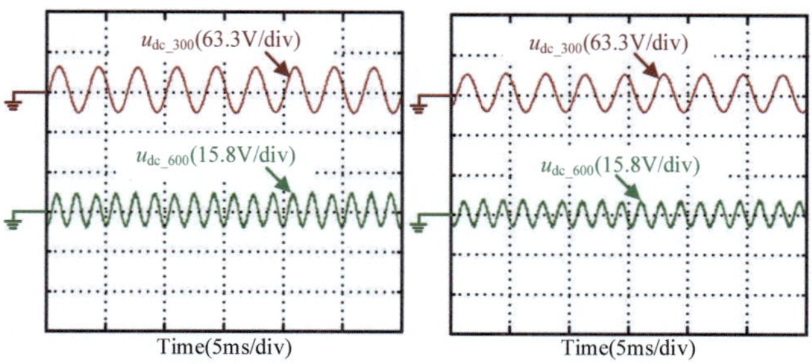

(d) Zoomed view with main harmonics prediction (e) Zoomed view with RCR-DVD

Fig. 3.32 Experimental results of DC-link voltage harmonic suppression, reprinted from Ref. [10], copyright 2022, with permission from Nature

Fig. 3.34b, c, respectively. After applying rectified current regulation, the THD of grid current decreases from 45.6% to 33.8%, and fluctuation of the DC-link voltage decreases from 145.7 to 107.9 V. With the harmonics prediction, the THD of grid currents decreases to 32.9%, and fluctuation of the DC-link voltage decreases from 107.9 to 81.1 V as shown in Fig. 3.34d. Figure 3.34e shows the waveforms after applying rectified current regulation with DC-link voltage decoupling. The suppression effect on the grid current and DC-link voltage harmonics is further improved.

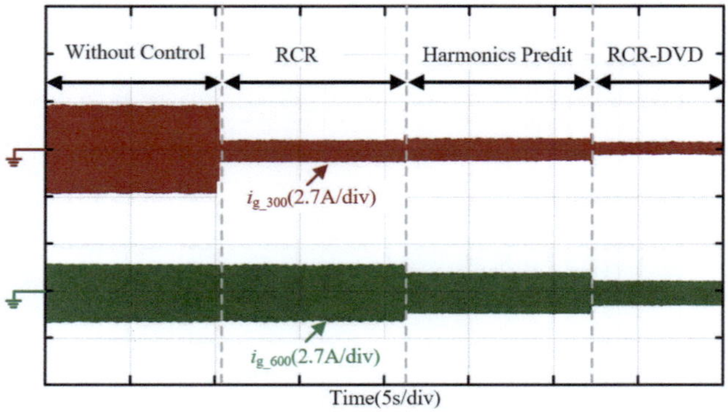

(a) Enable control strategies

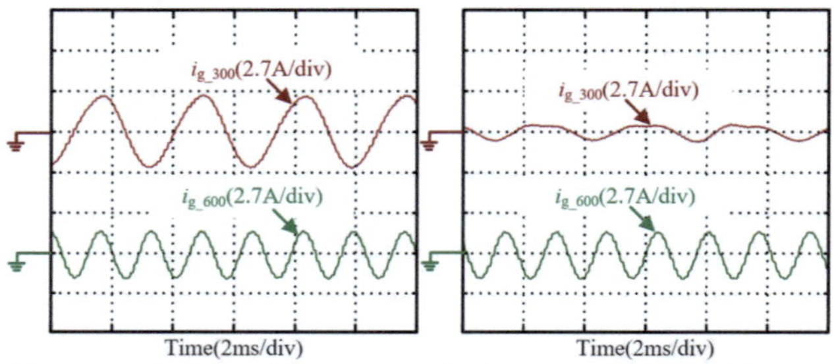

(b) Zoomed view without control (c) Zoomed view with RCR

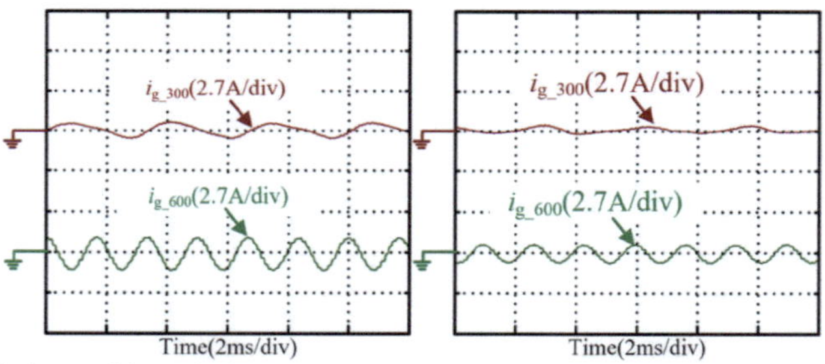

(d) Zoomed view with main harmonics prediction (e) Zoomed view with RCR-DVD.

Fig. 3.33 Experimental results of inductive current harmonic suppression, reprinted from Ref. [10], copyright 2022, with permission from Nature

The THD of the grid current is further decreased to 30.6% and the fluctuation of DC-link voltage decreases to 66.1 V. Fourier analysis shows that compared with the waveform before the control, the 5th, 11th, and 13th harmonics component in the grid current are reduced after applying rectified current regulation. With the proposed strategy, the 7th, 11th, and 13th harmonics component in the grid current are further reduced. After applying rectified current regulation with DC-link voltage decoupling, the fluctuation of the DC-link voltage vector and the THD of the gird current are significantly decreased, resulting in an improvement in the quality of grid currents.

Figure 3.35 shows the THD of the grid current after applying rectified current regulation with DC-link voltage decoupling at different power levels. As can be seen, applying the proposed control strategy can effectively reduce the THD of the grid-side current and improve the power quality of the grid-side input current under different working conditions.

3.4 Method for Grid Current Quality Improvement in Electrolytic Capacitorless PMSM Drives with Low DC-Link Inductance

3.4.1 Analysis of the Relationship Between Grid Current and LC Resonance Phenomenon

For the electrolytic capacitorless PMSM drive with low DC-link inductance, the increase of the resonant frequency and the complex coupling between the grid and DC-link current currents result in the performance reduction of active damping method. To improve the quality of the grid current in drives with low DC-link inductance, further analysis of the grid current and LC resonance is required based on the impedance model. By analyzing the ratio of the harmonic amplitude of the grid current to the harmonic amplitude of the rectified voltage near the resonant frequency, the impact of LC resonance on the grid current harmonics is quantitatively calculated. The small signal model of the drive, as shown in Fig. 3.36, is established to analyze the relationship between the rectified voltage (as the input) and the grid current harmonics (as the response). The rectified voltage u_{g} can be expressed as

$$u_{\mathrm{g}} = 3U_{\mathrm{g}}\left[1 - \sum_{n=1}^{\infty} 2\cos\left(6n\omega_{\mathrm{g}}t + 6n\varphi_{\mathrm{g}}\right) / \left(36n^2 - 1\right)\right] / \pi \tag{3.36}$$

where U_{g} and φ_{g} are the amplitude and phase of the line voltage, respectively.

The DC-link voltage u_{dc} and DC side current i_{g} can be expressed as

$$\begin{cases} u_{\mathrm{dc}} = U_{\mathrm{dc0}} + \sum\limits_{n=1}^{\infty} u_{\mathrm{dc\_n}} = U_{\mathrm{dc0}} + \sum\limits_{n=1}^{\infty} U_{\mathrm{dc\_n}} \sin\left(6n\omega_{\mathrm{g}}t + \varphi_{\mathrm{dc\_n}}\right) \\ i_{\mathrm{g}} = I_{\mathrm{g0}} + \sum\limits_{n=1}^{\infty} i_{\mathrm{g\_n}} = I_{\mathrm{g0}} + \sum\limits_{n=1}^{\infty} I_{\mathrm{g\_n}} \sin\left(6n\omega_{\mathrm{g}}t + \varphi_{\mathrm{g\_n}}\right) \end{cases} \tag{3.37}$$

where $U_{\mathrm{dc\_n}}$ and $\varphi_{\mathrm{dc\_n}}$ are the amplitude and phase of $u_{\mathrm{dc\_n}}$, respectively. I_{g0} and $i_{\mathrm{g\_n}}$ are the mean value and the nth harmonic of the DC-link current. $I_{\mathrm{g\_n}}$ and $\varphi_{\mathrm{g\_n}}$ are the amplitude and phase of $i_{\mathrm{g\_n}}$, respectively.

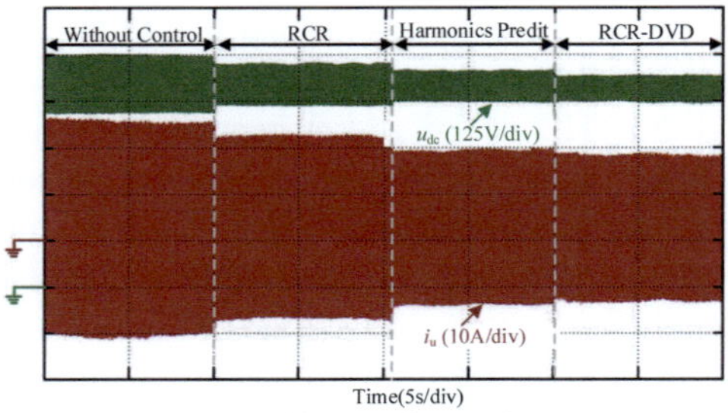

(a) Enable control strategies.

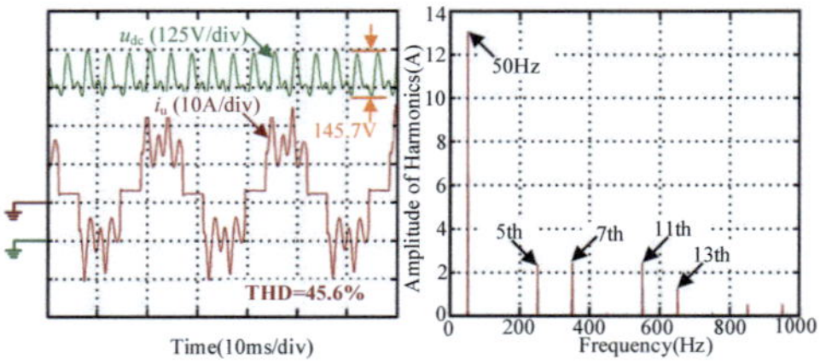

(b) Zoomed view without control and Fourier analysis of the grid current

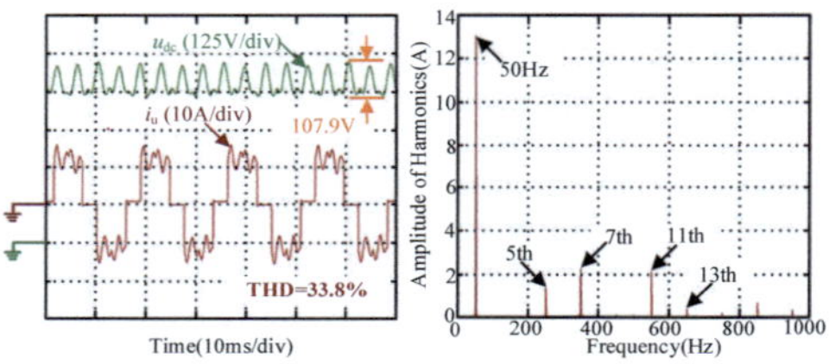

(c) Zoomed view with RCR and Fourier analysis of the grid current

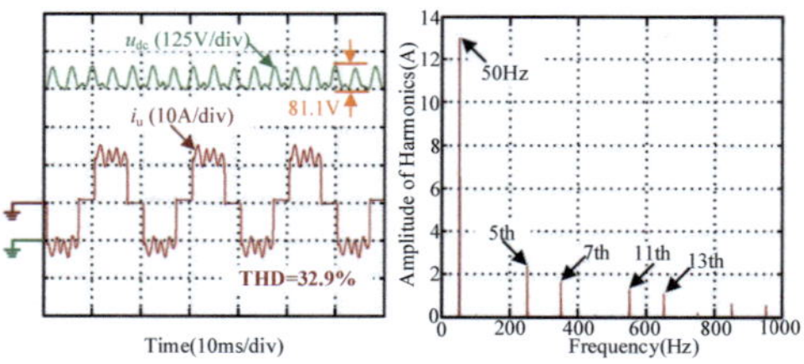

(d) Zoomed view with main harmonics prediction and Fourier analysis of the grid current

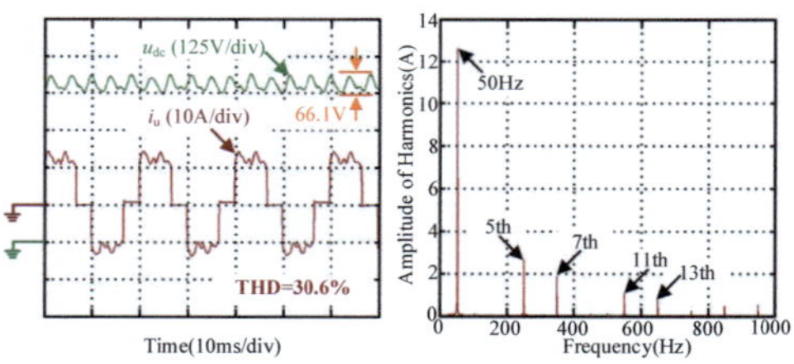

(e) Zoomed view with RCR-DVD and Fourier analysis of the grid current.

Fig. 3.34 Experimental results of DC-link voltage and grid current, reprinted from Ref. [10], copyright 2022, with permission from Nature

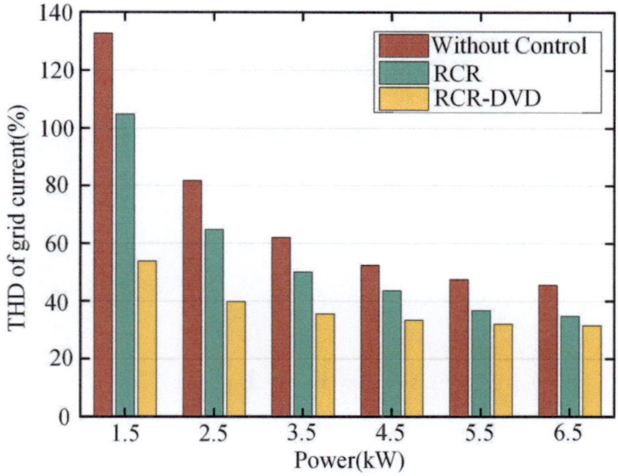

Fig. 3.35 THD of grid-side current before and after applying control strategies at different power levels

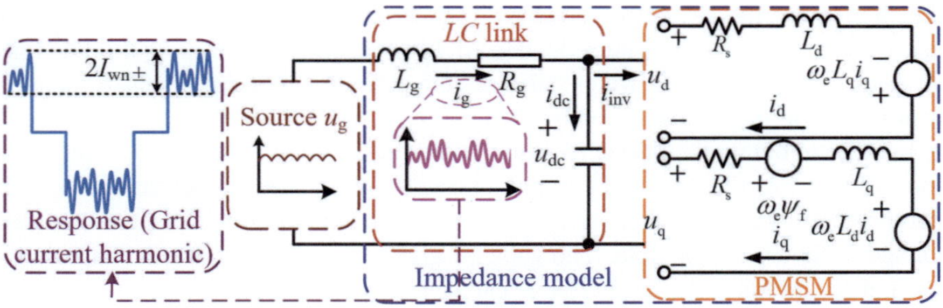

Fig. 3.36 Impedance model with the drive with small DC-link capacitor, reprinted from Ref. [11], copyright 2022, with permission from Nature

The DC-link current is directly related to the grid current, which means the grid current harmonics can be obtained by $\Delta i_g/\Delta u_g$. Δi_g and Δu_g are the small signals of DC-link current and rectified voltage. The amplitude and phase errors caused by the sampling process cannot be ignored under the high- frequency resonance, whose transfer function $\boldsymbol{G}_h$ is

$$\mathbf{G}_h = \left(1 - e^{-sT_s}\right)/sT_s \tag{3.38}$$

$\boldsymbol{K}_m$ is defined as the transfer function between the DC-link current and the rectified voltage, which is derived as

$$\mathbf{K}_m = \Delta i_g/\Delta u_g = 1/\left[1/(\mathbf{Y}_m + C_{dc}s) + \left(L_g s + R_g\right)\right] \tag{3.39}$$

$\Delta\boldsymbol{K}_m$ is defined as

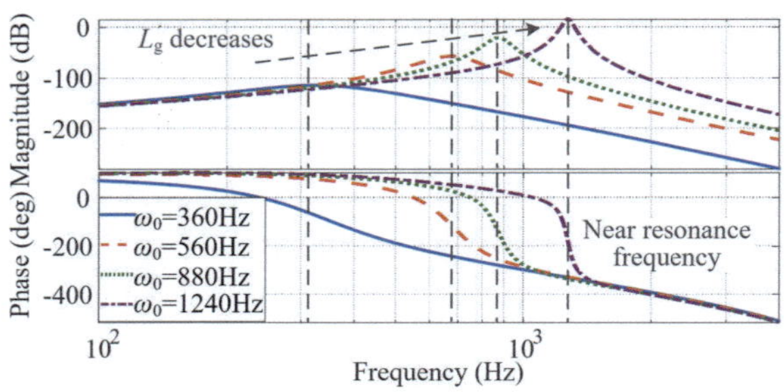

Fig. 3.37 Block diagram of $\Delta\boldsymbol{K}_m$ under different values of ω_0

$$\Delta\mathbf{K}_m = \mathbf{K}_m - \mathbf{K}_{m,id} \tag{3.40}$$

where $\boldsymbol{K}_{m,id}$ is the transfer function of $\Delta i_g/\Delta u_g$ under the ideal condition ($T_s = 0$).

From (3.40), the sampling and updating errors of the harmonics caused by LC resonance increase as ω_0 increases. $\Delta\boldsymbol{K}_m$ can be applied to analyze the deviations of the DC-link current caused by the digital algorithm, whose Bode diagrams are shown in Fig. 3.37. As ω_0 increases from 360 to 1240 Hz, the amplitude and phase deviations near the resonant frequency vary from − 114 dB and – 68° to 4 dB and – 184°, respectively. The reduction of L_g increases the errors of the DC-link current harmonics near ω_0, which enhances the difficulty of extracting the resonance information of grid current directly.

In order to obtain the grid current harmonics affected by LC resonance, the relationship between the grid and DC-link currents needs to be further analyzed. The phase relationship among the grid current, the DC-link current and the rectified voltage is shown in Fig. 3.38. The *w*-phase grid current i_w contains the harmonics i_{wn-} and i_{wn+} at $(6n - 1)\omega_g$ and $(6n + 1)\omega_g$

$$\begin{aligned} i_{wn\pm} &= a_n \cos(6n \pm 1)\left(\omega_g t + \varphi\right) + b_n \sin(6n \pm 1)\left(\omega_g t + \varphi\right) \\ &= I_{wn\pm} \sin\left[(6n \pm 1)\left(\omega_g t + \varphi\right) + \varphi_{wn\pm}\right] \end{aligned} \tag{3.41}$$

where $I_{wn\pm}$ is the amplitude of the grid current harmonics at the frequency of $(6n \pm 1)\omega_g$, $\varphi_{wn\pm}$ is the phase difference between the grid current $i_{wn\pm}$ and the rectified voltage harmonics related to $6n\omega_g$.

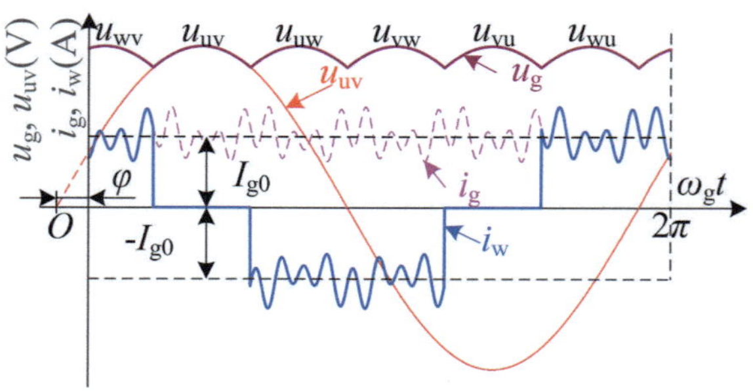

Fig. 3.38 Phase relationship among i_w, i_g and u_g

$$\begin{cases} a_{\rm n} = \pm\dfrac{2\sqrt{3}}{\pi}\left\{\dfrac{I_{\rm g0}}{6n\pm 1} - \displaystyle\sum_{n_1=1}^{\infty}\dfrac{(6n\pm 1)I_{{\rm g}\_n_1}\sin\varphi_{{\rm er}n_1}}{\left[(6n_1)^2-(6n\pm 1)^2\right]}\right\} \\ b_{\rm n} = \mp\displaystyle\sum_{n_1=1}^{\infty} 12\sqrt{3}n_1 I_{{\rm L}\_n_1}\cos\varphi_{{\rm er}n_1}\Big/\pi\left[(6n_1)^2-(6n\pm 1)^2\right] \end{cases} \tag{3.42}$$

In (3.42), $\varphi_{\rm ern}$ is the phase difference between the DC-link current $i_{\rm g\_n}$ and the rectified voltage harmonics at $6n\omega_{\rm g}$

$$\varphi_{\rm ern} = \varphi_{\rm g\_n} - 6n\varphi = \angle\mathbf{K}_{\rm mn} - \pi/2 \tag{3.43}$$

where $\boldsymbol{K}_{\rm mn}$ represents $\boldsymbol{K}_{\rm m}(j6n\omega_{\rm g})$.

From (3.42), the grid harmonics $i_{\rm wn-}$ and $i_{\rm wn+}$ are closely related to the mean value $I_{\rm g0}$ and harmonics $i_{\rm g\_n}$ of the DC-link current at $6n\omega_{\rm g}$, which means the harmonics of grid current is coupled with those of DC-link current. Hence, (3.41) can be rewritten as

$$\begin{aligned} i_{\rm wn\pm} = i_{\rm 0n\pm} + i_{\rm adn\pm} = 2\sqrt{3}\{\pm I_{\rm g0}\cos(6n\pm 1)(\omega_{\rm g}t+\varphi)/(6n\pm 1) \\ +M_{\rm 6n\pm}I_{\rm L\_n}\sin[(6n\pm 1)(\omega_{\rm g}t+\varphi)+\varphi_{\rm 6\_n\pm}]/(12n\pm 1)\}/\pi \end{aligned} \tag{3.44}$$

where $i_{\rm 0n\pm}$ and $i_{\rm adn\pm}$ are the grid current harmonics related to the mean value and harmonics of the DC-link current, respectively. The amplitude of $i_{\rm 0n\pm}$ is less influenced by the *LC* resonance.

In (3.44), $M_{\rm 6n\pm}$ is the amplitude ratio between $i_{\rm ad\pm}$ and $i_{\rm g\_n}$. $\varphi_{\rm 6n\pm}$is the phase difference between $i_{\rm ad\pm}$ and the rectified voltage harmonics at the frequency of $6n\omega_{\rm g}$

$$M_{\rm 6n\pm} = \sqrt{36n^2 + (\pm 12n+1)\sin^2\varphi_{\rm ern}},\ \varphi_{\rm 6n\pm} = \arcsin\left[(6n\pm 1)\sin\varphi_{\rm ern}/M_{\rm 6n\pm}\right] \tag{3.45}$$

From (3.44), the amplitude $I_{\rm wn\pm}$ and phase $\varphi_{\rm wn\pm}$ of the grid current harmonics can be expressed as

$$\begin{cases} I_{\rm wn\pm} = \dfrac{2\sqrt{3}}{\pi}\sqrt{\left[\dfrac{6M_{\rm 6n\pm}|\mathbf{K}_{\rm mn}|U_{\rm g}\sin\varphi_{\rm 6n\pm}}{\pi(36n^2-1)(12n\pm 1)} \pm \dfrac{I_{\rm g0}}{6n\pm 1}\right]^2 + \left[\dfrac{6M_{\rm 6n\pm}|\mathbf{K}_{\rm mn}|U_{\rm g}\cos\varphi_{\rm 6n\pm}}{\pi(36n^2-1)(12n\pm 1)}\right]^2} \\ \varphi_{\rm wn\pm} = \arcsin\dfrac{6M_{\rm 6n\pm}|\mathbf{K}_{\rm mn}|U_{\rm g}\cos\varphi_{\rm 6n\pm}}{\pi(36n^2-1)(12n\pm 1)I_{\rm wn\pm}} \end{cases} \tag{3.46}$$

The harmonic order $k_{\rm r}$ of the DC-link current which is severely affected by *LC* resonance is

$$k_{\rm r} = \mathbf{R}(1/6\omega_{\rm g}\sqrt{L_{\rm g}C_{\rm dc}}) \tag{3.47}$$

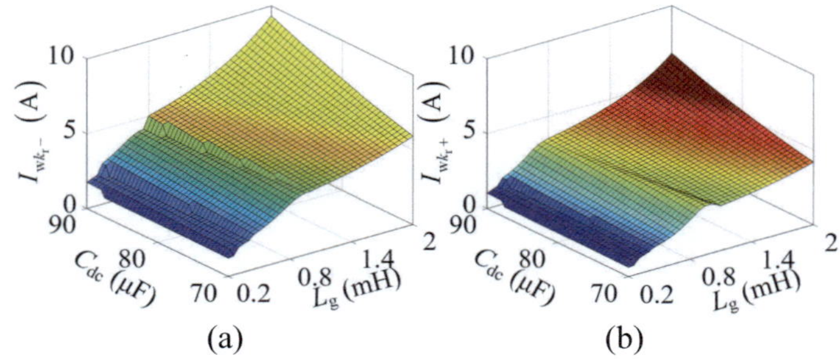

Fig. 3.39 Relationship between the grid current ripple and *LC* resonance under the output power of 5 kW. **a** At $(6k_{r-}1)\omega_g$. **b** At $(6k_r + 1)\omega_g$, reprinted from Ref. [11], copyright 2022, with permission from Nature

where $\boldsymbol{R}(\cdot)$ rounds off the input to the nearest integer. The harmonic orders of the grid current which are severely affected by *LC* resonance are $(6k_r$-1) and $(6k_r + 1)$.

The grid current harmonics near the resonant frequency are shown in Fig. 3.39, whose amplitudes decrease with the reduction of L_g and C_{dc}. As L_g and C_{dc} are lower than 0.6 mH and 75 μF respectively, the harmonic amplitudes become less than 2 A. Compared with the drive system with large DC-link inductance, the difficulty of extracting the resonant component at the grid further increases.

Besides, the coupling between the grid harmonics $i_{wn\pm}$ and the mean value I_{g0} of the DC-link current also affects the accuracy of the grid resonance extraction. If only suppressing the resonant harmonic $i_{g\_k_r}$ of the DC-link current, the harmonic amplitude I_{wk_r} at grid side may not decrease due to the change of the phase difference φ_{erk_r}. Taking the resonant frequency of 1258 Hz as an example for analysis. The harmonic amplitudes $I_{w4\pm}(k_r = 4)$ under the variation of i_{g_4} are shown in Fig. 3.40. Assuming $A_\pm$ and $B_\pm$ are the points without and with the method respectively, the dotted line shows operation trajectory of $I_{w4\pm}$. I_{w4-} and I_{w4+} are suppressed by 1.90 A and 0.60 A respectively. As for *B*-, I_{w4-} is lower than the amplitude of i_{m4-}, which indicates *LC* resonance suppresses I_{w4-} at this point. As I_{g_4} is further reduced to 0 A, I_{w4-} is increased from 0.75 A to 1 A, which worsens the grid quality. This means the DC-link current harmonics is not suitable to characterize the *LC* resonance information of the grid current in the drive systems with low inductance. Hence, the grid current improved method considering the influence of I_{g0} need to be further researched.

3.4.2 Feature Extraction Principle of Frequency Mapping

The block diagram of the resonance suppression strategy based on the closed loop of the feature signal is shown in Fig. 3.41. The DC-link current i_g is reconstructed with the motor current $\boldsymbol{i}_s$, duty cycle $D_{a,b,c}$, and DC-link voltage. The compensated feature signals $i_{fe(wk_r\pm)\_r}$ are generated by compensating the errors from the sampling

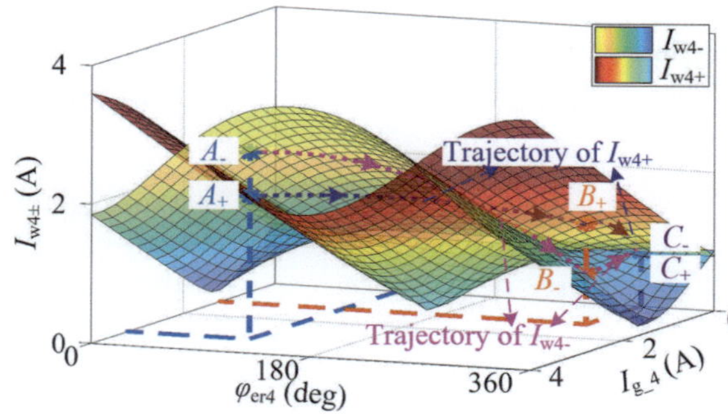

Fig. 3.40 Grid current amplitude near the resonant frequency under the variation of the amplitude and phase in the DC-link current, reprinted from Ref. [11], copyright 2022, with permission from Nature

and updating processes. x is the frequency coefficient, which adjusts the frequencies of $i_{\text{fe}(\text{w}k_r-)\_r}$ and $i_{\text{fe}(\text{w}k_r+)\_r}$ to $6\omega_g$ and $12\omega_g$ respectively. The output of control loop is the voltage adjusting angle $\Delta\theta_u$, which is superimposed on the motor estimated position $\hat{\theta}_e$. The adjusted voltage angle is θ_{un}, which is involve ed in the inverse Park transformation of the voltage reference. The advantage of the closed loop of grid feature signal is achieving the direct control of high-frequency resonance by decoupling the grid and the DC-link currents.

The vector diagram of the grid current and the DC-link current is shown in Fig. 3.42. From (3.44), the vector $I_{wn\pm}e^{j\varphi_{wn\pm}}$ of grid harmonics is the sum of the vectors $\boldsymbol{I}_{mn\pm}$ and $\boldsymbol{I}_{hn\pm}$ related to the mean value and harmonics of DC-link current respectively

$$I_{wn\pm}e^{j\varphi_{g\_n\pm}} = \frac{2\sqrt{3}}{\pi}\left[\pm j\frac{I_{g0}}{6n \pm 1} + \frac{M_{6n\pm}I_{g\_n}}{12n \pm 1}e^{j\varphi_{6n\pm}}\right] \tag{3.48}$$

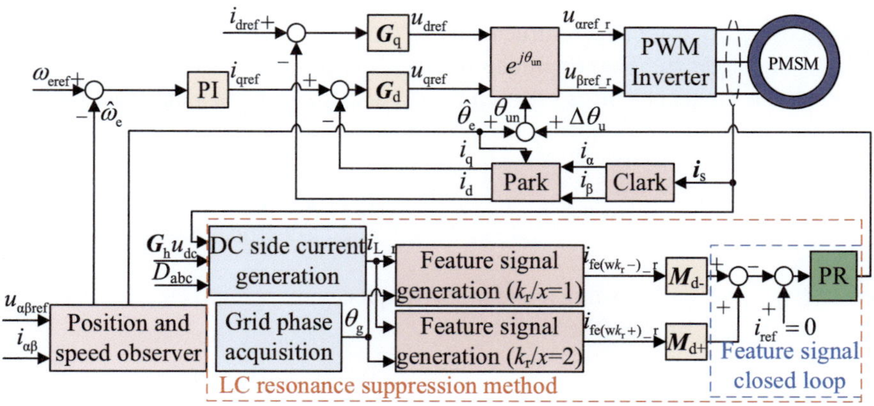

Fig. 3.41 Block diagram of resonance suppression strategy based on closed loop of feature signal at dominated frequencies, reprinted from Ref. [11], copyright 2022, with permission from Nature

Fig. 3.42 Vector diagram of the grid current and the DC-link current harmonics

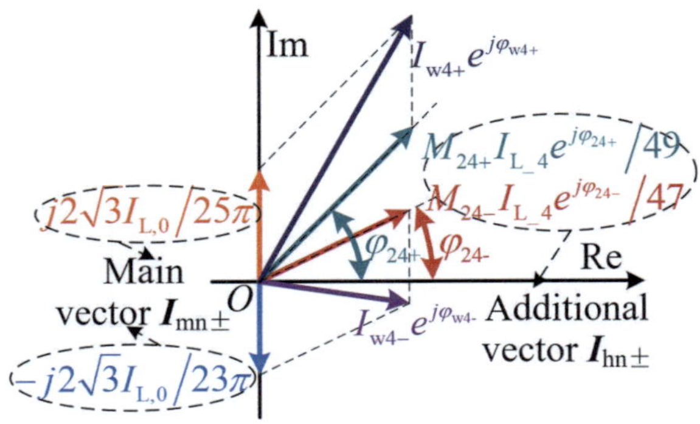

The ideal feature signals $i_{\mathrm{fe(wk\pm)}}$ are extracted according to the vector relationship in (3.48), whose amplitude and frequency are $I_{\mathrm{wn\pm}}$ and $6n\omega_{\mathrm{g}}/x(x > 1)$, respectively.

$$i_{\mathrm{fe(wn\pm)}} = I_{\mathrm{wn\pm}} \sin\left[6n\left(\omega_{\mathrm{g}}t + \varphi_{\mathrm{ad}}\right)/x + \varphi_{\mathrm{wn\pm}}\right] \tag{3.49}$$

where φ_{ad} is the additional phase to adjust the feature extraction.

In the system, the angular frequency of the grid line voltage, $\theta_{\mathrm{g}} = \omega_{\mathrm{g}}t + \varphi$, can be directly extracted, so the additional phase φ_{ad} can be set to φ. With the imaginary part of (3.48), $i_{\mathrm{fe(wn\pm)}}$ can be regarded as the sum of the ideal feature signals $i_{\mathrm{fe(mn\pm)}}$ and $i_{\mathrm{fe(hn\pm)}}$ related to the mean value and harmonics of i_{g} respectively

$$\begin{aligned} i_{\mathrm{fe(wn\pm)}} = \frac{2\sqrt{3}}{\pi}\left[i_{\mathrm{fe(mn\pm)}} + i_{\mathrm{fe(hn\pm)}}\right] = \frac{2\sqrt{3}}{\pi}\left\{\pm\frac{I_{\mathrm{g0}}}{6n \pm 1}\cos\left[6n\left(\omega_{\mathrm{g}}t + \varphi\right)/x\right]\right. \\ \left. + \frac{M_{\mathrm{6n\pm}}I_{\mathrm{L\_n}}}{12n \pm 1}\sin\left[6n\left(\omega_{\mathrm{g}}t + \varphi_{\mathrm{ad}}\right)/x + \varphi_{\mathrm{6n\pm}}\right]\right\} \end{aligned} \tag{3.50}$$

Hence, $i_{\mathrm{fe(mn\pm)}}$ can be reconstructed using the mean value of the DC-link current I_{g0} and the angle θ_{g}. The amplitude of $i_{\mathrm{fe(hn\pm)}}$ is proportional to that of $i_{\mathrm{g\_n}}$. To eliminate the frequency error between $i_{\mathrm{fe(hn\pm)}}$ and $i_{\mathrm{g\_n}}$, the mapping from the high frequency $6n\omega_{\mathrm{g}}$ to low frequency $6n\omega_{\mathrm{g}}/x$ can be expressed as

$$i_{\mathrm{fe(hn\pm)}} = \frac{M_{\mathrm{6n\pm}}}{12n \pm 1}\left[\cos\theta_{\mathrm{ch}(n/x)\pm}\ \sin\theta_{\mathrm{ch}(n/x)\pm}\right]\begin{bmatrix} I_{\mathrm{g\_n}}\sin\left(6n\omega_{\mathrm{g}}t + \varphi_{\mathrm{gn}}\right) \\ I_{\mathrm{g\_n}}\cos\left(6n\omega_{\mathrm{g}}t + \varphi_{\mathrm{gn}}\right) \end{bmatrix} \tag{3.51}$$

where $\theta_{\mathrm{ch}(n/x)\pm}$ is the adjusting angle of the frequency mapping.

Hence, $i_{\mathrm{fe(hn\pm)}}$ can be generated from $i_{\mathrm{g\_n}}$

$$i_{\mathrm{fe(hn\pm)}} = \frac{M_{\mathrm{6n\pm}}\left[\cos\theta_{\mathrm{ch}(n/x)\pm} + \sin\theta_{\mathrm{ch}(n/x)\pm}d(\cdot)/6k\omega_{\mathrm{g}}dt\right]i_{\mathrm{g\_n}}}{12n \pm 1} \tag{3.52}$$

And the adjusting angle $\theta_{\mathrm{ch}(n/x)\pm}$ can be expressed as

$$\begin{aligned}\theta_{\mathrm{ch}(n/x)\pm} &= 6n(\omega_g t + \varphi)/x + \varphi_{6n\pm} - (6n\omega_g t + \varphi_{gn}) \\ &= \arcsin\frac{(6n \pm 1)\sin\varphi_{\mathrm{ern}}}{\sqrt{36n^2 + (\mp 12n + 1)\sin^2\varphi_{\mathrm{ern}}}} - 6n\theta_g(1 - 1/x)\end{aligned} \tag{3.53}$$

According to (3.52), $i_{\mathrm{fe(hn}\pm)}$ can be extracted with the DC-link harmonic $i_{\mathrm{g\_n}}$, the grid angle θ_g and the phase error φ_{ern}. The comparison of the feature signals and resonant component of grid current is shown in Fig. 3.43. With the high-to-low frequency mapping in (3.52), $i_{\mathrm{fe}(wk_r\pm)}$ can characterize the high-frequency resonance at grid side, which decouples the grid and the DC-link currents. The frequency of $i_{\mathrm{fe}(wk_r\pm)}$ can be reduced to get away from the resonant frequency ω_0 through increasing x. From the dash dot line in Fig. 3.37, the amplitude and phase of $\Delta\boldsymbol{K}_{\mathrm{m}}$ are decreased with the input frequency, which means the frequency mapping can reduce the decay of the feature signal of the resonant component caused by the low DC-link inductance.

The block diagram of the generation of the feature signals is shown in Fig. 3.44. In order to increase the accuracy of the feature extraction, the phase delays of the applied variables ($\boldsymbol{i}_{\mathrm{s}}$, u_{dc} and D_{abc}) need to be the same, which are set as T_{s}

$$\begin{cases} \mathbf{i}_{\mathrm{s\_r}} = 0.5(1 + e^{-sT_{\mathrm{s}}})\mathbf{G}_{\mathrm{h}}\mathbf{i}_{\mathrm{s}} \approx e^{-0.5sT_{\mathrm{s}}}\mathbf{G}_{\mathrm{h}}\mathbf{i}_{\mathrm{s}} = e^{-sT_{\mathrm{s}}}|\boldsymbol{G}_{\mathrm{h}}|\mathbf{i}_{\mathrm{s}} \\ i_{\mathrm{dc\_r}} = C_{\mathrm{dc}}(1 - e^{-sT_{\mathrm{s}}})\mathbf{G}_{\mathrm{h}}u_{\mathrm{dc}}/T_{\mathrm{s}} \approx se^{-0.5sT_{\mathrm{s}}}C_{\mathrm{dc}}\mathbf{G}_{\mathrm{h}}u_{\mathrm{dc}} = e^{-sT_{\mathrm{s}}}|\boldsymbol{G}_{\mathrm{h}}|i_{\mathrm{dc}} \end{cases} \tag{3.54}$$

where $\boldsymbol{i}_{\mathrm{s\_r}}$ and $i_{\mathrm{dc\_r}}$ are the motor current and the adjusted DC-link current considering sampling delay, respectively.

The duty cycle generated in the FOC is applied at the next switching cycle. Hence, the duty cycle in the reconstruction process should be delayed two switching cycles. Hence, the reconstructed DC-link current $i_{\mathrm{g\_r}}$ can be presented as

$$i_{\mathrm{g\_r}} = e^{-0.5sT_{\mathrm{s}}}\mathbf{G}_{\mathrm{h}}sC_{\mathrm{dc}}u_{\mathrm{dc}} + \left(D_{\mathrm{a\_r}}i_{\mathrm{a\_r}} + D_{\mathrm{b\_r}}i_{\mathrm{b\_r}} + D_{\mathrm{c\_r}}i_{\mathrm{c\_r}}\right) = e^{-sT_{\mathrm{s}}}\mathbf{G}_{\mathrm{h}}i_{\mathrm{L}} \tag{3.55}$$

where $i_{\mathrm{(abc)\_r}}$ is the component of $\boldsymbol{i}_{\mathrm{s\_r}}$ on the *abc*-axis. $D_{\mathrm{(abc)\_r}}$ is the duty cycle with the delay of two switching cycles.

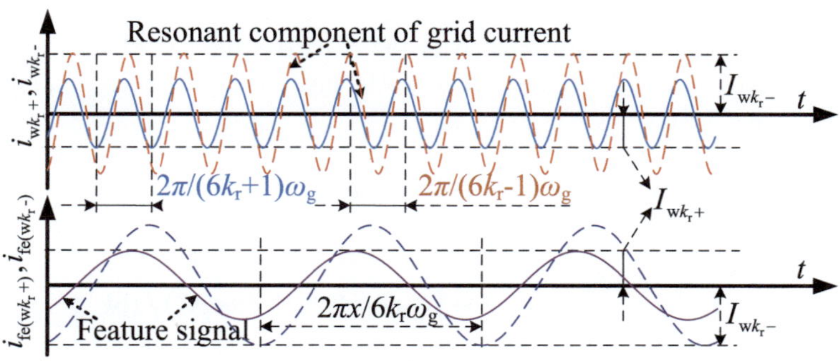

Fig. 3.43 Comparison of the grid current harmonics and the feature signals

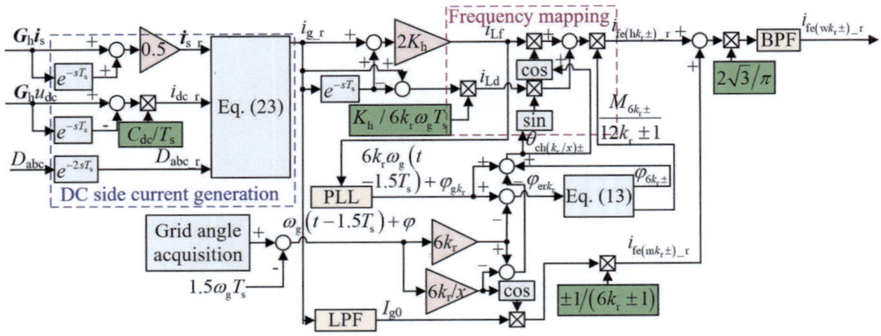

Fig. 3.44 Block diagram of the generation of the feature signals, reprinted from Ref. [11], copyright 2022, with permission from Nature

By compensating for the sampling errors of the DC-link current i_{gc}, the adjusted DC-link current i_{gf} and its quadrature signal i_{gd} are obtained. The harmonics $i_{Ld\_k_r}$ and $i_{Lf\_k_r}$ near the resonance frequency can be expressed as

$$\begin{cases} i_{Lf\_k_r} = \mathbf{L}^{-1}[0.5K_h(1+e^{-sT_s})i_{g\_r,k_r}] = I_{g\_k_r}\sin(6k_r\omega_g t + \varphi_{gk_r} - 9k_r\omega_g T_s) \\ i_{Ld\_k_r} = \mathbf{L}^{-1}[K_h(1-e^{-sT_s})i_{g\_r,k_r}/6k_r\omega_g T_s] = I_{g\_k_r}\cos(6k_r\omega_g(t-1.5T_s) + \varphi_{gk_r}) \end{cases} \tag{3.56}$$

where i_{g_r,k_r} is the resonant harmonic of i_{g_r}, $\boldsymbol{L}^{-1}(\cdot)$ is the inverse Laplace transformation, K_h is the compensation gain

$$K_h = 1/|\mathbf{G}_h(j6k_r\omega_g)| = 1/|(1-e^{-j6k_r\omega_g T_s})/j6k_r\omega_g T_s| \tag{3.57}$$

According to (3.56), $i_{gf\_k_r}$ is delayed by $1.5T_s$ compared to its actual value. The mean value of i_g can be extracted with the lowpass filter (LPF) and phase locked loop (PLL). With the delayed grid angle, the compensated feature signal $i_{fe(mn\pm)\_r}$ related to I_{g0} is extracted from (3.50). The compensated feature signal $i_{fe(hn\pm)\_r}$ related to i_{g_n} is extracted with the frequency mapping, and the compensated feature signals $i_{fe(wk_r-)\_r}$ and $i_{fe(wk_r+)\_r}$ are obtained with the bandpass filter (BPF) whose center frequencies are $6\omega_g$ and $12\omega_g$, respectively. The phase delays of $i_{fe(wk\pm)\_r}$, $i_{fe(mn\pm)\_r}$, and $i_{fe(hn\pm)\_r}$ are $1.5T_s$.

$$\begin{cases} i_{fe(wk_r-)\_r} = \mathbf{L}^{-1}[e^{-1.5sT_s}i_{fe(wk_r-)}] = \left[-I_{g0}\cos\left(6\theta_g - 9\omega_g T_s\right)/\left(6k_r-1\right)\right. \\ \quad \left.+M_{6k_r-}\left(\cos\theta_{ch1\pm}i_{Lf\_k_r} + \sin\theta_{ch1\pm}i_{Ld\_k_r}\right)/\left(12k_r-1\right)\right]\cdot 2\sqrt{3}\Big/\pi \\ i_{fe(wk_r+)\_r} = \mathbf{L}^{-1}[e^{-1.5sT_s}i_{fe(wk_r+)}] = \left[I_{g0}\cos\left(12\theta_g - 18\omega_g T_s\right)/\left(6k_r+1\right)\right. \\ \quad \left.+M_{6k_r+}\left(\cos\theta_{ch2\pm}i_{Lf\_k_r} + \sin\theta_{ch2\pm}i_{Ld\_k_r}\right)/\left(12k_r+1\right)\right]\cdot 2\sqrt{3}\Big/\pi \end{cases} \tag{3.58}$$

The amplitude of $i_{\mathrm{fe(w}k_\mathrm{r}\pm)\_r}$ is equal to the grid harmonics at $(6k_\mathrm{r} \pm 1)\omega_\mathrm{g}$. The amplitudes and phases of $i_{\mathrm{fe(w}k_\mathrm{r}-)\_r}$ and $i_{\mathrm{fe(w}k_\mathrm{r}-)\_r}$ need to be further adjusted to improve the suppression ability, which applies the additional components $\boldsymbol{M}_{\mathrm{d-}}$ and $\boldsymbol{M}_{\mathrm{d+}}$

$$\mathbf{M}_{\mathrm{d-}} = K_1(\cos\varphi_{\mathrm{d\_1}} + \sin\varphi_{\mathrm{d\_1}} \cdot s/6\omega_\mathrm{g}), \mathbf{M}_{\mathrm{d+}} = K_2(\cos\varphi_{\mathrm{d\_2}} + \sin\varphi_{\mathrm{d\_2}} \cdot s/12\omega_\mathrm{g}) \tag{3.59}$$

where K_1, K_2, $\varphi_{\mathrm{d\_1}}$, and $\varphi_{\mathrm{d\_2}}$ are the adjusting coefficient and phases for $i_{\mathrm{fe}k_\mathrm{r}-}$ and $i_{\mathrm{fe}k_\mathrm{r}+}$, respectively.

As seen in Fig. 3.43, the feature signal is a periodic sinusoidal signal. The closed loop of $i_{\mathrm{fe(w}k_\mathrm{r}\pm)\_r}$ applies the proportional resonant (PR) controller to increase the control performance. The resonant frequencies are set as $6\omega_\mathrm{g}$ and $12\omega_\mathrm{g}$.

$$\boldsymbol{G}_{\mathrm{PR\_C}} = K_{\mathrm{RP\_C}} + \frac{2K_{\mathrm{RR\_C1}}\omega_{\mathrm{c1}}s}{s^2 + 2\omega_{\mathrm{c1}}s + 36\omega_\mathrm{g}^2} - \frac{2K_{\mathrm{RR\_C2}}\omega_{\mathrm{c2}}s}{s^2 + 2\omega_{\mathrm{c2}}s + 144\omega_\mathrm{g}^2} \tag{3.60}$$

where $K_{\mathrm{RR\_C1}}$ and $K_{\mathrm{RR\_C2}}$ are the resonant gain of proportional resonant controller related to $6\omega_\mathrm{g}$ and $12\omega_\mathrm{g}$, respectively.

In the FOC, the duty cycle is obtained with the space vector pulse width modulation, whose reference is u_{dc}. The additional harmonics at motor side is generated by the DC-link ripples. Hence, the DC-link voltage can be seen as the input signal of the feature signals. The relationship between the adjusting angle $\Delta\theta_\mathrm{u}$ and DC-link voltage Δu_{dc} is expressed as

$$\Delta\theta_\mathrm{u} = -\boldsymbol{G}_{\mathrm{PR\_C}}(\mathbf{M}_{\mathrm{d-}}\, i_{\mathrm{fe(w}k_\mathrm{r}-)\_r} + \mathbf{M}_{\mathrm{d+}}\, i_{\mathrm{fe(w}k_\mathrm{r}+)\_r}) = \boldsymbol{G}_{\mathrm{mad}}\Delta u_{\mathrm{dc}} \tag{3.61}$$

where $\boldsymbol{G}_{\mathrm{mad}}$ is the transfer function from $\Delta\theta_\mathrm{u}$ to Δu_{dc}.

As shown in (3.61), the PR controller output depends on the difference between i_{ref} and $\mathbf{M}_{\mathrm{d-}}\, i_{\mathrm{fe(w}k_\mathrm{r}-)\_r} + \mathbf{M}_{\mathrm{d+}}\, i_{\mathrm{fe(w}k_\mathrm{r}+)\_r}$. The reference voltages $u_{\alpha\beta\mathrm{ref\_r}}$ in $\alpha\beta$-axes after control are derived as

$$\begin{bmatrix} u_{\alpha\mathrm{ref\_r}} \\ u_{\beta\mathrm{ref\_r}} \end{bmatrix} = \begin{bmatrix} \cos(\hat{\theta}_\mathrm{e} + \Delta\theta_\mathrm{u}) & \sin(\hat{\theta}_\mathrm{e} + \Delta\theta_\mathrm{u}) \\ -\sin(\hat{\theta}_\mathrm{e} + \Delta\theta_\mathrm{u}) & \cos(\hat{\theta}_\mathrm{e} + \Delta\theta_\mathrm{u}) \end{bmatrix} \begin{bmatrix} u_{\mathrm{dref}} \\ u_{\mathrm{qref}} \end{bmatrix} \tag{3.62}$$

As the position and speed observer converges, the estimated rotor position $\hat{\theta}_\mathrm{e}$ is equal to the equivalent sampled position θ_{ee}. The equivalent voltage references $u_{\mathrm{dref\_r}}$ and $u_{\mathrm{qref\_r}}$ in dq-axes can be expressed as

$$\begin{bmatrix} U_{\mathrm{dref\_r,0}} + \Delta u_{\mathrm{dref\_r}} \\ U_{\mathrm{qref\_r,0}} + \Delta u_{\mathrm{qref\_r}} \end{bmatrix} = \begin{bmatrix} U_{\mathrm{dref,0}} + \Delta u_{\mathrm{dref}} - U_{\mathrm{qref,0}}\Delta\theta_\mathrm{u} \\ U_{\mathrm{qref,0}} + \Delta u_{\mathrm{qref}} + U_{\mathrm{dref,0}}\Delta\theta_\mathrm{u} \end{bmatrix} \tag{3.63}$$

where $U_{\mathrm{d,qref\_r,0}}$, $U_{\mathrm{d,qref,0}}$, $\Delta u_{\mathrm{d,qref\_r}}$, and $\Delta u_{\mathrm{d,qref}}$ are the averages and the variations of the reference voltages in dq-axes with and without control, respectively.

According to (3.63), the regulation of voltage angle is equivalent to inject additional small variations $-U_{\mathrm{qref},0}\Delta\theta_{\mathrm{u}}$ and $U_{\mathrm{dref},0}\Delta\theta_{\mathrm{u}}$ into the reference voltage. The introduced voltages contain the harmonics at the frequencies of $6\omega_{\mathrm{g}}$ and $12\omega_{\mathrm{g}}$, which are adopted to reshape the impedance relationship between the grid resonance and rectified voltage.

3.4.3 Parameter Design and Performance Analysis

In order to obtain the optimal control effect, the parameters of adjusting components need to be analyzed. According to (2.14), after applying the proposed strategy, the actual d-q axis voltage variation can be expressed as

$$\begin{bmatrix}\Delta u_{\mathrm{d}}\\ \Delta u_{\mathrm{q}}\end{bmatrix}=C_0\left\{e^{-sT_{\mathrm{s}}}\begin{bmatrix}\Delta u_{\mathrm{dref}}-U_{\mathrm{qref},0}\Delta\theta_{\mathrm{u}}\\ \Delta u_{\mathrm{qref}}+U_{\mathrm{dref},0}\Delta\theta_{\mathrm{u}}\end{bmatrix}+\frac{\left(1-e^{-sT_{\mathrm{s}}}\mathbf{G}_{\mathrm{h}}\right)\Delta u_{\mathrm{dc}}}{U_{\mathrm{dc},0}}\begin{bmatrix}U_{\mathrm{dref},0}\\ U_{\mathrm{qref},0}\end{bmatrix}\right\} \tag{3.64}$$

According to (3.64), the transfer functions $\boldsymbol{W}_{\mathrm{dmad}}$ and $\boldsymbol{W}_{\mathrm{qmad}}$ between the DC-link voltage and stator current in dq-axes are reshaped as

$$\begin{cases}\mathbf{W}_{\mathrm{d,qmad}}=\mathbf{W}_{\mathrm{d,q}}+e^{-sT_{\mathrm{s}}}\mathbf{G}_{\mathrm{mad}}\mathbf{W}_{\mathrm{d,qre}}\\ \begin{bmatrix}\mathbf{W}_{\mathrm{dre}}\\ \mathbf{W}_{\mathrm{qre}}\end{bmatrix}=\dfrac{e^{-sT_{\mathrm{s}}}}{\mathbf{A}_{2\mathrm{q}}\mathbf{A}_{1\mathrm{d}}-\mathbf{A}_{2\mathrm{d}}\mathbf{A}_{1\mathrm{q}}}\begin{bmatrix}-U_{\mathrm{qref},0}\mathbf{A}_{2\mathrm{q}}-U_{\mathrm{dref},0}\mathbf{A}_{1\mathrm{q}}\\ U_{\mathrm{dref},0}\mathbf{A}_{1\mathrm{d}}+U_{\mathrm{qref},0}\mathbf{A}_{2\mathrm{d}}\end{bmatrix}\end{cases} \tag{3.65}$$

The transfer function of the inverter input admittance $\boldsymbol{Y}_{\mathrm{mad}}$ can be denoted as (3.66) after applying impedance reshaping

$$\begin{aligned}\mathbf{Y}_{\mathrm{mad}}=\mathbf{Y}_{\mathrm{m}}+\mathbf{Y}_{\mathrm{mre}}=\frac{3e^{-sT_{\mathrm{s}}}\mathbf{G}_{\mathrm{mad}}}{2U_{\mathrm{dc0}}}\big[\mathbf{W}_{\mathrm{dre}}\big(U_{\mathrm{d0}}+\omega_{\mathrm{e}}L_{\mathrm{d}}I_{\mathrm{q0}}+L_{\mathrm{d}}I_{\mathrm{d0}}s\\ +I_{\mathrm{d0}}R_{s}\big)+\mathbf{W}_{\mathrm{qre}}\big(U_{\mathrm{q0}}-\omega_{\mathrm{e}}L_{\mathrm{q}}I_{\mathrm{d0}}+L_{\mathrm{q}}I_{\mathrm{q0}}s+I_{\mathrm{q0}}R_{\mathrm{s}}\big)\big]+\mathbf{Y}_{\mathrm{m}}\end{aligned} \tag{3.66}$$

where $\boldsymbol{Y}_{\mathrm{mre}}$ is the virtual admittance introduced by the suppression strategy. The transfer function $\boldsymbol{G}_{\mathrm{mad}}$ is hard to realize, which means $\boldsymbol{Y}_{\mathrm{mre}}$ should be further simplified to reduce the difficulty of parameter design.

Based on the analysis, $\boldsymbol{L}^{-1}(\Delta\theta_{\mathrm{u}})$ only contains the harmonics at the frequencies of $6n\omega_{\mathrm{g}}$ $(n=1, 2)$. Different from designing $\boldsymbol{M}_{\mathrm{d}-}$ and $\boldsymbol{M}_{\mathrm{d}+}$ in a full frequency band, the amplitude and phase characteristics can be designed separately at specific frequency, which can bring the better harmonic suppression effect. Hence, $\boldsymbol{Y}_{\mathrm{mre}}$ can be simplified as

$$\begin{aligned}\mathbf{Y}_{\mathrm{mre}}=&\ (Y_{\mathrm{re1}}\cos\varphi_{\mathrm{re1}}+Y_{\mathrm{re1}}\sin\varphi_{\mathrm{re1}}s/6\omega_{\mathrm{g}})\mathbf{G}_{\mathrm{BPF1}}\\ &+(Y_{\mathrm{re2}}\cos\varphi_{\mathrm{re2}}+Y_{\mathrm{re2}}\sin\varphi_{\mathrm{re2}}s/12\omega_{\mathrm{g}})\mathbf{G}_{\mathrm{BPF2}}\end{aligned} \tag{3.67}$$

where $\boldsymbol{G}_{\mathrm{BPF1}}$ and $\boldsymbol{G}_{\mathrm{BPF2}}$ are the transfer functions of BPFs for $i_{\mathrm{fe}k_{\mathrm{r}}-}$ and $i_{\mathrm{fe}k_{\mathrm{r}}+}$, respectively. Y_{re1}, Y_{re2}, φ_{re1} and φ_{re2} are the amplitudes and phases of $\boldsymbol{Y}_{\mathrm{mre}}$ at the frequencies of $6\omega_{\mathrm{g}}$ and $12\omega_{\mathrm{g}}$, respectively.

The characteristics of $\boldsymbol{Y}_{\mathrm{mre}}$ at the frequencies of $6\omega_{\mathrm{g}}$ and $12\omega_{\mathrm{g}}$ should be the same before and after the design. Under $i_{\mathrm{d}} = 0$ control, Y_{ren} and φ_{ren} should be set as

$$\begin{cases} Y_{\mathrm{ren}} = 1.5|\mathbf{G}_{\mathrm{madn}}\mathbf{M}_{\mathrm{ren}}|/U_{\mathrm{dc,0}},\ \varphi_{\mathrm{ren}} = \angle\mathbf{G}_{\mathrm{madn}}\mathbf{M}_{\mathrm{ren}} - 6n\omega_{\mathrm{g}}T_{\mathrm{s}} \\ \mathbf{M}_{\mathrm{ren}} = \mathbf{W}_{\mathrm{dren}}(U_{\mathrm{d,0}} + \omega_{\mathrm{e}}L_{\mathrm{d}}I_{\mathrm{q,0}}) + \mathbf{W}_{\mathrm{qren}}(U_{\mathrm{q,0}} + I_{\mathrm{q,0}}R_{\mathrm{s}} + j6L_{\mathrm{q}}I_{\mathrm{q,0}}n\omega_{\mathrm{g}}) \end{cases} \tag{3.68}$$

In (3.68), $\boldsymbol{M}_{\mathrm{ren}}$ and $\boldsymbol{G}_{\mathrm{madn}}$ represent $\boldsymbol{M}_{\mathrm{re}}(j6n\omega_{\mathrm{g}})$ and $\boldsymbol{G}_{\mathrm{mad}}(j6n\omega_{\mathrm{g}})$. From (3.68), $\boldsymbol{W}_{\mathrm{dre}}$ and $\boldsymbol{W}_{\mathrm{qre}}$ are related to the system parameters and operation conditions, which means the design of $\boldsymbol{G}_{\mathrm{madn}}$ is essential to reduce the grid current amplitude caused by the resonance

$$\begin{cases} \boldsymbol{G}_{\mathrm{mad1}}U_{\mathrm{dc1}}e^{j\varphi_{\mathrm{dc\_1}}} = -(K_{\mathrm{p}} + K_{\mathrm{r1}})K_1 I_{\mathrm{w}k_{\mathrm{r}},\mathrm{mad-}}\, e^{j(\varphi_{\mathrm{d\_1}} + \varphi_{\mathrm{w}k_{\mathrm{r}},\mathrm{mad-}} + 6\varphi - 9\omega_{\mathrm{g}}T_{\mathrm{s}})} \\ \boldsymbol{G}_{\mathrm{mad2}}U_{\mathrm{dc2}}e^{j\varphi_{\mathrm{dc\_2}}} = -(K_{\mathrm{p}} + K_{\mathrm{r2}})K_2 I_{\mathrm{w}k_{\mathrm{r}},\mathrm{mad+}}\, e^{j(\varphi_{\mathrm{d\_2}} + \varphi_{\mathrm{w}k_{\mathrm{r}},\mathrm{mad+}} + 12\varphi - 18\omega_{\mathrm{g}}T_{\mathrm{s}})} \end{cases} \tag{3.69}$$

where $I_{\mathrm{wn,mad}\pm}$ is the amplitude of the grid current with control at $(6n \pm 1)\omega_{\mathrm{g}}$. $\varphi_{\mathrm{wn,mad}\pm}$ is the phase difference between the grid current and the rectified voltage related to $6n\omega_{\mathrm{g}}$ with control.

From (3.66), the transfer functions among the DC-link current, DC-link voltage and rectified voltage $\boldsymbol{K}_{\mathrm{mad}}$ and $\boldsymbol{K}_{\mathrm{umad}}$ after control are expressed as

$$\begin{cases} \mathbf{K}_{\mathrm{mad}} = \Delta i_{\mathrm{L}}/\Delta u_{\mathrm{grec}} = 1/\left[1/(\mathbf{Y}_{\mathrm{m}} + \mathbf{Y}_{\mathrm{mre}} + C_{\mathrm{dc}}s) + \left(L_{\mathrm{dc}}s + R_{\mathrm{g}}\right)\right] \\ \mathbf{K}_{\mathrm{umad}} = \Delta u_{\mathrm{dc}}/\Delta u_{\mathrm{grec}} = 1/\left[1 + \left(L_{\mathrm{dc}}s + R_{\mathrm{g}}\right)(\mathbf{Y}_{\mathrm{m}} + \mathbf{Y}_{\mathrm{mre}} + C_{\mathrm{dc}}s)\right] \end{cases} \tag{3.70}$$

Hence, $I_{\mathrm{wn,mad}\pm}$ and $\varphi_{\mathrm{wn,mad}\pm}$ can be expressed as

$$\begin{cases} I_{\mathrm{wn,mad}\pm} = \frac{2\sqrt{3}}{\pi}\sqrt{\left[\frac{6M_{\mathrm{6n,mad}\pm}|\mathbf{K}_{\mathrm{madn}}|U_{\mathrm{g}}\sin\varphi_{\mathrm{6n,mad}\pm}}{\pi(36n^2-1)(12n\pm1)} \pm \frac{I_{\mathrm{g0}}}{6n\pm1}\right]^2 + \left[\frac{6M_{\mathrm{6n}\pm}|\mathbf{K}_{\mathrm{madn}}|U_{\mathrm{g}}\cos\varphi_{\mathrm{6n,mad}\pm}}{\pi(36n^2-1)(12n\pm1)}\right]^2} \\ \varphi_{\mathrm{wn,mad}\pm} = \angle\left(\pm j\frac{I_{\mathrm{g0}}}{6n\pm1} + \frac{6M_{\mathrm{6n,mad}\pm}U_{\mathrm{g}}|\mathbf{K}_{\mathrm{madn}}|}{\pi(36n^2-1)(12n\pm1)}e^{j\varphi_{\mathrm{6n,mad}\pm}}\right) \end{cases} \tag{3.71}$$

where $\boldsymbol{K}_{\mathrm{madn}}$ represents $\boldsymbol{K}_{\mathrm{mad}}(j6n\omega_{\mathrm{g}})$

$$\begin{cases} M_{\mathrm{6n,mad}\pm} = \sqrt{36n^2 + (\pm12n + 1)\sin^2\left(\angle\mathbf{K}_{\mathrm{madn}} - \pi/2\right)} \\ \varphi_{\mathrm{6n,mad}\pm} = \arcsin\left[(6n \pm 1)\sin\left(\angle\mathbf{K}_{\mathrm{madn}} - \pi/2\right)/M_{\mathrm{6n,mad}\pm}\right] \end{cases} \tag{3.72}$$

The amplitudes and phases of the harmonics of the DC-link voltage at the frequency of $6\omega_{\mathrm{g}}$ and $12\omega_{\mathrm{g}}$ after control can be expressed as

$$U_{\mathrm{dcn}} = 6U_{\mathrm{g}}|\mathbf{K}_{\mathrm{umadn}}|/\left(36n^2 - 1\right)\pi,\ \varphi_{\mathrm{dcn}} = 6n\varphi - \pi/2 + \angle\mathbf{K}_{\mathrm{umadn}} \tag{3.73}$$

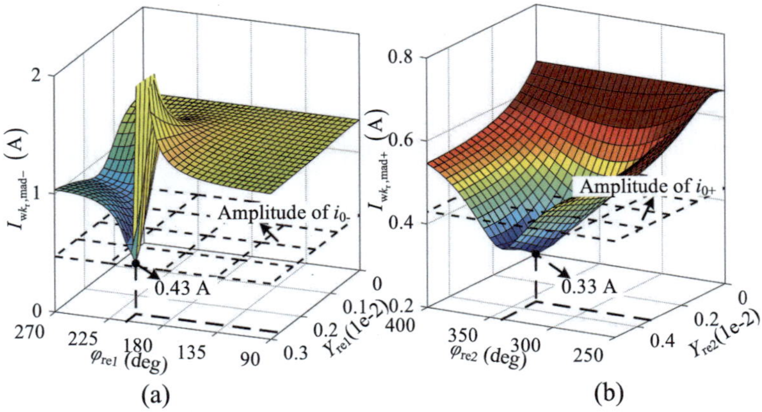

Fig. 3.45 Grid harmonic amplitudes near LC resonant frequency with different values of Y_{ren} and φ_{ren} ($n = 1, 2$). **a** At $(6k_{\text{r}} - 1)\omega_{\text{g}}$. **b** At $(6k_{\text{r}} + 1)\omega_{\text{g}}$

where $\boldsymbol{K}_{\text{umadn}}$ represents $\boldsymbol{K}_{\text{umad}}(j6n\omega_{\text{g}})$ ($n = 1, 2$).

From (3.71), the adjusting coefficient and phases K_{n} and $\varphi_{\text{d\_n}}$ can be designed through analyzing the amplitudes of resonant components $I_{wk_r,\text{mad -}}$ and $I_{wk_r,\text{mad +}}$ at grid side. The harmonic amplitude $I_{wk_r,\text{mad -}}$ under different values of Y_{re1} and φ_{re1} is shown in Fig. 3.45a. $I_{wk_r,\text{mad -}}$ is reduced from 1.14 A to 0.43 A when Y_{re1} and φ_{re1} are set as 2.7e−3 and 210°, respectively. The harmonic amplitude $I_{wk_r,\text{mad +}}$ with different amplitudes and phases of $\boldsymbol{Y}_{\text{mre}}$ at $12\omega_{\text{g}}$ are demonstrated in Fig. 3.45b. $I_{wk_r,\text{mad +}}$ is decreased from 0.75 A to 0.33 A when Y_{re2} and φ_{re2} are set as 3.5e−3 and 335 deg, respectively, which illustrates that the control method can reduce the harmonic of the grid current at $(6k_{\text{r}} - 1)\omega_{\text{g}}$ and $(6k_{\text{r}} + 1)\omega_{\text{g}}$, effectively.

With the designed Y_{ren} and φ_{ren}, the amplitudes and phases of $\boldsymbol{K}_{\text{mad}}$ and $\boldsymbol{K}_{\text{umad}}$ at $6\omega_{\text{g}}$ and $12\omega_{\text{g}}$ can be obtained. From (3.68), (3.69), and (3.73), the adjusting coefficient and phases of $\boldsymbol{M}_{\text{d}-}$ and $\boldsymbol{M}_{\text{d}+}$ can be derived as

$$\begin{cases} K_1 = 4U_{\text{dc},0}Y_{\text{re1}}U_{\text{g}}|\mathbf{K}_{\text{umad1}}|/35\pi\left[I_{wk_r,\text{mad}-}|\mathbf{M}_{\text{re1}}|\left(K_{\text{p}} + K_{\text{r1}}\right)\right] \\ \varphi_{\text{d\_1}} = \varphi_{\text{re1}} + \angle\mathbf{K}_{\text{umad1}} + \pi/2 + 15\omega_{\text{g}}T_{\text{s}} - \angle\mathbf{M}_{\text{re1}} - \varphi_{wk_r,\text{mad -}} \\ K_2 = 4U_{\text{dc},0}Y_{\text{re2}}U_{\text{g}}|\mathbf{K}_{\text{umad1}}|/143\pi\left[I_{wk_r,\text{mad}+}|\mathbf{M}_{\text{re2}}|\left(K_{\text{p}} + K_{\text{r2}}\right)\right] \\ \varphi_{\text{d\_2}} = \varphi_{\text{re2}} + \angle\mathbf{K}_{\text{umad2}} + \pi/2 + 30\omega_{\text{g}}T_{\text{s}} - \angle\mathbf{M}_{\text{re2}} - \varphi_{wk_r,\text{mad +}} \end{cases} \quad (3.74)$$

Figure 3.46 shows the harmonic amplitudes of grid current under different operation conditions with high and low frequency coefficients. With the designed coefficients ($Y_{\text{re1}} = 2.7\text{e}{-3}$, $\varphi_{\text{re1}} = 210°$), $I_{wk_r,\text{mad -}}$ can be reduced effectively under the frequency coefficient of 4 and 2. Specifically, the harmonic amplitude at $(6k_{\text{r}} - 1)\omega_{\text{g}}$ can be reduced from 1.13 A to 0.52 A under the speed and torque of 75 Hz and 30 N·m, which is lower 0.30 A than the result under the frequency coefficient of 1 and 0.8. As shown in Fig. 3.46b, the amplitudes of the grid current at $(6k_{\text{r}} + 1)\omega_{\text{g}}$ can also be reduced more effectively with the high frequency coefficient. Meanwhile, the

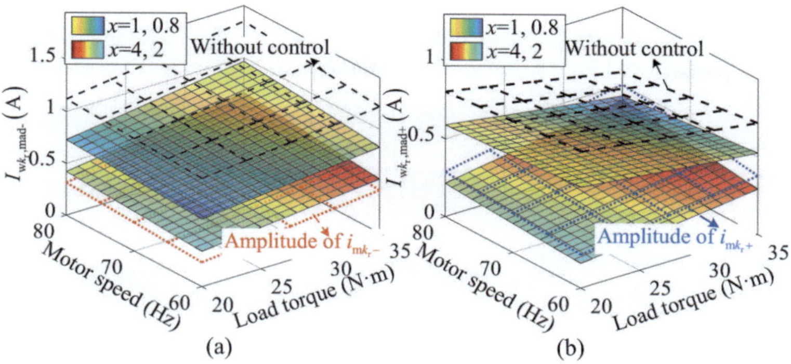

Fig. 3.46 Resonant component of the grid current under different operating conditions with high and low frequency coefficients of the proposed resonance suppression strategy. **a** At (6kr − 1)ωg. **b** At (6kr + 1)ωg

difference between $I_{w4,mad\pm}$ and the amplitude of $i_{04\pm}$ does not exceed 0.03 A. The deterioration of high-frequency resonance on grid current is almost eliminated under the coupling between the grid and DC-link currents. Meanwhile, the proposed resonance suppression method can work effectively through improving the frequency coefficient in a wide speed and torque region.

3.4.4 Experimental Results

Figure 3.47 shows the experimental results at 75 Hz. The waveforms from the top to the bottom are the DC-link voltage, the feature signals $i_{fe(w4\pm)\_r}$, and the *w*-phase grid current. $i_{fe(w4-)\_r}$ and $i_{fe(w4+)\_r}$ are the feature signals of the 23th and 25th harmonics in the grid current. With the control of $i_{fe(w4-)\_r}$, the peak-to-peak values of the DC-link voltage and grid current fall from 115.6 V and 53.3 A to 103.1 V and 40.0 A, respectively. After the further control of $i_{fe(w4+)\_r}$, the peak-to-peak values can be reduced to 96.9 V and 34.6 A, respectively. The harmonic amplitudes of the grid current at 1150 Hz and 1250 Hz are decreased from 1.1 A and 0.7 A to 0.5 A and 0.3 A, respectively. Figure 3.47b–d are the zoomed views of the experimental results without and with the proposed strategy. The distortion of the DC-link voltage and grid current can be suppressed with the method. The frequency coefficient *x* is set as 4 and 2 for the harmonic suppression at $23\omega_g$ and $25\omega_g$, respectively. Hence, the frequencies of $i_{fe(w4-)\_r}$ and $i_{fe(w4+)\_r}$ are set as $6\omega_g$ ($k_r/x = 1$) and $12\omega_g$ ($k_r/x = 2$) respectively, and their amplitudes are equal to the amplitudes of the 23th and 25th harmonics in the grid current, respectively.

Figure 3.48 shows the results with the conventional method. The peak-to-peak values of the DC-link voltage and grid current are reduced from 115.6 V and 53.3 A to 104.3 V and 46.7 A, respectively. The control performance is lower compared to the large inductance drive system. Further increasing the gain results in the 1250 Hz

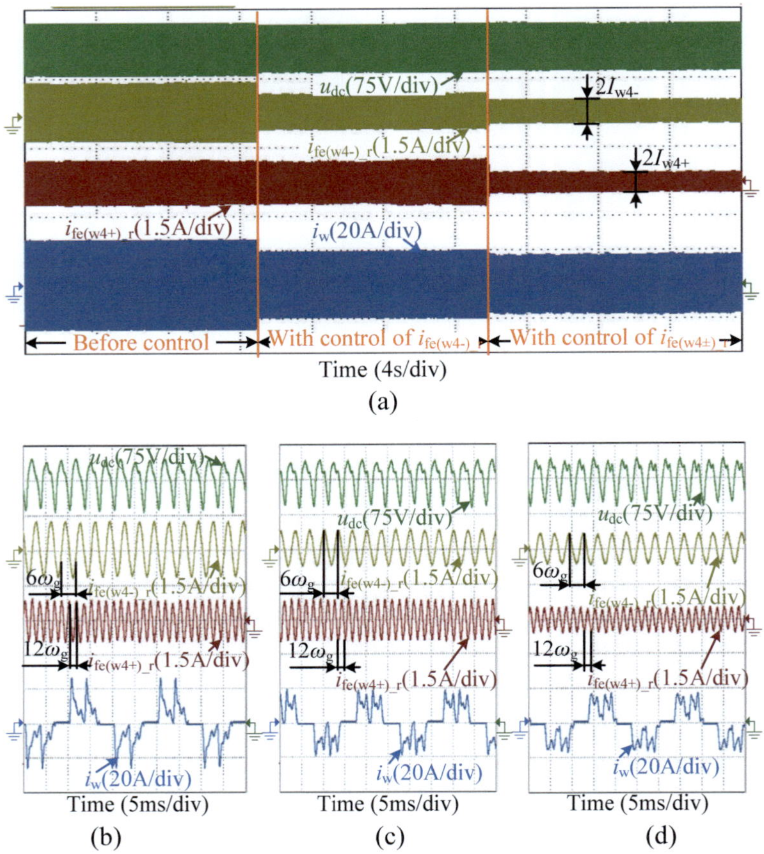

Fig. 3.47 Experimental results with the proposed resonance suppression method at the motor frequency of 75 Hz, reprinted from Ref. [11], copyright 2022, with permission from Nature. **a** Enable the proposed strategy. **b** Zoomed view without the proposed strategy. **c** Zoomed view with control of $i_{fe(w4-)\_r}$. **d** Zoomed view with control of $i_{fe(w4\pm)\_r}$

harmonic amplitude of the grid current decreasing from 0.7 A to 0.4 A, while the 1150 Hz harmonic amplitude increases from 0.7 A to 1.1 A. This indicates that directly extracting LC resonance information does not yield significant results in low-inductance drive systems.

Figure 3.49 shows the fast Fourier transformation (FFT) analysis of the grid current at the motor frequency of 75 Hz. By comparing Fig. 3.49a and Fig. 3.49b, the total harmonic distortion (THD) of the grid current can be reduced by 2.4% with the general parameters of the conventional method. As can be seen in Fig. 3.49c, the resonant components can be effectively suppressed with the proposed method, and the THD can be reduced from 77.2% to 60.3%. Besides, the harmonic amplitudes at 250 Hz and 350 Hz are also decreased from 5.3 A and 3.2A to 4.2A and 0.6A, respectively. Figure 3.49d shows the FFT of grid current with the control of $i_{fe(w4-)\_r}$

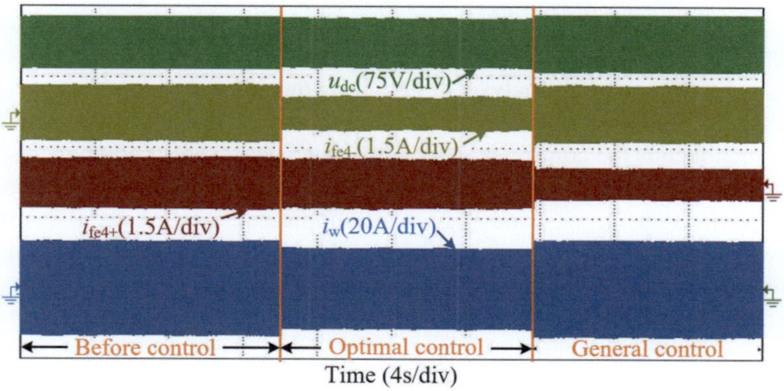

Fig. 3.48 Experimental results with the conventional method at the motor frequency of 75 Hz

and $i_{fe(w4+)\_r}$, and the THD can be further reduced to 45.0%. The harmonics at 550 Hz and 650 Hz are reduced to 1.5 A and 1.4A, respectively.

Figure 3.50a shows the influence of the proposed strategy on the grid current harmonics related to the mean value and harmonics of the DC-link current, whose amplitudes are equal to those of $i_{fe(m4\pm)\_r}$ and $i_{fe(h4\pm)\_r}$, respectively. The grid harmonic amplitudes related to the DC-link current at $6k_r\omega_g$ are decreased from

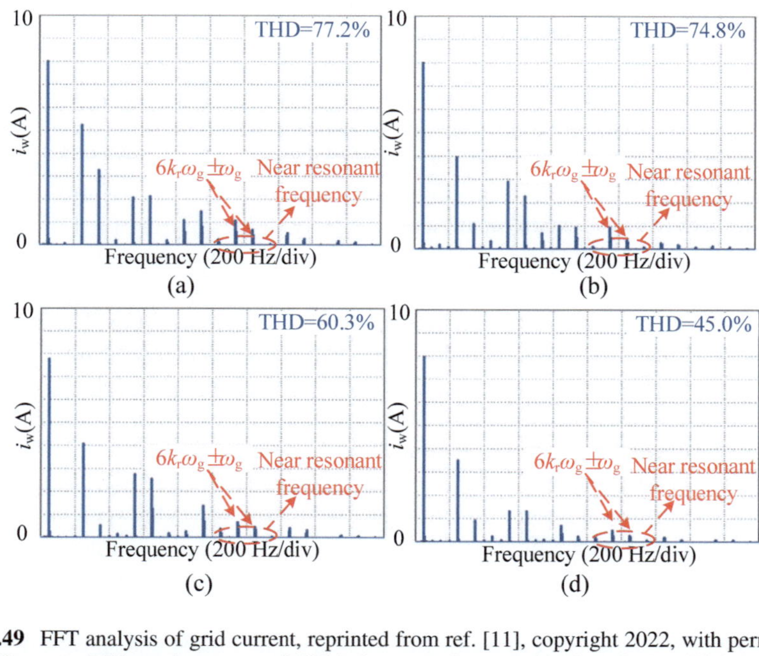

Fig. 3.49 FFT analysis of grid current, reprinted from ref. [11], copyright 2022, with permission from Nature. **a** Without the strategy. **b** With the general conventional method. **c** With control of $i_{fe(w4-)\_r}$. **d** With control of $i_{fe(w4\pm)\_r}$

1.1 A and 1.0A to 0.7 A and 0.8 A, respectively. Meanwhile, the phase difference between $i_{fe(m4\pm)\_r}$ and $i_{fe(h4\pm)\_r}$ changes from 79.8° and − 96.7° to 123.0 deg and − 154.4°. From the vector analysis in Fig. 3.42, the deterioration effect of $i_{fe(h4\pm)\_r}$ on the grid harmonic amplitude $I_{wk\pm}$ is increased as the phase difference get close to 0°, which means the control strategy can reduce the resonant components of the grid current. Figure 3.50b shows the effectiveness of the conventional method. With optimal parameters, $i_{fe(h4-)\_r}$ and $i_{fe(h4+)\_r}$ are suppressed to 0.9 A and 0.8 A respectively, which results in the decrease of 0.4 A and 0.1 A in the resonant components of the grid current. Increasing the control gain, the amplitudes of $i_{fe(h4-)\_r}$ and $i_{fe(h4+)\_r}$ are further reduced to 0.7 A and 0.6 A, respectively. The suppression effect of *LC* resonance in the DC-link current is improved. However, the phase difference between $i_{fe(m4-)\_r}$ and $i_{fe(h4-)\_r}$ varies from 123.0 to 66.3°, which causes the increase of the deterioration effect of $i_{L,4}$ on the grid current harmonic i_{w4-}. The current coupling between the grid and DC-link worsens the grid harmonic amplitude at 1150 Hz under the conventional method.

To further verify the effectiveness of the strategy, Fig. 3.51 shows the experimental results at the motor frequency of 60 Hz. The peak-to-peak values of the grid current are reduced from 45.2 A to 30.4 A, respectively.

To analyze the implication of the proposed method over DC-link, the DC-link voltage, the actual DC-link current, the reconstructed DC-link current, and the current harmonic at 1200 Hz are shown in Fig. 3.51b, c. The amplitude of the resonant component in the DC-link current is reduced from 1.4 A to 0.9 A. The ripple amplitude of the reconstructed DC-link current i_{Lf} is equal to that of the actual value i_L. And the phase delay between i_{Lf} and i_L is delay of $1.5T_s$.

In order to verify the robustness of the strategy, the grid-line impedance is set as 0.1mH, and the resonance frequency is 880 Hz. Figure 3.52 shows the experimental results at 75 Hz. As can be seen in Fig. 3.52a, the harmonic order k_r of the DC-link current which is severely affected by *LC* resonance varies from 4 to 3. The frequencies of compensated feature signals $i_{fe(w3-)\_r}$ and $i_{fe(w3+)\_r}$ are also set as $6\omega_g$ and $12\omega_g$ respectively, which means $x = 3$, 1.5. With the control of i_{fe3-}, the peak-to-peak values of the DC-link voltage and grid current fall from 96.9 V and 58.3 A to 71.8 V and 36.7 A, respectively. After the further control of $i_{fe(w3+)\_r}$, the peak-to-peak values can be reduced to 64.6 V and 29.8 A, respectively. The harmonic amplitudes of the grid current at 850 Hz and 950 Hz are decreased from 5.1 A and 3.6 A to 2.1 A and 0.9 A, respectively. This means the proposed method maintains the effectiveness under the variations of the AC-side (grid side) impedance.

The experimental waveforms of variable speed operation are shown in Fig. 3.53a when the operation frequency varies from 60 to 75 Hz and back to 60 Hz. Figure 3.53b shows the experimental waveform when the load torque varies between 30 N·m and 37 N m. The disturbance of motor speed from the torque step variation is lower than 2 Hz. Meanwhile, the amplitude of the resonant component in the grid current remains below 0.7 A during the torque and speed changes. Hence the proposed method has the tolerance of dynamic performance, and torque and speed ripple.

Figure 3.54 shows the THD of grid current under different output power conditions. As the output power increases, the control performance of the conventional

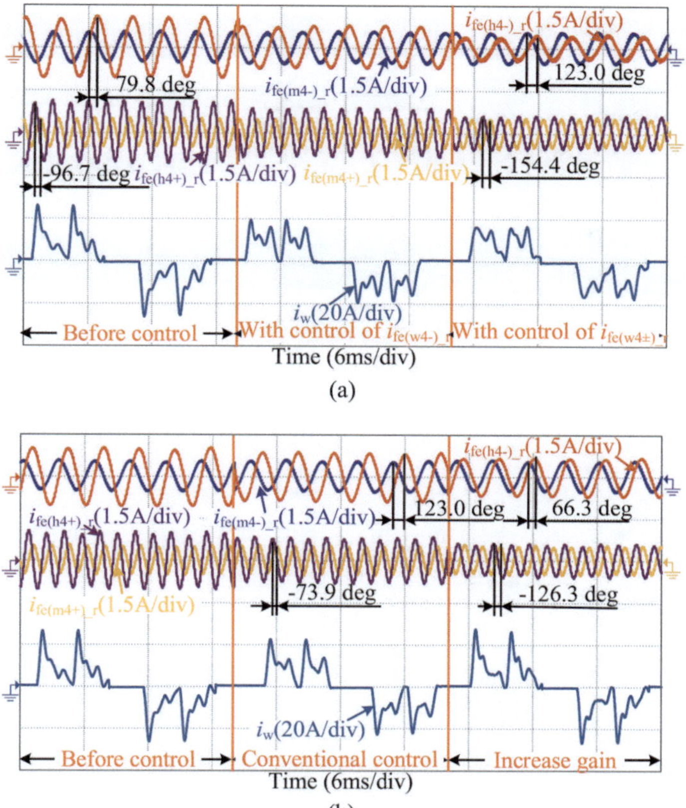

Fig. 3.50 Experimental results of grid current harmonics, reprinted from ref. [11], copyright 2022, with permission from Nature. **a** With the proposed method. **b** With the conventional method

method with optimal parameters reduces significantly. With the proposed method, the THD of the grid current can be reduced by more than 30%.

3.5 Summary

To improve the grid current quality of electrolytic capacitorless drives, this chapter investigates various suppression strategies. The main contributions and conclusions are as follows:

(1) A stability control method based on inherent harmonics impedance shaping is proposed, bandpass filters are used to extract the inherent harmonics of the DC-link voltage. Virtual impedance then designed for each harmonic. Considering the impact of DC-link voltage sampling delay, the angle of the virtual admittance

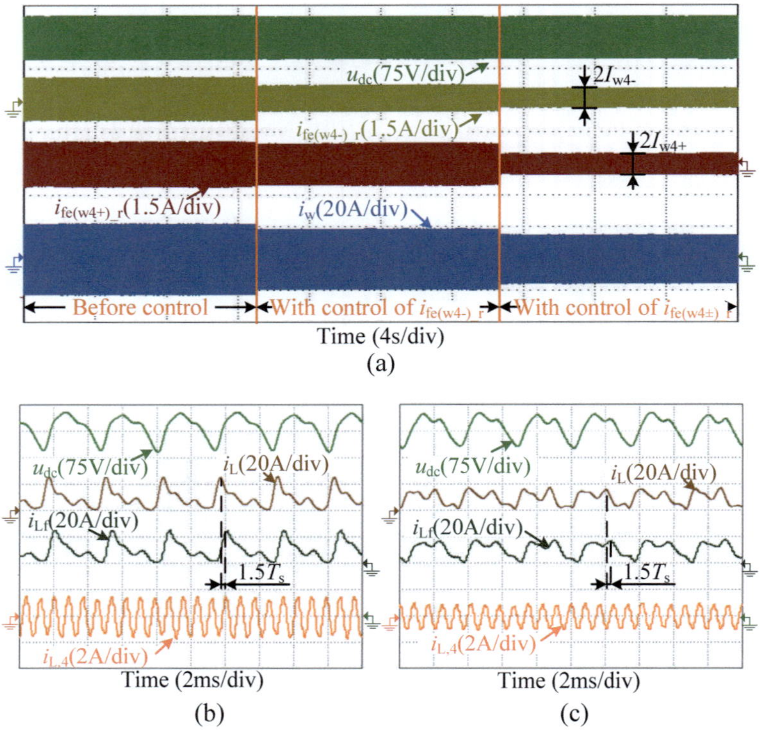

Fig. 3.51 Experimental results with the proposed resonance suppression method at the motor frequency of 60 Hz. **a** Zoomed view. **b** Implication of the method over DC-link current without the strategy. **c** Implication of the method over DC-link current with the strategy

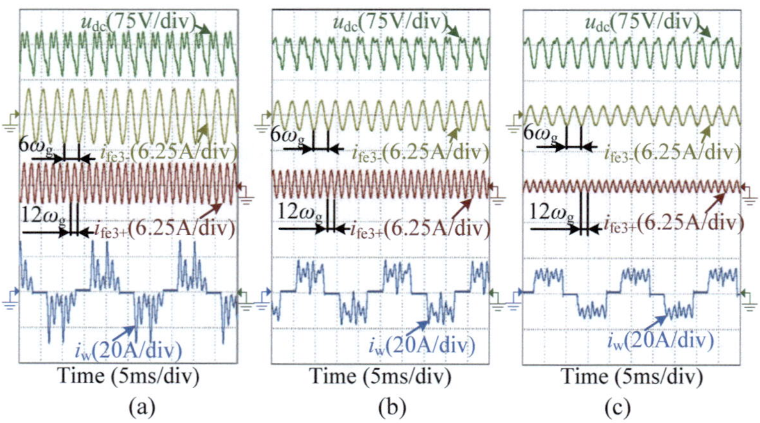

Fig. 3.52 Experimental results with the proposed resonance suppression method under the grid-line impedance of 0.1mH, reprinted from Ref. [11], copyright 2022, with permission from Nature. **a** Zoomed view without the proposed strategy. **b** Zoomed view with control of $i_{fe(w3-)\_r}$. **c** Zoomed view with control of $i_{fe(w3\pm)\_r}$

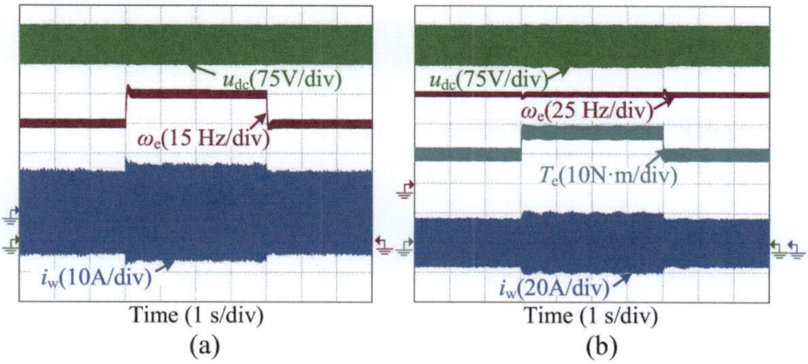

Fig. 3.53 Dynamic experimental waveforms of the proposed method. **a** Speed dynamic process. **b** Load torque dynamic process

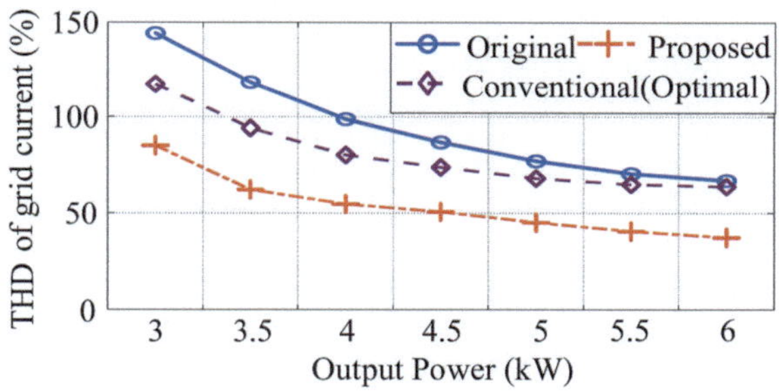

Fig. 3.54 THD of grid current under different output power conditions, reprinted from Ref. [11], copyright 2022, with permission from Nature

is compensated to improve the accuracy of the damping control. Experimental results show that the proposed method can effectively suppress LC resonance and reduce harmonics in the DC-link voltage and grid current.

(2) A method based on rectified current regulation with DC-link voltage decoupling (RCR-DVD) is proposed to enhance the harmonics suppression performance of grid currents. The method predicts the main harmonics in the DC-link voltage periodically and establishes a closed-loop control for DC-link voltage harmonics to achieve harmonic suppression. At the same time, the disturbance terms of the DC-link voltage in the small-signal control loop of the harmonic suppression method are decoupled, further suppressing the DC-side current harmonics and improving the grid current quality. Experimental results demonstrate that the proposed method can simultaneously suppress both DC-link and grid current harmonics.

(3) A suppression strategy based on the feature extraction of the grid current harmonics from frequency mapping is proposed. With the high-to-low frequency mapping, the low-frequency feature signals of grid current harmonics are applied to characterize the high-frequency resonance, which reduce the sampling and update errors caused by the higher resonant frequency. On the basis, the closed-loop control of the low-frequency feature signal is adopted to reshape the

impedance relationship between the grid current and rectified voltage. Therefore, the harmonics of the grid current can be decoupled from the DC side current, and the high-frequency resonance of the grid current can be directly controlled, which realizes resonance suppression in the low DC-link inductance drives. Experimental results verify the effectiveness of proposed suppression strategy.

References

1. W. Lee and S. K. Sul. DC-Link voltage stabilization for reduced DC-link capacitor inverter [J]. IEEE Transaction on Industry Application, 2014, 50(1): 404–414.
2. Y. A. I. Mohamed, A. A. A. Radwan, and T. K. Lee. Decoupled reference voltage based active DC link stabilization for PMSM drives with tight-speed regulation [J]. IEEE Transactions on Industrial Electronics, 2012, 59(12): 4523–4536.
3. P. Magne, D. Marx, B. N. Mobarakeh and S. Pierfederici. Large-signal stabilization of a DC-link supplying a constant power load using a virtual capacitor: impact on the domain of attraction [J]. IEEE Transaction on Industry Application, 2012, 48(3): 878–887.
4. M. Wu, and D. D. C. Lu. A novel stabilization method of LC input filter with constant power loads without load performance compromise in DC microgrids [J]. IEEE Transactions on Industrial Electronics, 2015, 62(7): 4552–4562.
5. D. Ding et al., Impedance reshaping for inherent harmonics in PMSM driveswith small DC-link capacitor, IEEE Transactions on Power Electronics, vol. 37, no. 12, pp. 14265–14279, Dec. 2022.
6. G. Zhang, G. Wang, D. Xu, and N. Zhao. ADALINE-network-based PLL for position sensorless interior permanent magnet synchronous motor drives [J]. IEEE Transactions on Power Electronics, 2016, 31(2): 1450–1460.
7. H. Mosskull. Optimal DC-link stabilization design [J]. IEEE Transactions on Industrial Electronics, 2015, 62(8): 5031–5044.
8. L. Mathe, L. Török, D. Wang and D. Sera, "Resonance reduction for ac drives with small capacitance in the DC-.Link," IEEE Trans. Ind. Appl., vol. 53, no. 4, pp. 3814–3820, Jul-Aug. 2017.
9. K. Li, T. Sun, H. Li, Y. Wang, W. Li and L. Gong, "Grid current quality improvement for three-phase diode rectifier-fed small DC-Link capacitance IPMSM drives," IEEE Trans. Energy. Convers., vol. 37, no. 4, pp. 2310–2320, Dec. 2022.
10. D. Ding, H. Xie, B. Li, G. Wang, R. Gao, Z. Ren, W. Yue, D. Xu, "Harmonic Suppression Based on Rectified Current Regulation With DC-Link Voltage Decoupling for Electrolytic Capacitorless PMSM Drives," IEEE Trans. Power Electron., vol. 39, no. 2, pp. 2213–2225, Feb. 2024.
11. R. Gao, D. Ding, G. Wang, Q. Wang, G. Zhang and D. Xu, "Feature Extraction of Frequency-Mapping-Based Resonance Suppression in PMSM Drives With Low DC-Link Capacitance and Inductance," IEEE Trans. Power Electron., vol. 40, no. 4, pp. 4848–4861, Apr. 2025.

Chapter 4
Grid Current Improvement Strategy Under Non-ideal Grid Conditions

4.1 Introduction

Chapter 3 proposes a grid-side power quality enhancement method under ideal grid, with experimental results validating the effectiveness of the proposed method. However, in practical applications, power grids often exhibit varying degrees of non-ideal characteristics, such as grid background harmonics and variations in grid impedance. These factors can further interfere with the stability of the drive and grid current quality. This chapter improves grid-side power quality under non-ideal grid conditions from two perspectives to enhancing the robustness and anti-interference capability of the drive in non-ideal grid.

4.2 Harmonic Energy Regulation for Grid Current Quality Improvement

4.2.1 Robustness of the Drive to Grid Impedance

The active damping strategy with parallel virtual impedance has a clear physical significance and is widely applied in grid-connected inverters, effectively suppressing LCL resonance and improving the stability of the drive [1, 2]. Similarly, the active damping method with parallel virtual impedance is also suitable for electrolytic capacitorless drives. A common active damping strategy is to add virtual resistors in series or parallel within the drive to enhance its damping characteristics. To minimize the disturbance of damping power on the motor power, an active damping strategy where the virtual admittance Y_{damp} is connected in parallel across the C_{dc} is used as an example for stability analysis [3]. The corresponding topology is shown in Fig. 4.1.

D. Ding et al., *Electrolytic Capacitorless Permanent Magnet Synchronous Motor Drives*, https://doi.org/10.1007/978-981-95-5601-4_4

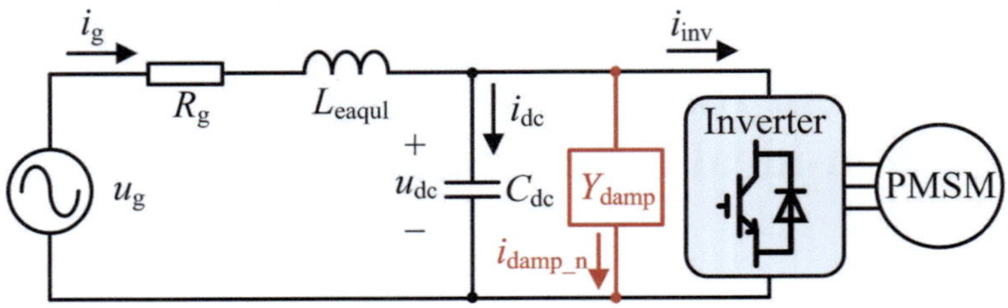

Fig. 4.1 Topology of the cascaded system

After the parallel virtual admittance is applied, the transfer function at the system's inherent frequency can be expressed as

$$\begin{aligned} G(s) = \frac{u_{dc}(s)}{u_g(s)} &= \frac{1}{(L_{eaqul}s + R_g)(C_{dc}s + Y_{m\_in} + Y_{damp}) + 1} \\ &= \frac{1}{(L_{eaqul}s + R_g)(C_{dc}s + Y_{m\_in\_damp}) + 1} \end{aligned} \quad (4.1)$$

where $Y_{m\_in\_damp}$ is the inverter input admittance after applying the parallel virtual admittance. According to Fig. 4.1, the block diagram of the drive is shown in Fig. 4.2. To reduce the power disturbance caused by active damping strategy, the virtual admittance is constructed by considering only the harmonics at the system's inherent frequency.

The Nyquist plots of $Y_{m\_in\_damp}(s)/Y_{LC\_out}(s)$ under different L_{eaqul} is shown in Fig. 4.3. When L_{eaqul} is 2.5mH, the intersection is located at the right of (− 1, 0), and the stability of the drive system is enhanced compared with the original topology. However, with the increase of L_{eaqul}, the intersection gradually moves away from (− 1, 0), and the stability of the drive gradually decreases. So, the robustness of the active damping strategy to the grid impedance is not satisfactory.

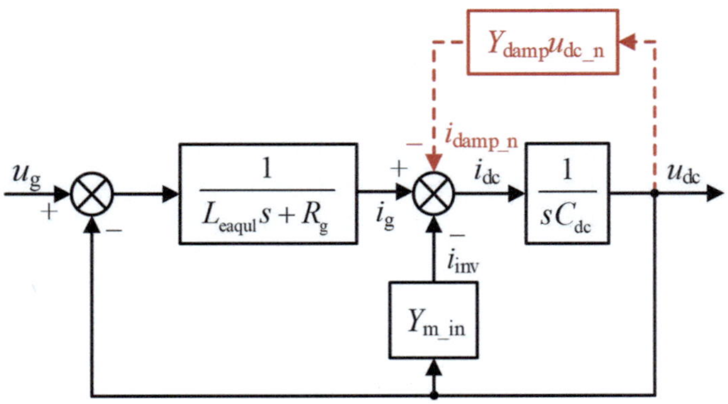

Fig. 4.2 Block diagram of the drive after parallelling virtual impedance

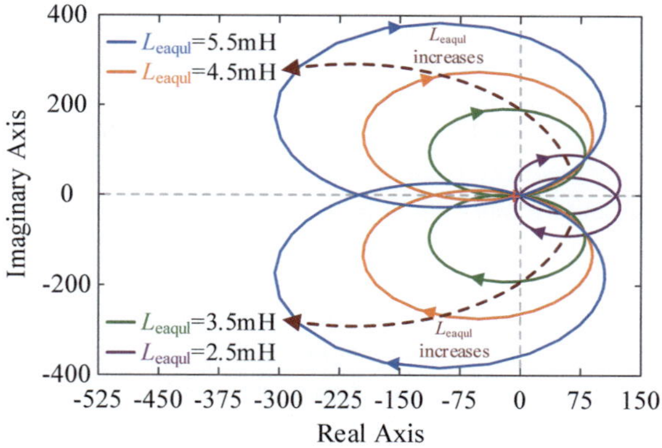

Fig. 4.3 Nyquist plots of $Y_{m\_in\_damp}(s)/Y_{LC\_out}(s)$

4.2.2 *Grid Current Improvement Strategy Based on Harmonic Energy Regulation*

The analysis of how grid impedance changes affect the active damping strategy shows that current- or voltage-based active damping methods provide system damping by extracting relevant harmonics from the DC link. The harmonic energy inject ed is exchanged with the energy stored in the energy storage elements of drive. This section examines the harmonic energy exchange mechanism between these storage elements, focusing on the energy exchange modes of the drive. It also explores a robust harmonic energy regulation method for the grid impedance. The small-signal model of the drive is shown in Fig. 4.4, where the energy relationship between the DC-link and the motor winding is represented by a controlled current source.

The rectified voltage u_g can be expressed as [3]

$$u_g = u_{dN}\left[1 - \sum_{n=1}^{\infty} \frac{2}{(6n)^2 - 1} cos\left(6\omega_g nt\right)\right] \tag{4.2}$$

where $n = 1, 2, 3\ldots$, $u_{dN} = 3\sqrt{3}u_{peak}/\pi$, u_{peak} is the peak value of the phase voltage.

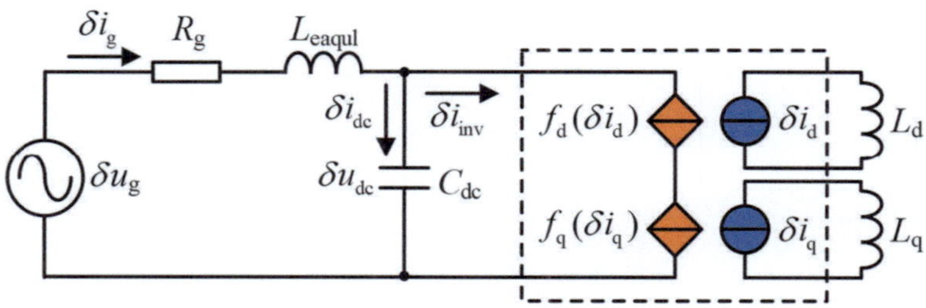

Fig. 4.4 Small signal model of the drive

According to (4.2), the rectified voltage not only contains DC components, but also contains harmonics at integer multiples of $6\omega_g$. When the drive operates, the harmonic energy at $6n\omega_g$ frequency will be exchanged between the energy storage elements, which affects the grid current quality as well as the stability of the drive In the electrolytic capacitorless permanent magnet motor drive, the energy storage elements mainly include the DC-link capacitance C_{dc}, the equivalent inductance L_{eaqul} and the motor inductance $L_{d,q}$.

From the perspective of harmonic energy, the vector control ensures motor performance through the effective speed and current control. As a result, harmonic energy primarily flows between C_{dc} and L_{eaqul}, leading to poor quality of grid current. Existing active damping control strategies mitigate grid current harmonics by injecting harmonic energy into motor windings, thereby reducing the harmonic energy stored in C_{dc} and L_{eaqul}. However, this method of harmonic injection does not account for the optimal regulation of harmonic energy among energy storage elements. If the control strategy focuses on reducing harmonic energy in L_{eaqul} and better regulating harmonic energy regulation across all energy storage elements, it could effectively suppress grid current harmonics and improve the robustness to grid impedance variations of the system.

When flux-weakening control is not implemented, the d-axis current is significantly smaller than the q-axis current. Therefore, the following analysis focuses only on the harmonic energy in the q-axis inductance L_q [4]. When C_{dc} and L_q are the main energy storage components, as illustrated in Fig. 4.5, dark arrows indicate higher harmonic energy flow, while light arrows represent lower harmonic energy flow. In this case, harmonic energy predominantly exchanges between C_{dc} and L_q, while the harmonic energy in L_{eaqul} is minimized.

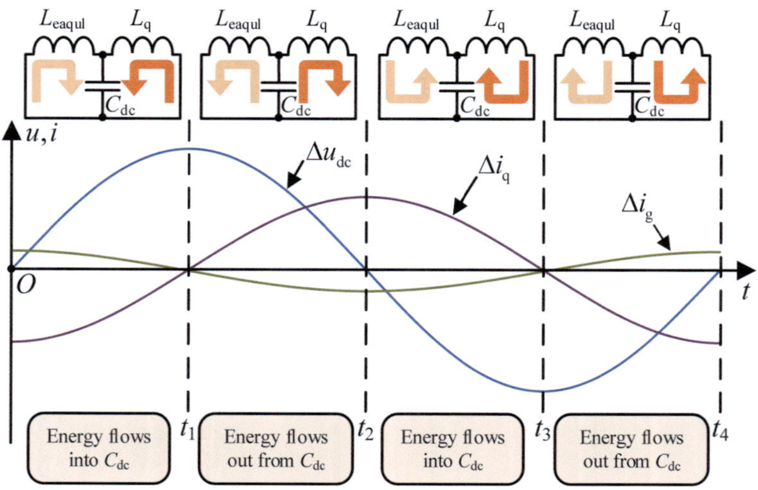

Fig. 4.5 Energy exchange state when Cdc and Lq act as the main energy storage elements

In summary, if the relationship between motor current and DC-link voltage can be effectively controlled to achieve the energy exchange mode depicted in Fig. 4.5, the grid current quality can be significantly improved, and the robustness of the system to grid impedance can be enhanced. To realize this energy exchange mode, it is essential to control the harmonic components of the motor current. Specifically, in terms of amplitude, increasing the harmonic amplitudes of the motor current at the frequencies of $6\omega_g$ and $12\omega_g$ can equalize the harmonic energy stored in C_{dc} and L_q, as expressed in (4.3).

$$\begin{cases} \frac{1}{2}C_{dc}\left|u_{dc}(j6\omega_g)\right|^2 = \frac{1}{2}L_q\left|i_q(j6\omega_g)\right|^2 \\ \frac{1}{2}C_{dc}\left|u_{dc}(j12\omega_g)\right|^2 = \frac{1}{2}L_q\left|i_q(j12\omega_g)\right|^2 \end{cases}. \tag{4.3}$$

In terms of phase, the control strategy should adjust the phase of the motor current components at the frequencies of $6\omega_g$ and $12\omega_g$, so that they can lag the harmonics at the same frequency in DC-link voltage by 90°, as expressed in (4.4).

$$\begin{cases} \angle i_q(j6\omega_g) - \angle u_{dc}(j6\omega_g) = -\frac{\pi}{2} + 2k\pi \\ \angle i_q(j12\omega_g) - \angle u_{dc}(j12\omega_g) = -\frac{\pi}{2} + 2k\pi \end{cases} \tag{4.4}$$

where k is an integer.

Considering that the control performance of active damping strategies applied to electrolytic capacitorless permanent magnet motor drives may reduce when grid impedance changes, a grid current quality improvement strategy based on harmonic energy regulation is proposed. This approach enhances the robustness of the control strategy against variations in grid impedance. The block diagram is shown in Fig. 4.6. The proposed strategy mainly includes two parts: the regulation parameters matching and the harmonic energy regulation. The amplitude processing unit and the phase processing unit regulate the harmonics at the frequencies of $6\omega_g$ and $12\omega_g$, which are extracted from the DC-link voltage u_{dc} by the bandpass filters. The strategy considers only the inherent harmonics at the frequencies of $6\omega_g$ and $12\omega_g$ in u_{dc} [3]. Specifically, $BPF_1(s)$ is a bandpass filter with the center frequency of $6\omega_g$. $BPF_2(s)$ is a bandpass filter with the center frequency of $12\omega_g$. LPF(s) is a low-pass filter. To achieve harmonic energy regulation, the reconstructed DC-link voltage sampling values are applied to the SVPWM.

Based on the theoretical energy exchange mode, the gain coefficients K_{gain_i}(i = 1, 2) and delay coefficients α_i(i = 1, 2) are designed to regulate harmonic energy. Before regulation, C_{dc} and L_{eaqul} are considered the main energy storage elements. In this energy exchange mode, the phase of the inherent harmonics in the DC-link voltage aligns with that of the inherent harmonics in the q-axis current. Both phases lag the inherent harmonics of the DC-link current by 90deg. After regulation, C_{dc} and L_q become the main energy storage elements. In this regulated mode, the phase

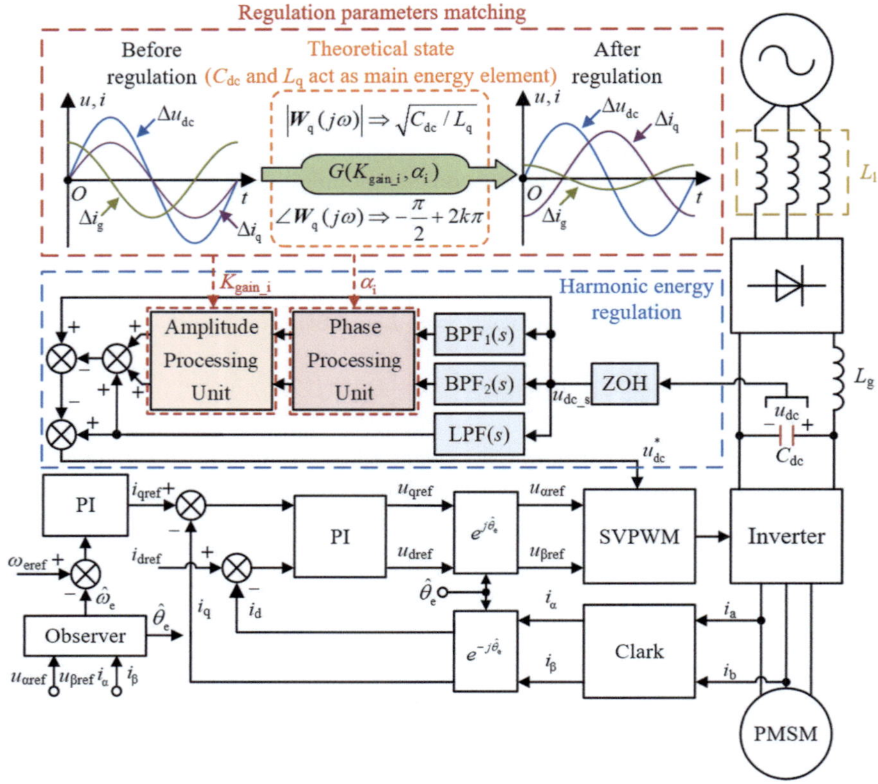

Fig. 4.6 Block diagram of grid current improvement strategy based on harmonic energy regulation, reprinted from Ref. [7], copyright 2022, with permission from Nature

of the inherent harmonics in the DC-link current leads the inherent harmonics in the DC-link voltage by 90°. The phase of the inherent harmonics in the DC-link voltage leads the inherent harmonics in the q-axis current by 90°. The rotor position for this study is obtained using a sensorless observer [5, 6].

As discussed in (4.3) and (4.4), the harmonics of the motor current require adjustments in both amplitude and phase. Δi_q can be expressed as follows

$$\Delta i_q = \frac{\left(\Delta u_{dc} - e^{-sT_s}\Delta u_{dc\_samp}\right)\left(U_{qref0}A_{1d} - U_{dref0}A_{2d}\right)}{U_{dc0}\left(A_{2q}A_{1d} - A_{2d}A_{1q}\right)}. \tag{4.5}$$

The phase and amplitude of Δi_q can be adjusted by changing u_{dc_samp}. Consequently, (2.23) can be simplified to

$$i_q(s) = f\left[G(s)u_{dc\_samp}(s)\right] \tag{4.6}$$

where f represents the function operator, $G(s)u_{dc\_samp}(s)$ serves as the independent variable of the function, and $i_q(s)$ is the dependent variable. Here, $G(s)$ denotes the processing unit for the sampled DC-link voltage u_{dc_samp}.

To satisfy the conditions specified in (4.3) and (4.4), $G(s)$ must possess the ability to adjust both the amplitude and phase of the motor current harmonics. $G(s)$ can be expressed as

$$G(s) = G_{\mathrm{BPF}_1} K_{\mathrm{gain\_1}} e^{-\alpha_1 T_s s} + G_{\mathrm{BPF}_2} K_{\mathrm{gain\_2}} e^{-\alpha_2 T_s s} - 1 \tag{4.7}$$

where G_{BPF1} is the bandpass filter with the center frequency of $6\omega_g$, and G_{BPF2} is the bandpass filter with the center frequency of $12\omega_g$. The gain coefficients K_{gain_1} and K_{gain_2} are designed for harmonic adjustment at the frequencies of $6\omega_g$ and $12\omega_g$, respectively. Similarly, the delay coefficients α_1 and α_2 correspond to the harmonic phase adjustments at $6\omega_g$ and $12\omega_g$, respectively.

After adjusting the amplitude and phase of the motor current harmonics, the transfer function between the reconstructed DC-link voltage sampling value and the actual motor d-q axis currents, denoted as $W_{d,qreg}(s)$, can be defined. According to (4.3), to minimize the harmonic energy stored in L_{eaqul}, the harmonic energy stored in C_{dc} and L_q must be balanced. This condition can be expressed as

$$\begin{cases} \left|W_{\mathrm{qreg}}(j6\omega_g)\right| = \sqrt{\frac{C_{dc}}{L_q}} \\ \left|W_{\mathrm{qreg}}(j12\omega_g)\right| = \sqrt{\frac{C_{dc}}{L_q}} \end{cases}. \tag{4.8}$$

According to (4.4), the q-axis current should lag the DC-link voltage by 90°. This relationship can be expressed as

$$\begin{cases} \angle W_{\mathrm{qreg}}(j6\omega_g) = -\frac{\pi}{2} + 2k\pi \\ \angle W_{\mathrm{qreg}}(j12\omega_g) = -\frac{\pi}{2} + 2k\pi \end{cases}. \tag{4.9}$$

Based on Eqs. (4.8) and (4.9), the characteristic surface is illustrated in Fig. 4.7. From Fig. 4.7, a set of solutions meeting the control requirements can be obtained: $W_{qreg}(j6\omega_g)| = |W_{qreg}(j12\omega_g)| = -27.6$ dB, $\angle W_{qreg}(j6\omega_g) = \angle W_{qreg}(j12\omega_g) =$ -90 deg. Accordingly, the parameters can be set as follows: $K_{gain\_1} = 1.9$, $\alpha_1 = 11.6$; $K_{gain\_2} = 6.4$, $\alpha_2 = 5.1$.

After applying the control strategy, the Nyquist diagram of $Y_{m\_in}(s)/Y_{LC\_out}(s)$ under different equivalent inductances is shown in Fig. 4.8. As L_{eaqul} varies from 2.5 mH to 5.5 mH, the intersection remains at the right of (− 1, 0), and the drive remains stable. It can be seen that compared with the original system, the control strategy is able to reduce the sensitivity of the drive to changes in the grid impedance.

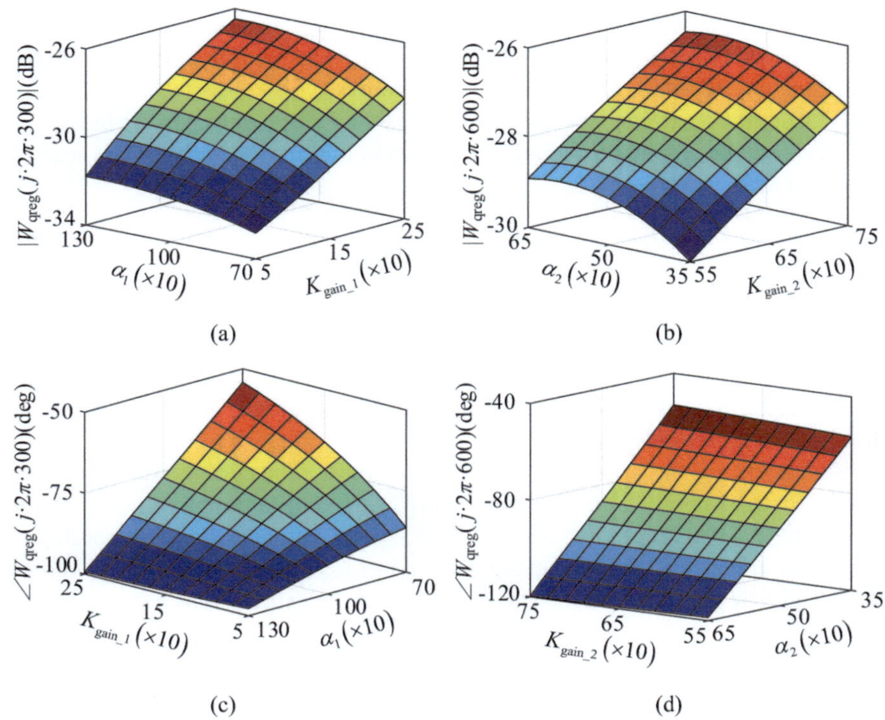

Fig. 4.7 Characteristic surface of $|W_{qreg}(j\omega)|$ and $\angle W_{qreg}(j\omega)$, reprinted from Ref. [7], copyright 2022, with permission from Nature. **a** $|W_{qreg}(j6\omega_g)|$. **b** $|W_{qreg}(j12\omega_g)|$. **c** $\angle W_{qreg}(j6\omega_g)$. **d** $\angle W_{qreg}(j12\omega_g)$

4.2.3 Experimental Verification

Figure 4.9 shows the experimental results of the grid current, DC-link voltage, and q-axis current before and after adopting the proposed control strategy under the motor operating frequency 70 Hz and rated torque conditions. It can be seen from Fig. 4.9a–c that the grid current is significantly improved with the proposed control strategy. As shown in Fig. 4.9d, e, the THD of the grid current is reduced from the original 53 to 32%, and the 5th, 7th, 11th, and 13th harmonic components of the grid current are all reduced.

Moreover, the amplitude and phase relationship between the q-axis current and DC-link voltage at the specific frequency are all adjusted with the proposed control strategy. At the frequency of $6\omega_g$, $|W_q(j6\omega_g)|$ increases from 0.02 dB to 0.05 dB, and $\angle W_q(j6\omega_g)$ changes from 45.36° to − 95.04°, as shown in Fig. 4.9f, g. At the frequency of $12\omega_g$, $|W_q(j12\omega_g)|$ increases from 0.02 to 0.05 dB, and $\angle W_q(j12\omega_g)$ changes from − 41.04° to − 88.56°, as shown in Fig. 4.9h, i. It should be noted that $|W_{q\_new}(j\omega)|$ with the proposed control is slightly larger than the theoretical value of formula (4.8), which is caused by the variation of L_q with current during the motor

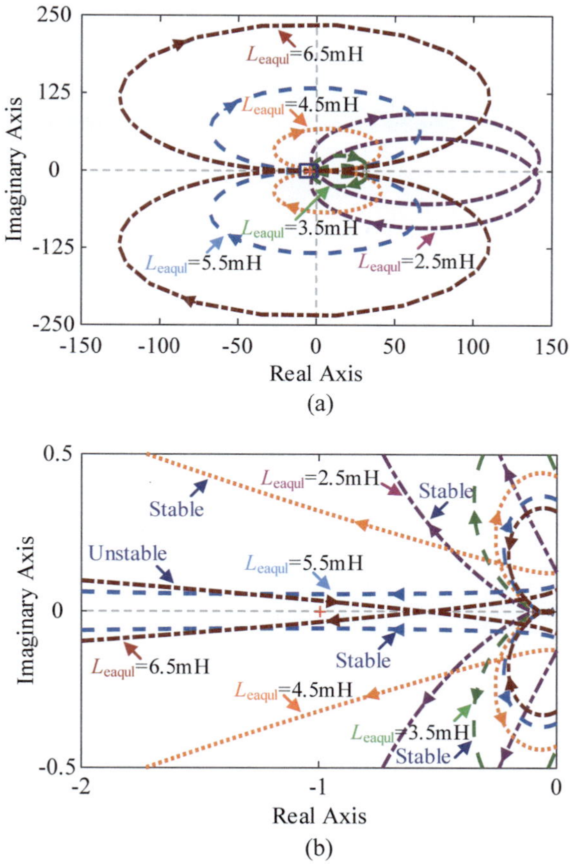

Fig. 4.8 Nyquist plots of $Y_{m\_in}(s)/Y_{LC\_out}(s)$ under different L_{eaqul} after applying control. **a** Whole diagra. **b** Zoomed view

operation [8]. Therefore, it can be seen that the proposed control strategy can reduce the THD of the grid current, the fluctuation components of the grid current, and the fluctuation amplitude of the DC-link voltage which is reduced from 146 to 104 V.

In order to further verify the effectiveness of harmonic energy regulation after applying the proposed control strategy, Fig. 4.10 shows the experimental results of the DC-link voltage u_{dc} and the DC-link current i_g before and after applying the control strategy. It can be seen that the proposed control strategy can regulate harmonic energy and the fluctuation amplitude of i_g can reduce from 9.65 A to 6.56 A.

In order to verify the robustness of the proposed control strategy against the grid impedance variations, experiments are conducted under three conditions with L_{eaqul} set to 2.5 mH, 4.4 mH, and 5.6 mH while the motor operated at 70 Hz and rated torque. The results are compared with the active damping scheme using parallel virtual resistance [3], as shown in Fig. 4.11.

When L_{eaqul} set to 2.5 mH, the THD of the grid current was 31.8% with the proposed control strategy and 33.7% with the active damping scheme as shown in Fig. 4.11a, b. When L_{eaqul} set to 4.4 mH, the THD of the grid current was 33.2% with

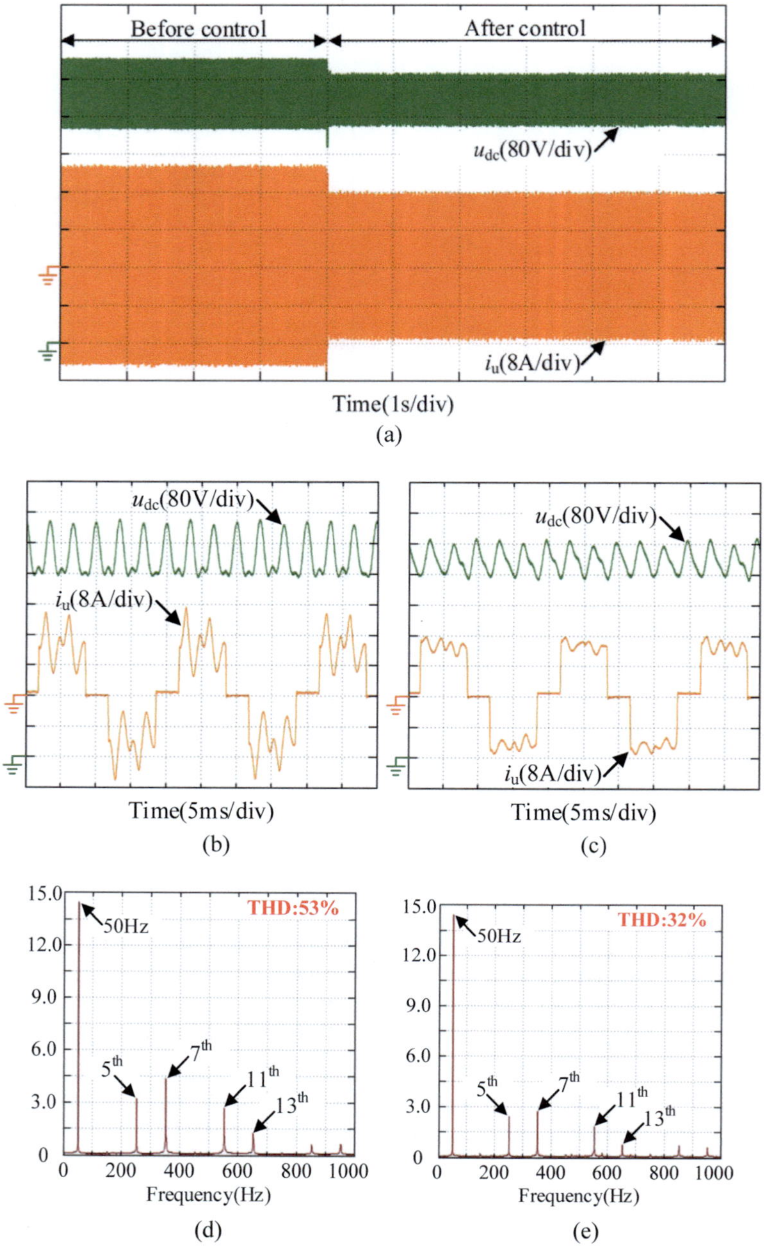

Fig. 4.9 Experimental results at the motor operating frequency of 70 Hz, reprinted from Ref. [7], copyright 2022, with permission from Nature. **a** Variations of DC-link voltage and grid-side current before and after applying the proposed control strategy. **b** Before applying the proposed control strategy. **c** After applying the proposed control strategy. **d** FFT analysis before applying the proposed control strategy **e** FFT analysis before applying the proposed control strategy **f** Harmonic component before applying the proposed control strategy ($6\omega_g$) **g** Harmonic component after applying the proposed control strategy ($6\omega_g$) **h** Harmonic component before applying the proposed control strategy ($12\omega_g$) **i** Harmonic component after applying the proposed control strategy ($12\omega_g$)

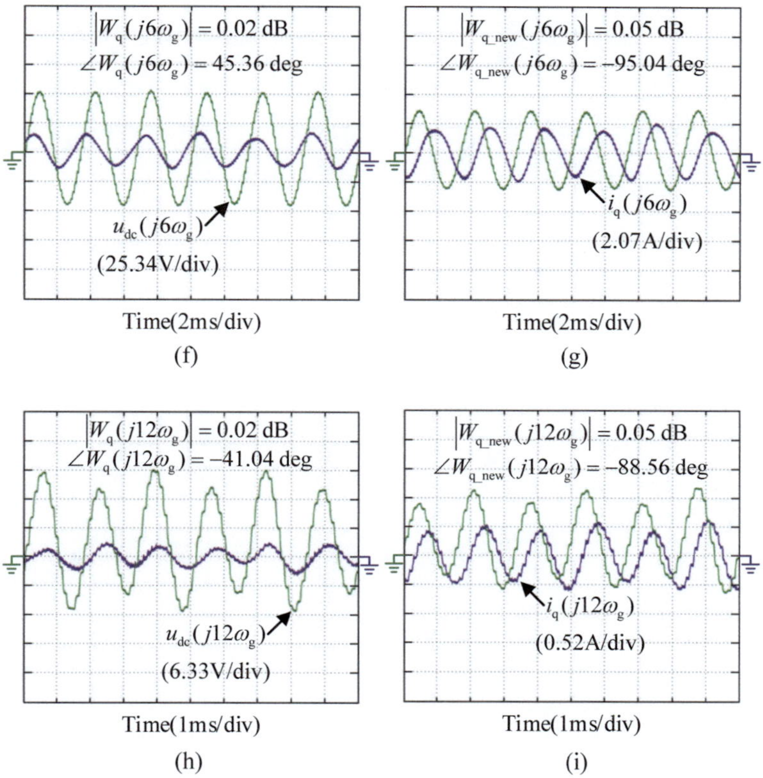

Fig. 4.9 (continued)

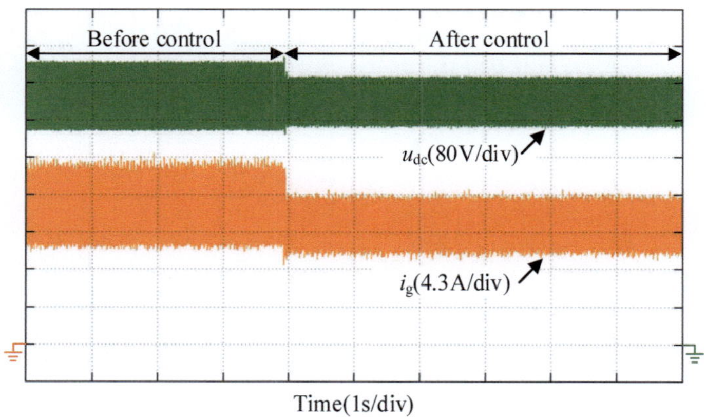

Fig. 4.10 The i_g variation before and after applying the proposed control strategy

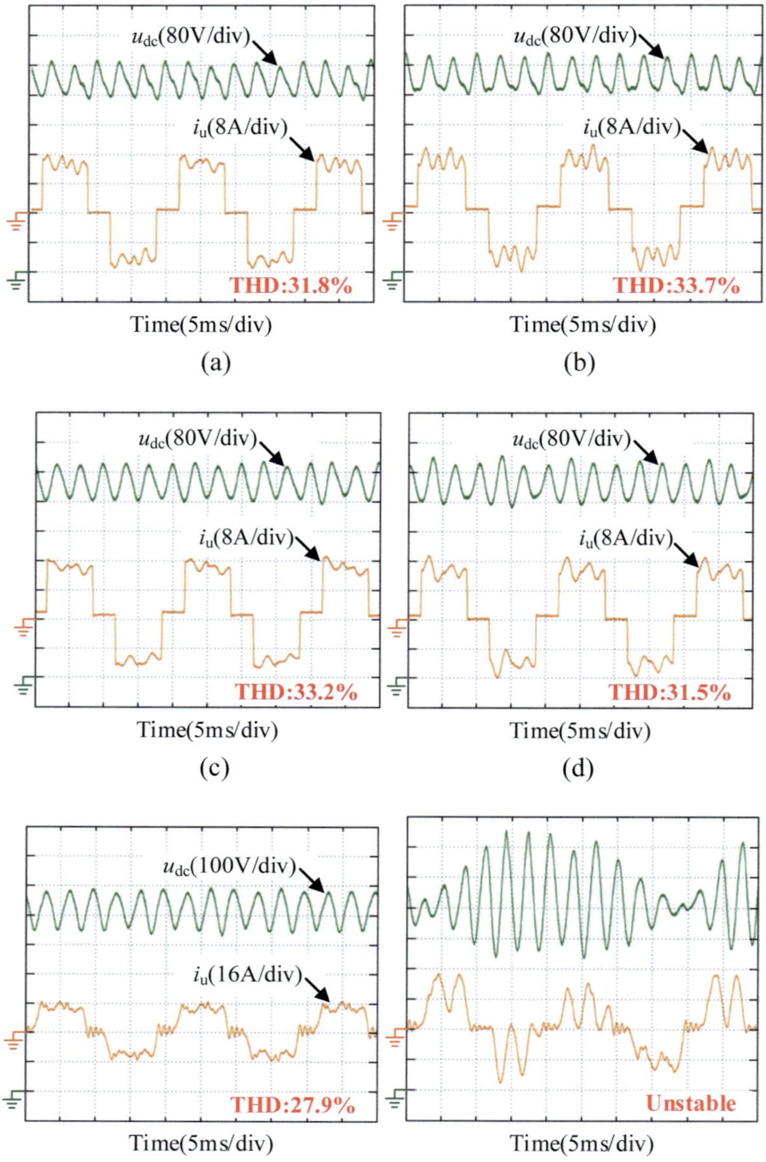

Fig. 4.11 The experimental results against the grid-side impedance variations, reprinted from Ref. [7], copyright 2022, with permission from Nature. **a** Applying the proposed control strategy when L_{eaqul} is 2.5 mH. **b** Applying the active damping scheme when L_{eaqul} is 2.5 mH. **c** Applying the proposed control strategy when L_{eaqul} is 4.4 mH. **d** Applying the active damping scheme when L_{eaqul} is 4.4 mH. **e** Applying the proposed control strategy when L_{eaqul} is 5.6 mH. **f** Applying the active damping scheme when L_{eaqul} is 5.6 mH

the proposed strategy and 31.5% with the active damping strategy, as illustrated in Fig. 4.11c, d. When L_{eaqul} further increased to 5.6 mH, the THD of the grid current was reduced to 27.9% with the proposed strategy and unstable with the active damping strategy, as shown in Fig. 4.11e, f. These results demonstrate that, compared with the active damping strategy, the proposed control strategy exhibits better adaptability against the grid impedance variations.

Figure 4.12 presents the experimental results of the u_{dc} and motor current i_{a} before and after applying the proposed control strategy under the rated torque and motor frequency of 70 Hz when L_{eaqul} is 5.6 mH. After applying the control strategy, the fluctuation of the DC-link voltage decreased from 230 to 99 V while the THD of the motor current increased slightly from 4.51 to 5.01%. The motor phase current primarily contained harmonics at 230, 370, 530, and 670 Hz, with the respective harmonic amplitudes changing from 0.88 A, 0.49 A, 0.08 A, and 0.05 A to 0.86 A, 0.51 A, 0.24 A, and 0.19 A. The experimental results indicate that the proposed control strategy can introduce several minimal additional harmonics into the motor current.

Figure 4.13 shows the experimental results at the different motor frequencies of 50 and 60 Hz when L_{eaqul} is 5.6 mH. The results indicate that the drive system becomes unstable with the active damping scheme but the proposed control strategy can maintain the stability of the system and the grid current quality.

Figure 4.14 shows the experimental results when the motor speed ω_{e} dynamically changes at the same load. The results indicate that the fluctuation of grid current is reduced by 15% and the DC-link voltage fluctuation was reduced by 7% with the proposed control strategy when the motor speed ω_{e} dynamically changes.

Figure 4.15 illustrates the variations of the THD of the grid current before and after applying the proposed control strategy under different motor power. The results indicate that the THD of the grid current is improved to varying degrees across all power points, validating the effectiveness of the proposed control strategy.

4.3 Grid Side Current Harmonics Suppression Strategy Based on Rectified Voltage Full-Feedforward

4.3.1 Harmonic Characteristics Analysis of the Drive Under Unbalanced Grid Voltage

Voltage drops at single-phase is one of the common operating conditions of unbalanced gird. The connection of high-power single-phase loads, the switching of transformer equipment, and single-phase grounding short circuit faults can all cause a drop in single-phase voltage. Therefore, this article focuses on analyzing the operating condition where the amplitude of the input u-phase voltage drops. At this moment, the three-phase voltage on the grid side can be expressed as

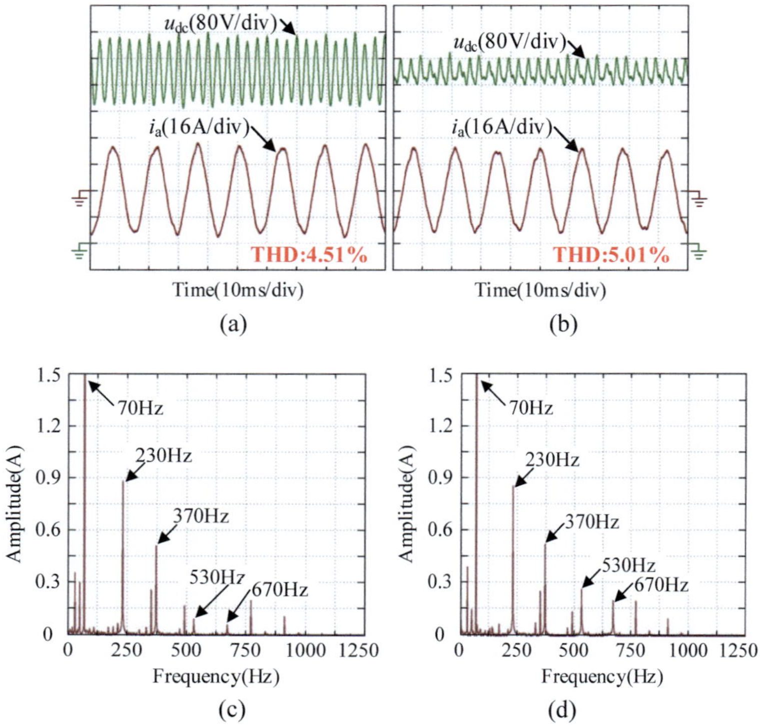

Fig. 4.12 The variations of motor current before and after applying the proposed control strategy, reprinted from Ref. [7], copyright 2022, with permission from Nature. **a** The experimental results before applying the proposed control strategy. **b** The experimental results after applying the proposed control strategy. **c** The FFT analysis results before applying the proposed control strategy. **d** The FFT analysis results after applying the proposed control strategy

$$\begin{cases} u_{un} = \sqrt{2}k_{dr}U\sin(\omega_g t) \\ u_{vn} = \sqrt{2}k_{dr}\sin(\omega_g t - \frac{2\pi}{3}) \\ u_{wn} = \sqrt{2}k_{dr}\sin(\omega_g t + \frac{2\pi}{3}) \end{cases} \tag{4.10}$$

where k_{dr} is the drop coefficient, and u_{un}, u_{vn}, and u_{wn} are the three-phase input voltages.

When the u phase voltage drops, the rectifier bridge diode conduction angle changes as shown in Fig. 4.16. At this time, waveforms of the grid current of u phase i_u and the rectified voltage u_g are distortional. The change of diode conduction angle $\Delta\theta$ after the drop of u-phase voltage can be expressed as

$$\Delta\theta = \left|\phi - \frac{\pi}{6}\right| \tag{4.11}$$

where ϕ can be expressed as

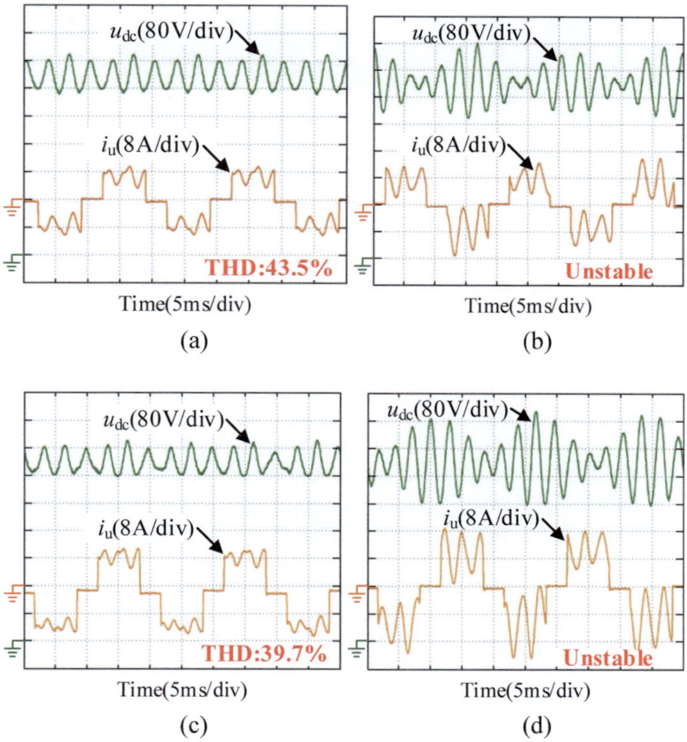

Fig. 4.13 The experimental results at the different operating frequencies, reprinted from Ref. [7], copyright 2022, with permission from Nature. **a** Applying the proposed control strategy at the operating frequency of 50 Hz. **b** Applying the active damping scheme at the operating frequency of 50 Hz. **c** Applying the proposed control strategy at the operating frequency of 60 Hz. **d** Applying the active damping scheme at the operating frequency of 60 Hz

$$\phi = \tan^{-1}\left(\frac{\sqrt{3}}{2k_{\mathrm{dr}}+1}\right). \tag{4.12}$$

Therefore, the rectified voltage u_{g} under unbalanced grid conditions over a period can be expressed as [9, 10]

$$\begin{cases} u_{\mathrm{un}} - u_{\mathrm{wn}} & \frac{\pi}{2} \le \omega_{\mathrm{g}}\mathrm{t} \le \frac{5\pi}{6} - \Delta\theta \\ u_{\mathrm{vn}} - u_{\mathrm{wn}} & \frac{5\pi}{6} - \Delta\theta \le \omega_{\mathrm{g}}\mathrm{t} \le \frac{7\pi}{6} + \Delta\theta \\ u_{\mathrm{vn}} - u_{\mathrm{un}} & \frac{5\pi}{6} - \Delta\theta \le \omega_{\mathrm{g}}\mathrm{t} \le \frac{7\pi}{6} + \Delta\theta \end{cases}. \tag{4.13}$$

Performing Fourier series expansion at (4.13)

$$u_{\mathrm{g}} = \frac{a_0}{2} + \sum_{n=1}^{\infty}\left[a_{\mathrm{n}}\cos(n\omega_{\mathrm{g}}t) + b_{\mathrm{n}}\sin(n\omega_{\mathrm{g}}t)\right]. \tag{4.14}$$

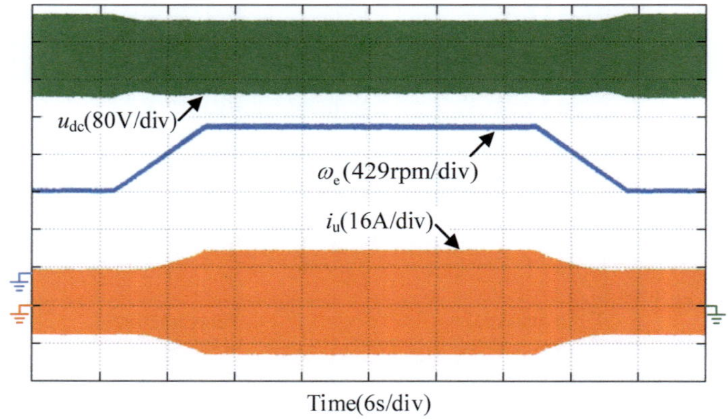

(a)

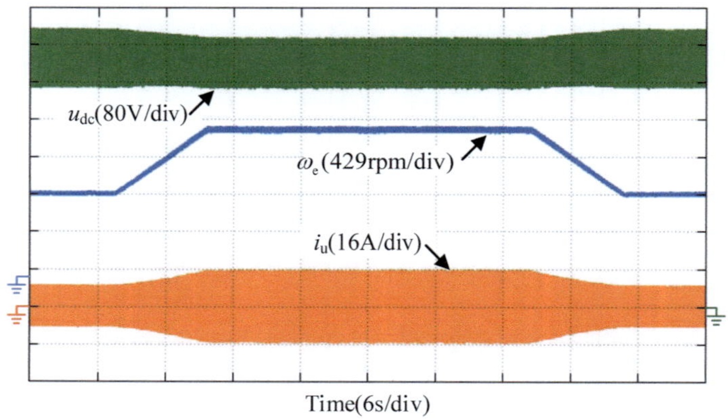

(b)

Fig. 4.14 The experimental results with the dynamic changes of the motor speed, reprinted from Ref. [7], copyright 2022, with permission from Nature. **a** The experimental results before applying the proposed control strategy. **b** The experimental results after applying the propsed control strategy

This article analyzes the condition that the amplitude of the *u*-phase voltage drops by 20%. At this time, the unbalance factor is approximately 7.2%. The remaining drop scenarios are similar. Since u_g is an even function, b_n is zero. a_0 and a_n can be expressed as (4.15).

Under balanced grid conditions, the effective value of the grid line voltage is 380 V, with a frequency of 50 Hz. As can be seen from (4.15), under balanced grid condition, the rectified voltage mainly contains harmonics that are integer multiples of 6 times and 12 times the grid voltage, as shown in Fig. 4.17a. When the amplitude of the grid-side input u-phase voltage drops by 20%, the amplitude of each harmonic in the rectified voltage is shown in Fig. 4.17b according to Eq. (4.15). Compared to balanced grid condition, significant even harmonics appear in the rectified voltage.

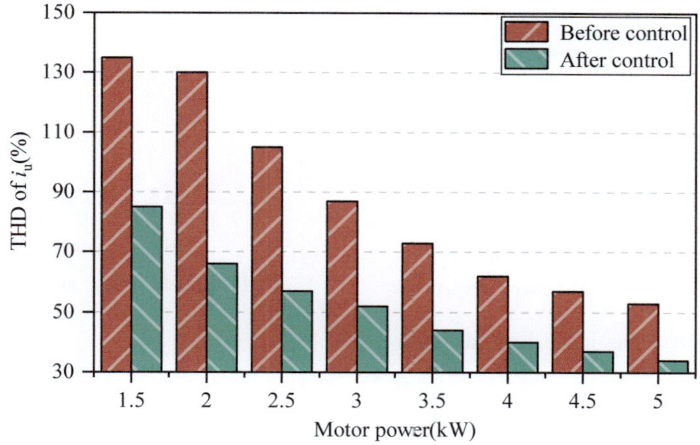

Fig. 4.15 The statistical results of the THD of the grid-side current before and after applying the proposed control strategy

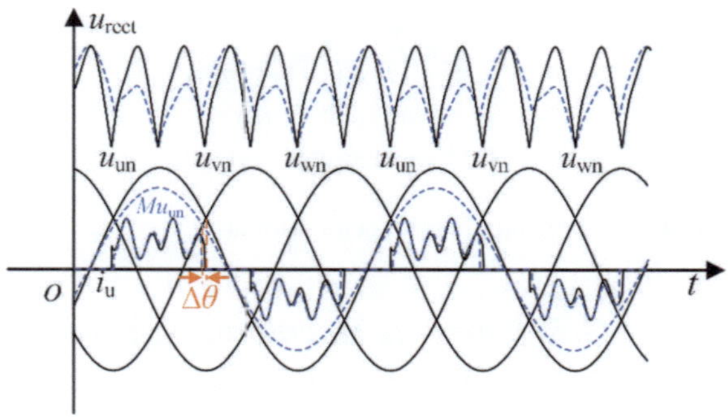

Fig. 4.16 Conduction angle of rectifier bridge after u phase voltage drop

$$\begin{cases} a_0 = \frac{22(5\sqrt{6}+2\sqrt{122})\omega_g}{25\pi^2} \\ a_n = -\frac{110\sqrt{6}\omega_g \cos[\frac{n\pi}{2}]\cos[100n\pi t]\cos[\frac{n\pi}{2}]}{25(-1+n^2)\pi^2} \\ -\frac{44\sqrt{122}\omega_g \cos[\frac{n\pi}{2}]\cos[100n\pi t]\cos[\frac{n\pi}{2}]}{25(-1+n^2)\pi^2} \end{cases}. \tag{4.15}$$

At this time, the inductor current will contain harmonic components with the same frequency as the rectified voltage. The Fourier expansion of grid current under unbalanced grid conditions obtained through analysis methods under balanced grid conditions. Under the operating condition of the motor with a power of 4 kW, the harmonics in the u-phase current before and after the u-phase voltage drops by 20% are shown in Fig. 4.18. Under balanced grid conditions, the grid currents mainly contain harmonics of order 5, 7, 11, and 13, as shown in Fig. 4.18a. Under the condition of u-phase voltage drops, the odd-order harmonics in the grid current increase as shown in Fig. 4.18b.

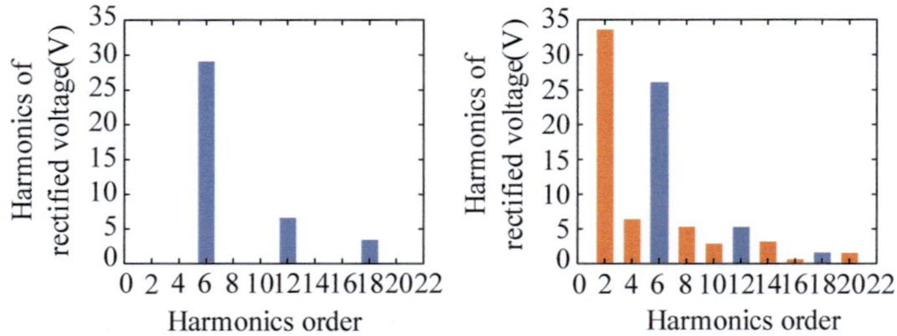

(a) balanced grid voltage (b) *u* phase voltage grop

Fig. 4.17 Harmonics in the rectified voltage

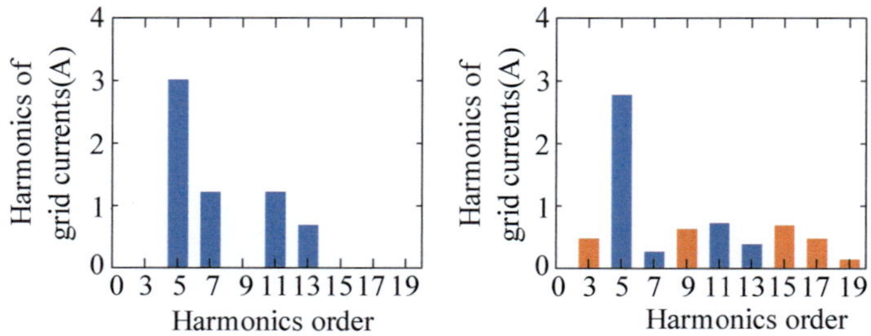

(a) balanced grid voltage (b) *u* phase voltage grop

Fig. 4.18 Harmonics in the u phase current

4.3.2 Grid Currents Harmonic Suppression Strategy Based on Rectified Voltage Full-Feedforward

As analyzed previously, harmonics of the 2nd, 4th, 8th, and other higher orders with large amplitudes appear in the rectifier voltage under unbalanced gird conditions. At this point, the effectiveness of the control strategies that targeting specific harmonic orders exhibits is reduced. Therefore, in order to address the issues of increased harmonics in the inductor current and degraded grid current quality under unbalanced gird conditions, this paper proposes a grid current harmonics suppression strategy based on full-feedforward of the rectifier voltage. The control block diagram is shown in Fig. 4.19. The inverter input current is reconstructed from the motor currents to obtain the rectifier voltage. Then, the harmonic information in the rectified voltage is extracted. Based on the small-signal model of the inductor current control loop, the full feedforward function of the rectifier voltage is designed. Then the damping voltage is injected into the q-axis voltage, which increasing the equivalent output impedance on the grid side and reducing grid current harmonics during unbalanced grid condition.

When the power loss of the inverter is ignored, the small signal variation of inverter current can be expressed as

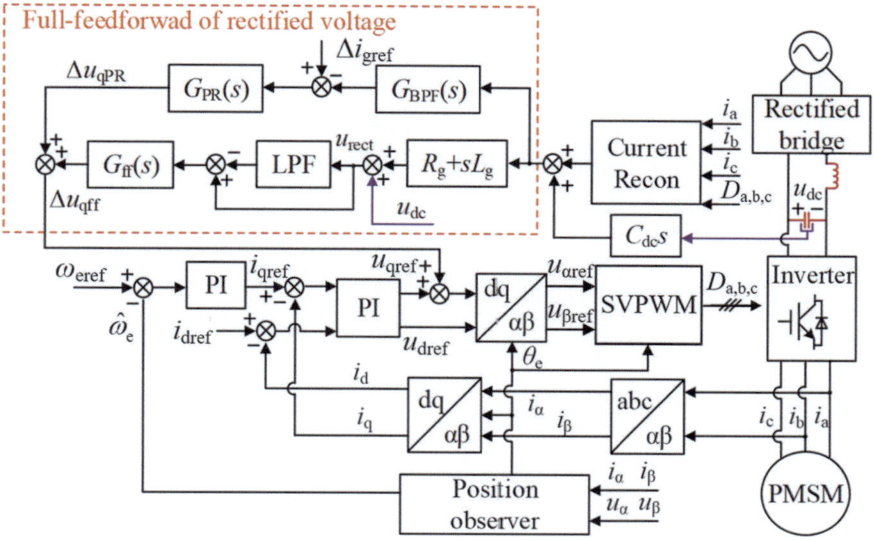

Fig. 4.19 Block diagram of the gird harmonics suppression method based on the rectified voltage full-feedforward [11]

$$\Delta i_{\text{inv}} = \frac{1.5\left(\Delta u_{\text{q}} I_{\text{q}0} + U_{\text{q}0} \Delta i_{\text{q}}\right) - \Delta u_{\text{dc}} I_{\text{inv}0}}{U_{\text{dc}0}}. \tag{4.16}$$

Combining (4.16) with the block diagram shown in Fig. 4.19, the small-signal block diagram for inductor current can be constructed as shown in Fig. 4.20 [11]·Performing equivalent transformations on the block diagram shown in Fig. 4.20. The transformation process is depicted in Fig. 4.21, where dashed lines represent the original form of the model and solid lines represent the transformed form. Firstly, the feedforward of the small signal variation of q-axis voltage Δu_q in Fig. 4.21 is modified to the feedforward of the small signal variation of q-axis current Δi_q. Then, the feedback point of the small signal variations of inductor current Δi_g is shifted from the output of the small signal variations of inverter current Δi_{inv} to the output of the small signal variations of motor side output power ΔP_L. The equivalent model can be obtained which is shown in Fig. 4.21a. The simplification process is illustrated in Fig. 4.21b. Based on the Fig. 4.21b, the feedforward and feedback branches of Δi_q are merged and the forward path of the model, and the feedback branch of the small signal variations of DC-link voltage Δu_{dc} is also merged with the forward path. Then, the simplified model is shown in Fig. 4.21c. The feedback point of Δi_g in Fig. 4.21c is then shifted backwards, and the forward path is simplified and merged, which resulting in the equivalent model shown in Fig. 4.21d. In Fig. 4.21d, G1(s) and G2(s) are the transfer functions derived through equivalent transformations, where G_1(s) represents the transfer function between Δu_q and Δu_{dc}, and G_2(s) represents the transfer function between the small signal variations of voltage on DC side equivalent impedance and the Δi_g, respectively.

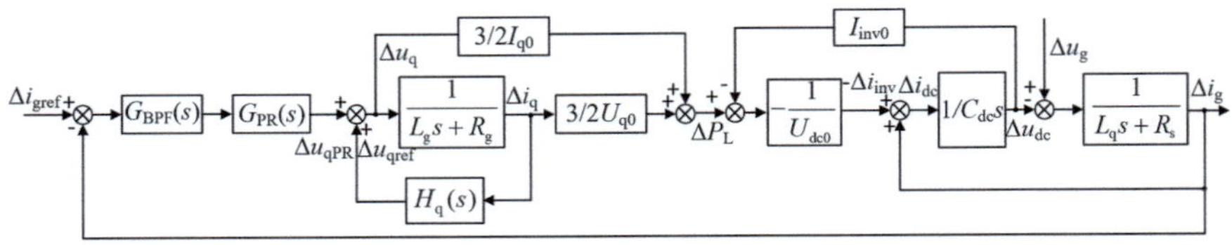

Fig. 4.20 Control block diagram of the small signal variations of inductor current

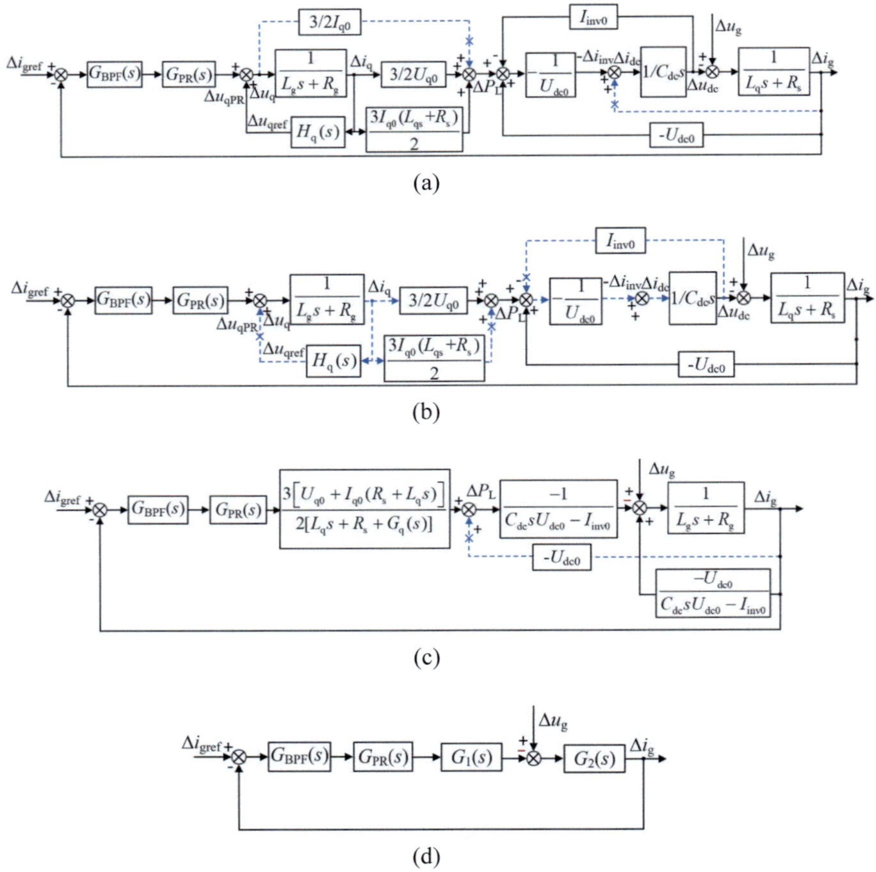

Fig. 4.21 Control block diagram of the small signal variations of inductor current

The transfer functions of $G_1(s)$ and $G_2(s)$ can be expressed as

$$G_1(s) = -\frac{3\left[I_{q0}(L_q s + R_s) + U_{q0}\right]}{2(U_{dc0}C_{dc}s - I_{inv0})\left[L_q s + R_s + H_q(s)\right]} \tag{4.17}$$

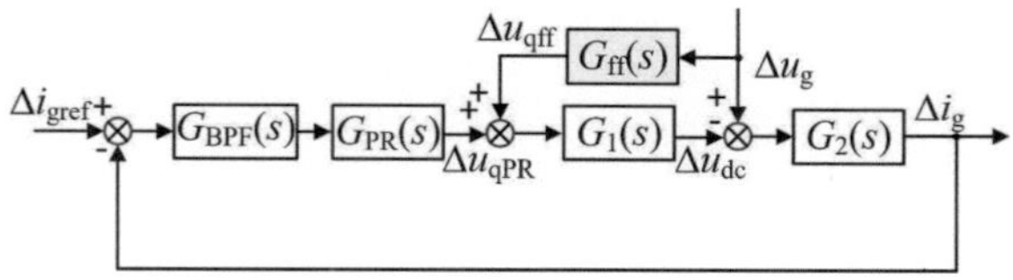

Fig. 4.22 Simplified block diagram of the inductor current based on the rectified voltage full-feedforward

$$G_2(s) = \frac{U_{dc0}C_{dc}s - I_{inv0}}{U_{dc0} + (L_g s + R_g)(U_{dc0}C_{dc}s - I_{inv0})}. \tag{4.18}$$

From Fig. 4.21d, the open loop transfer function of the closed control loop of the inductor current $T_A(s)$ can be expressed as

$$T_A(s) = -G_{BPF}(s)G_{PR}(s)G_1(s)G_2(s). \tag{4.19}$$

Therefore, Δi_g can be expressed as

$$\frac{T_A(s)}{1+T_A(s)}\Delta i_{gref} + \frac{G_2(s)}{1+T_A(s)}\Delta u_g. \tag{4.20}$$

As analyzed previously, under unbalanced grid conditions, the rectifier voltage contains significant even-order harmonic components. Combining the Eq. (4.20), it can be deduced that in order to reduce the impact of rectifier voltage harmonics on the inductor current control loop, a feedforward branch of the small signal variations of rectifier voltage can be added as shown in Fig. 4.22. At this point, the Δi_g can be expressed as.

$$\Delta i_g = \frac{T_A(s)}{1+T_A(s)}\Delta i_{gref} + \frac{G_2(s)}{1+T_A(s)}\Delta u_g - \frac{G_{ff}(s)G_1(s)G_2(s)}{1+T_A(s)}\Delta u_g \tag{4.21}$$

where the G_{ff}(s) is feedforward function of rectified voltage.

Combining (4.21), it can be deduced that in order to reduce the impact of rectifier voltage harmonics on the inductor current control loop, the full feedforward function of the rectifier voltage can be derived as

$$G_{ff}(s) = \frac{1}{G_1(s)} = \frac{-2(U_{dc0}C_{dc}s - I_{inv0})\left[L_q s + R_s + H_q(s)\right]}{3\left[I_{q0}(L_q s + R_s) + U_{q0}\right]}. \tag{4.22}$$

From (4.21), it can be seen that when the full feedforward function satisfies (4.22), the rectifier voltage disturbance branch in the inductor current control loop is decoupled. Based on the topology of the motor system, the small signal variations of the rectifier voltage can be expressed as

$$\Delta u_g = \Delta u_{dc} + \Delta i_g(R_g + sL_g). \tag{4.23}$$

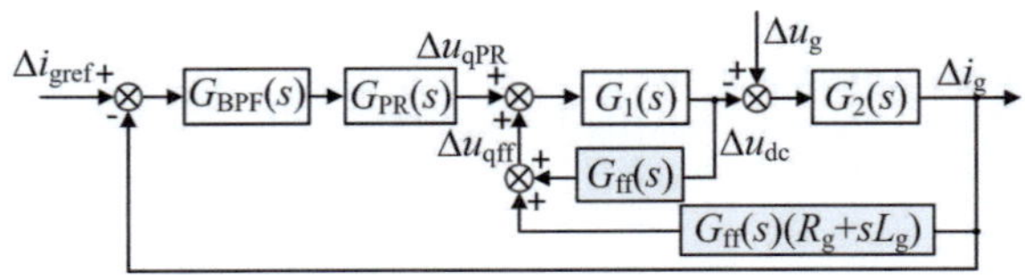

Fig. 4.23 Equivalent block diagram of the rectified voltage full-feedforward control

Therefore, the equivalent block diagram shown in Fig. 4.22 can be converted to model shown in Fig. 4.23. The full feedforward of the rectifier voltage can be equivalently considered as the feedforward of the DC-link voltage and inductor current, which eliminating the need for an additional rectifier voltage sampling circuit.

When the full-feedforward function satisfies (4.22), theoretically, the harmonic disturbance of the rectifier voltage in the inductor current control loop is eliminated. However, there are errors in the reconstructed inductor current in practical control due to the sampling and digital control delays. Therefore, there is also a digital control delay in the feedforward signal. After applying full-feedforward control of the rectified voltage, the actual small signal variations of the inductor current can be expressed as

$$\Delta i_g = \frac{T_A(s)}{1+T_A(s)}\Delta i_{gref} + \frac{G_2(s)[1-T_d(s)]}{1+T_A(s)}\Delta u_g. \tag{4.24}$$

Therefore, the Bode diagram of the transfer function between the rectified voltage and the inductor current before and after applying full-feedforward control of the rectified voltage is shown in Fig. 4.24. It can be seen that when the conventional inductor current control is applied, the inductor current control loop exhibits weak suppression capability against harmonic disturbances in the rectified voltage within the mid-frequency band ($2\omega_g$–$12\omega_g$). After applying full- feedforward control of the rectified voltage, the magnitude of the transfer function between the rectified voltage and the inductor current decreases. The harmonic disturbance suppression capability is enhanced.

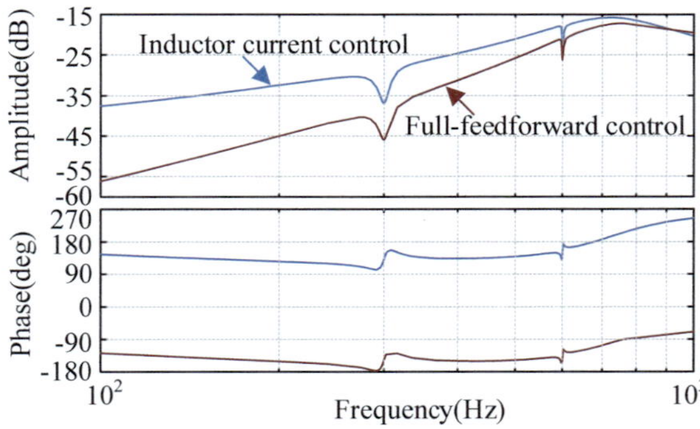

Fig. 4.24 Bode diagram between the Δu_g and Δi_g under different control strategies

4.3.3 Stability Analysis of Grid Side Current Harmonic Suppression Strategy Based on Full Feedforward of Rectified Voltage

To ensure the stable operation of the system, it is necessary to analyze the stability of the system before and after applying full feedforward control of the rectified voltage. From (4.21), the small signal variation of the inductor current can be derived as

$$\Delta i_g = i_0 - \frac{\Delta u_g}{Z_0(s)} \tag{4.25}$$

where i_0 and $Z_0(s)$ can be expressed as

$$i_0 = \frac{T_A(s)}{1 + T_A(s)} \Delta i_{gref} \tag{4.26}$$

$$Z_0(s) = \frac{1 + T_A(s)}{G_2(s) - G_{ff}(s)G_1(s)G_2(s)}. \tag{4.27}$$

Substituting (4.23) into (4.25), the small signal variations of inductor current can be expressed as

$$\Delta i_L = \left[i_0 - \frac{\Delta u_{dc}}{Z_0(s)} \right] \frac{1}{1 + \left[\frac{Z_g(s)}{Z_0(s)} \right]} \tag{4.28}$$

where $Z_g(s) = R_g + sL_g$

Through Norton equivalent circuit, the motor system can be equivalently represented by the impedance model shown in Fig. 4.25, where $Y_{LC}(s)$ is the output admittance of the LC section, $Z_g(s)$ is the equivalent impedance on the DC side, and $Z_0(s)$ is the equivalent output impedance on the grid side. At this point, the stability of the motor system can be judged by analyzing the stability of the two terms in (4.28).

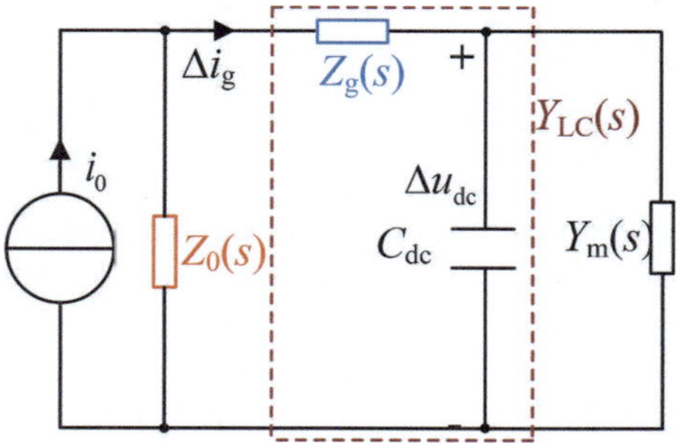

Fig. 4.25 Norton equivalent circuit of the motor system

The first term in (4.28), $[i_0 - [\Delta u_{dc}/Z_0(s)]$, can be regarded as the expression of small signal variation of the inductor current when the equivalent impedance on the DC side $Z_g(s)$ is equal to zero. At this point, the motor system can be considered as a cascaded system consisting of the equivalent impedance on the inverter side and the output impedance of the grid-side LC section. The output impedance of the grid-side LC section $Y_{LC}(s) = sC_{dc}$. After applying the full-feedforward strategy of the rectified voltage, the small signal variations of the given voltages on the dq axes of the motor are

$$\begin{bmatrix} \Delta u_{dref} \\ \Delta u_{qref} \end{bmatrix} = \begin{bmatrix} -H_d(s) & -\omega_e L_q \\ \omega_e L_d & -H_q(s) \end{bmatrix} \begin{bmatrix} \Delta i_d \\ \Delta i_q \end{bmatrix} + \begin{bmatrix} 0 \\ -G_{BPF}(s) G_{PR\_C}(s) i_{g\_r} \end{bmatrix} + \begin{bmatrix} 0 \\ G_{ff}(s)\Delta u_g \end{bmatrix}. \quad (4.29)$$

Therefore, the relationship between the small signal variations of the dq-axis currents and the small signal variations of the DC-link voltage can be expressed as

$$\begin{bmatrix} N_{d\_rect}(s) \\ N_{d\_rect}(s) \end{bmatrix} = \begin{bmatrix} N_{dmad}(s) \\ N_{qmad}(s) \end{bmatrix} + G_{mad3}(s)\frac{\Delta u_g}{\Delta u_{dc}} \begin{bmatrix} M_q(s) \\ -K_d(s) \end{bmatrix} \quad (4.30)$$

where $G_{mad3}(s)$ can be expressed as

$$G_{mad3}(s) = \frac{G_{ff}(s)}{K_q(s)K_d(s) - M_d(s)M_q(s)}. \quad (4.31)$$

From the topology of the electrolytic capacitorless drive, Δu_g can be expressed as

$$\Delta u_g = (C_{dc} s \Delta u_{dc} + \Delta i_{inv}) Z_g(s) + \Delta u_{dc}. \quad (4.32)$$

Therefore, after applying the full-feedforward control strategy of the rectified voltage, the output admittance of the motor side can be expressed as

$$\begin{aligned} Y_{\_rect}(s) = &-\frac{I_{inv0}}{U_{dc0}} + \frac{3N_{d\_rect}(s)\left(U_{d0} + \omega_e L_d I_{q0} + L_d I_{d0} s + I_{d0} R_s\right)}{2U_{dc0}} \\ &+ \frac{3N_{q\_rect}(s)\left(U_{q0} - \omega_e L_q I_{d0} + L_q I_{q0} s + I_{q0} R_s\right)}{2U_{dc0}} \end{aligned} \quad (4.33)$$

The stability of the cascaded system can be characterized by the Nyquist diagram of $Y_{\_rect}(s)/Y_{LC}(s)$. As shown in Fig. 4.26, when the equivalent impedance on the DC side $Z_g(s)$ is equal to zero, the Nyquist contour of $Y_{\_rect}(s)/Y_{LC}(s)$ does not enclose the point (– 1, j0), indicating that the motor system remains stable.

Therefore, the stability of the system depends on the second term in (4.28), which is $1/[1 + Z_g(s)/Z_0(s)]$. This term can be regarded as a negative feedback closed-loop control system with a forward gain of 1 and a feedback loop gain of $[Z_g(s)/Z_0(s)]$. To ensure the stability of the motor system, the open-loop transfer function $[Z_g(s)/$

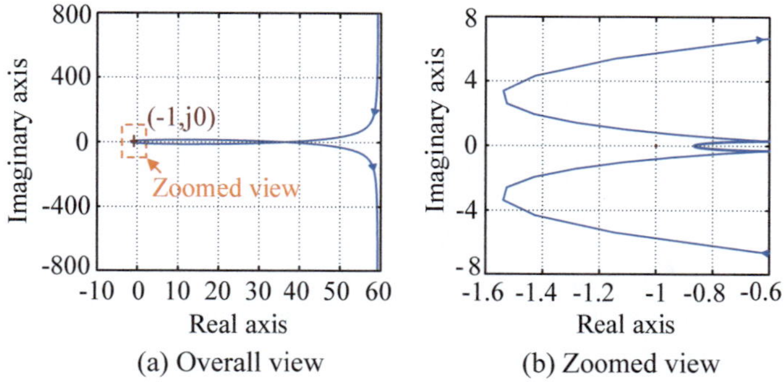

Fig. 4.26 Nyquist diagram of $Y_{\_rect}(s)/Y_{LC}(s)$ when the impedance of the DC side is zero

$Z_0(s)]$ of the control system needs to satisfy the Nyquist stability criterion. When the phase margin (PM) of the open-loop transfer function of the control system is greater than zero, the system remains stable; otherwise, the motor system becomes unstable. For the equivalent negative feedback closed-loop control loop corresponding to the second term in (4.28), its phase margin depends on the phase difference between $Z_g(s)$ and $Z_0(s)$ at the intersection frequency f_{int}

$$\begin{aligned} PM &= 180^\circ - [\angle Z_g(j2\pi f_{int}) - \angle Z_0(j2\pi f_{int})] \\ &= 90^\circ + \angle Z_0(j2\pi f_{int}) \end{aligned} \tag{4.34}$$

where $\angle Z_g(j2\pi f_{int})$ represents the phase of the equivalent impedance on the DC side at the intersection frequency f_{int}, and $\angle Z_0(j2\pi f_{int})$ represents the phase of the equivalent output impedance on the grid side at the intersection frequency f_{int}.

When the DC side impedance $Z_g(s)$ is considered purely inductive, $\angle Z_g(j2\pi f_{int}) = 90^\circ$, the phase margin of the closed-loop control system is minimized, resulting in the worst system stability. According to (4.34), the system remains stable when $\angle Z_0(j2\pi f_{int}) > -90^\circ$. Before applying full-feedforward of the rectified voltage, the equivalent output impedance on the grid side $Z_{ori}(s)$ can be expressed as

$$Z_{ori}(s) = \frac{1 + T_A(s)}{G_2(s)}. \tag{4.35}$$

In practical control, due to the presence of sampling and digital control delays, the equivalent output impedance on the grid side after applying full-feedforward control of the rectified voltage can be expressed as

$$Z_{ff}(s) = \frac{1}{1 - T_d(s)} Z_{ori}(s). \tag{4.36}$$

Before and after applying full-feedforward control of the rectified voltage, the Bode diagram of $Z_g(s)$, $Z_{ori}(s)$, and $Z_{ff}(s)$ are shown in Fig. 4.27. From the magnitude-frequency response curves, it can be seen that after applying full- feedforward control of the rectified voltage, the magnitude of the equivalent output impedance on the grid side increases, enhancing the harmonics suppression ability for the rectified voltage. Meanwhile, from the phase-frequency response curves, it can be known that although there is a phase lag in the equivalent output impedance on the grid side after applying the control, the system remains stable.

When the equivalent inductance on the DC side changes, the Bode diagram of $Z_{ff}(s)$ and $Z_g(s)$ are shown in Fig. 4.28. As the equivalent inductance on the DC side increases, the phase margin of the motor system decreases. However, the system remains stable in the certain range. The proposed control method exhibits robustness against variations in line distribution parameters.

Figure 4.29 shows the Bode diagram of $Z_{ff}(s)$ and $Z_g(s)$ when the q-axis inductance of the motor varies. As shown in the figure, when there is a deviation between the q-axis inductance value in the full feedforward function and the actual inductance value, the magnitude of the equivalent output impedance on the grid side decreases, resulting in a weakened harmonic suppression capability of the motor system. From

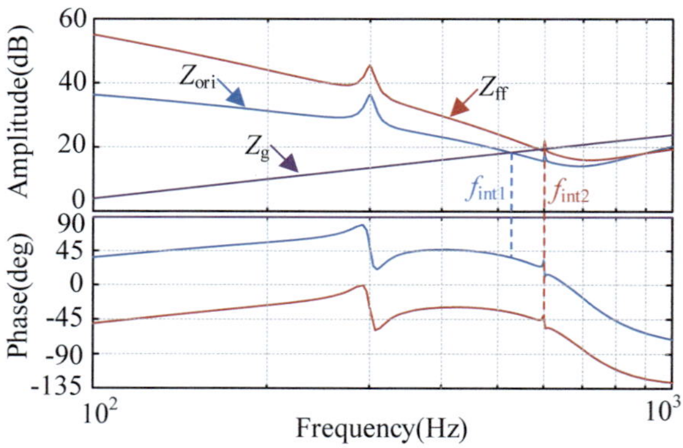

Fig. 4.27 Bode diagram of the impedance of DC-side and grid side

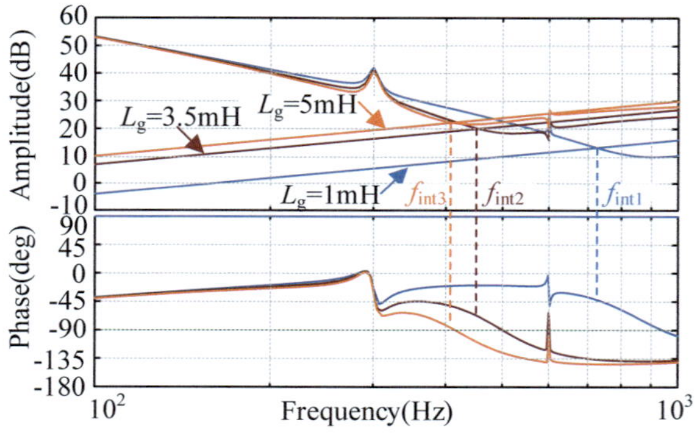

Fig. 4.28 Phase margin of the drive under the variations of equivalent inductance on the DC side

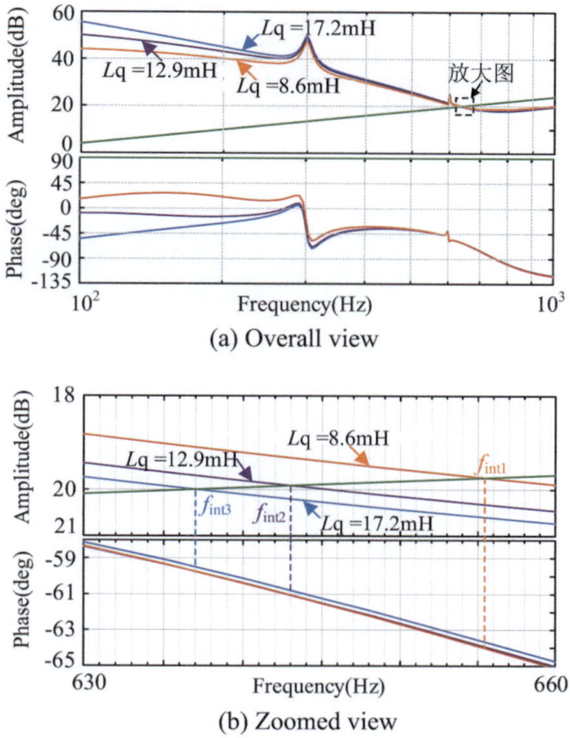

Fig. 4.29 Phase margin of the drive under the variation of q-axis inductance of the motor

the phase-frequency response, it can be observed that as the deviation increases, the steady-state margin of the motor system decreases. However, when the q-axis inductance varies within a certain range, the motor system remains stable.

4.3.4 Experiment Result

Figure 4.30 shows the experimental waveforms of DC-link voltage and inductor current before and after applying control. It can be seen from the Fig. 4.30 that when the *u*-phase voltage drops, the peak-to-peak value of the DC-link voltage increases from 119.9 to 203.7 V, and the peak-to-peak value of the inductor current increases from 10.8 to 14.0 A. After applying inductor current control, the peak-to-peak value of the DC-link voltage is 169.8 V, and the peak-to-peak value of the inductor current is 10.4 A. After applying full-feedforward control of the rectifier voltage, the peak-to-peak value of the DC-link voltage is 160.4 V, and the peak-to-peak value of the inductor current is 7.9 A. This indicates that applying full-feedforward control of the rectifier voltage can further reduce the fluctuations of DC-link voltage and inductor current.

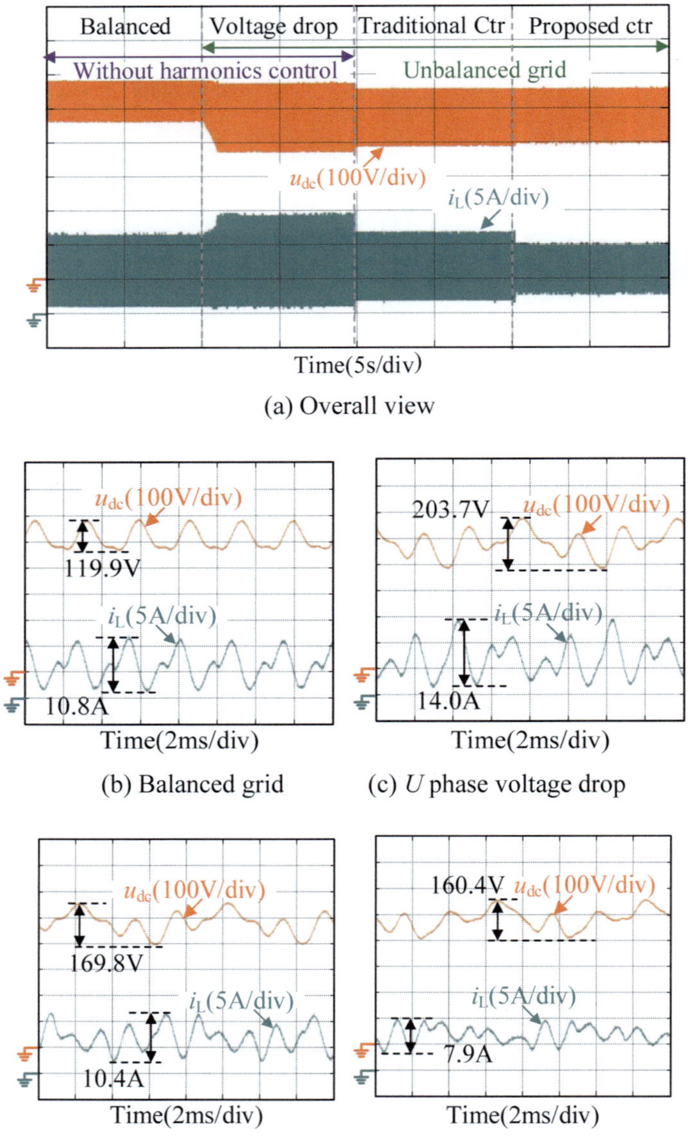

(a) Overall view

(b) Balanced grid

(c) *U* phase voltage drop

(d) Inductor current control

(e) Full-feedforward control

Fig. 4.30 Experimental results of DC-link voltage and inductor current

Figure 4.31 presents the FFT waveform of the inductor current. Under balanced grid conditions, the inductor current mainly contains harmonics at $6\omega_g$ and $12\ \omega_g$, with amplitudes of 3.5 A and 1.9 A, respectively. After the *u*-phase voltage drops, the inductor current mainly contains harmonics at $2\ \omega_g$, $4\ \omega_g$, $6\ \omega_g$, $8\ \omega_g$, $10\ \omega_g$, and $12\ \omega_g$, with amplitudes of 0.9 A, 0.4 A, 3.0 A, 2.1 A, 1.7 A, and 1.4 A, respectively. After applying inductor current control, the corresponding harmonic amplitudes are 1.0 A, 0.4 A, 1.1A, 2.0 A, 1.5 A, and 1.2 A. After further applying full feedforward control of the rectifier voltage, the corresponding harmonic amplitudes are 0.8 A, 0.1 A, 0.7 A, 0.7 A, 0.9 A, and 1.1 A. Therefore, after applying full feedforward control of the rectifier voltage, the harmonics in the inductor current are further suppressed.

Figure 4.32 shows the three-phase grid currents waveforms before and after applying control. When the *u*-phase voltage drops, the fluctuations of the *u*-phase and *w*-phase currents increase, and the waveforms experience significant distortion. After applying inductor current control, the fluctuations of the grid currents decrease. After further applying full feedforward control of the rectifier voltage, the fluctuations of the grid currents are further suppressed, and the distortion is improved. From the Fourier analysis of the *u*-phase current in Fig. 4.32b–e, it can be seen that harmonics such as 150 Hz and 450 Hz are newly introduced into the grid-side current after the *u*-phase voltage drops, which is consistent with theoretical analysis. After the *u*-phase voltage drops, the amplitudes of the harmonics at 150 Hz, 250 Hz, 350 Hz, 450 Hz, 550 Hz, and 650 Hz in the *u*-phase current are 0.6A, 1.3 A, 2.4 A, 2.0 A, 0.8

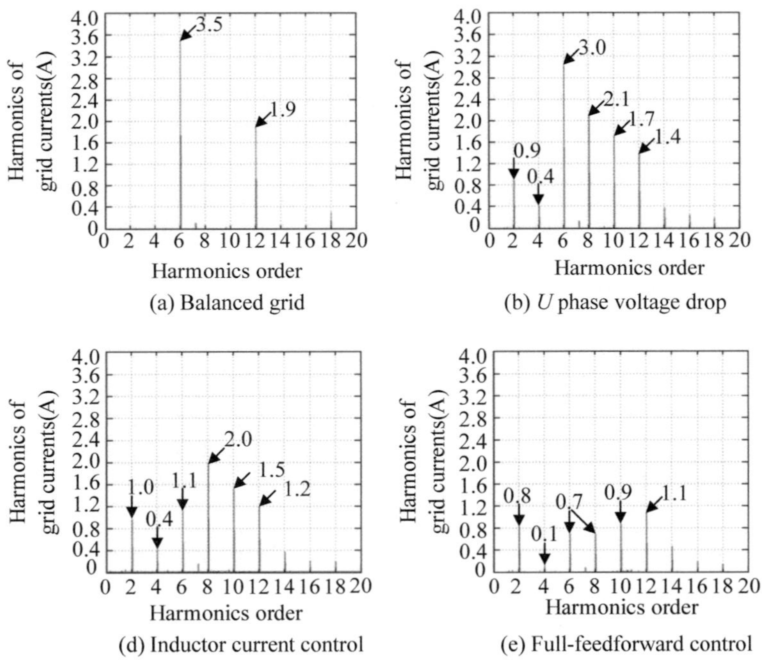

Fig. 4.31 Fourier analysis of the inductor current

A, and 0.7 A, respectively. At this point, the Total Harmonic Distortion (THD) of the u-phase current increases from 57.5% to 64.3%. After applying inductor current control, the corresponding harmonic amplitudes are 0.7A, 1.0A, 1.6A, 1.8A, 0.7A, and 0.8A, respectively, and the THD of the u-phase current decreases to 51.7%. After further applying full-feedforward control of the rectifier voltage, the corresponding harmonic amplitudes are 0.7 A, 1.2 A, 1.0 A, 1.0 A, 0.7 A, and 0.7 A, respectively, and the THD of the u-phase current further decreases to 39.8%. Therefore, compared to conventional inductor current control, the amplitudes of harmonics at frequencies such as 350 Hz, 450 Hz, and 650 Hz in the u-phase current and the THD are further reduced, and the quality of the grid-side input current is improved after applying the full feedforward strategy of the rectifier voltage.

Figure 4.33 shows the experimental results of DC-link voltage, inductor current, and u-phase current waveforms under the background of voltage distortion with a 5th harmonic component possessing 3% of the fundamental wave's amplitude in the three-phase voltage. As shown in Fig. 4.33a, after applying inductor current control, the fluctuations in DC-link voltage, inductor current, and u-phase current decrease. When full-feedforward control of the rectifier voltage is further applied, the corresponding waveform fluctuations are further reduced, and the distortion is improved. From Fig. 4.33b–d, it can be observed that after the grid side input voltage is distorted, the u phase current mainly contains harmonics at 180 Hz, 250 Hz, 350 Hz, 380 Hz, 480 Hz, 550 Hz, and 650 Hz, with corresponding amplitudes of 0.7 A, 1.9 A, 2.5 A, 1.0 A, 0.8 A, 1.8 A, and 0.6 A, respectively. At this point, the THD of the u phase current is 62.6%. After applying inductor current control, the corresponding harmonic amplitudes are 0.7 A, 1.2 A, 1.6 A, 1.3 A, 1.0 A, 1.4 A, and 0.7 A, respectively, and the THD of the u phase current decreases to 46.6%. When full-feedforward control of the rectifier voltage is further applied, the corresponding harmonic amplitudes are 0.6 A, 1.2 A, 1.5 A, 0.6 A, 0.5 A, 1.2 A, and 0.7 A, respectively, and the THD of the u phase current further decreases to 40.9%. Therefore, under the condition of voltage distortion, the implementation of full-feedforward control of the rectifier voltage strategy can still effectively suppress grid-side input current harmonics and reduce THD.

Figure 4.34 shows the THD of the grid currents before and after applying control under various operating conditions under unbalanced grid. Across different operating conditions, applying the grid current harmonics suppression strategy based on full-feedforward control of the rectifier voltage can effectively reduce the THD of the grid currents and enhance the power quality of the grid currents. However, due to limitations such as DC-link voltage sampling accuracy, digital control delay, and motor system stability, the THD of the grid-side current cannot reach its theoretical limit.

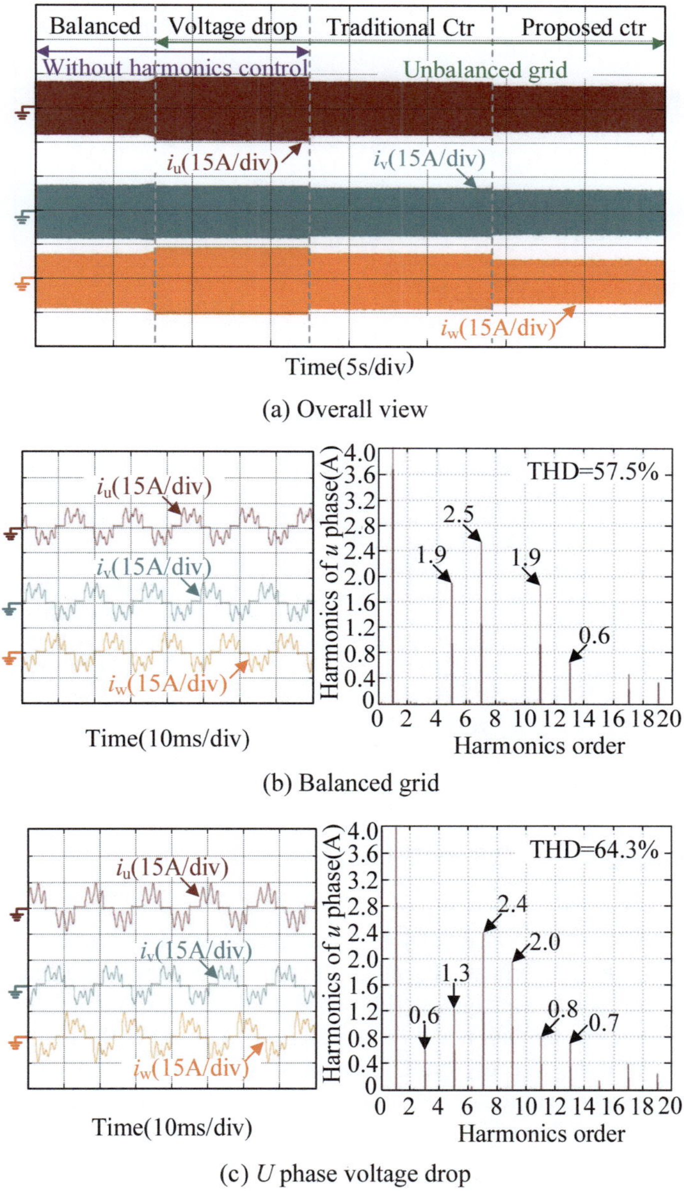

(a) Overall view

(b) Balanced grid

(c) U phase voltage drop

Fig. 4.32 Experimental results of three phase input currents and Fourier analysis of the u phase current

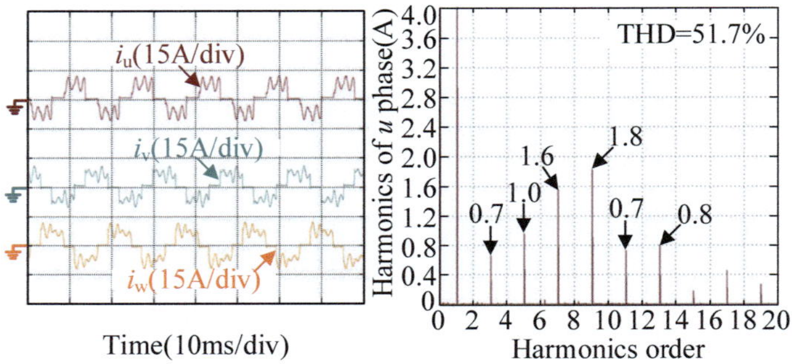

(d) Inductor current control

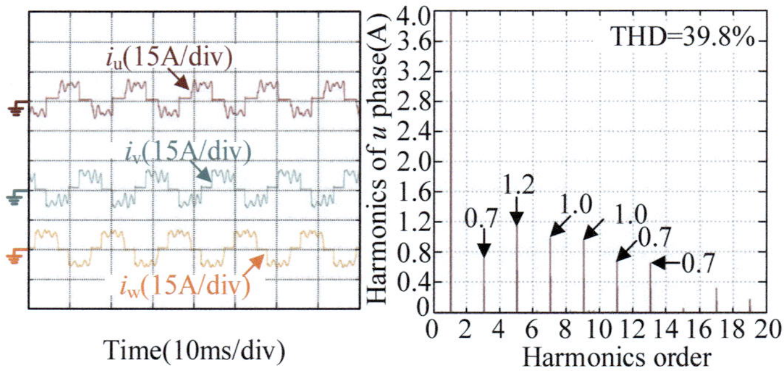

(e) Full-feedforward control

Fig. 4.32 (continued)

4.4 Summary

To improve the grid current quality in drive under non-ideal grid conditions, this chapter focuses on two typical non-ideal grid conditions: grid impedance variation and unbalanced grid. It investigates grid current quality improvement strategies based on harmonic energy regulation and full-feedforward control of the rectifier voltage, effectively addressing the key issue of decreased control performance of conventional active damping strategies under non-ideal grid conditions. The main work and conclusions are as follows:

(1) To address the critical issue of decreases in the performance of active damping strategies and system stability in electrolytic capacitorless PMSM drive when grid impedance varies, a grid current quality improvement strategy based on harmonic energy regulation is proposed to enhance the robustness of the drive to grid impedance. Firstly, based on harmonic energy regulation theory, when the DC-link capacitor and motor inductance serve as the main harmonic energy storage elements, the quality of the grid current can be improved. Secondly, using the derived small-signal model of the system, the DC-link voltage command

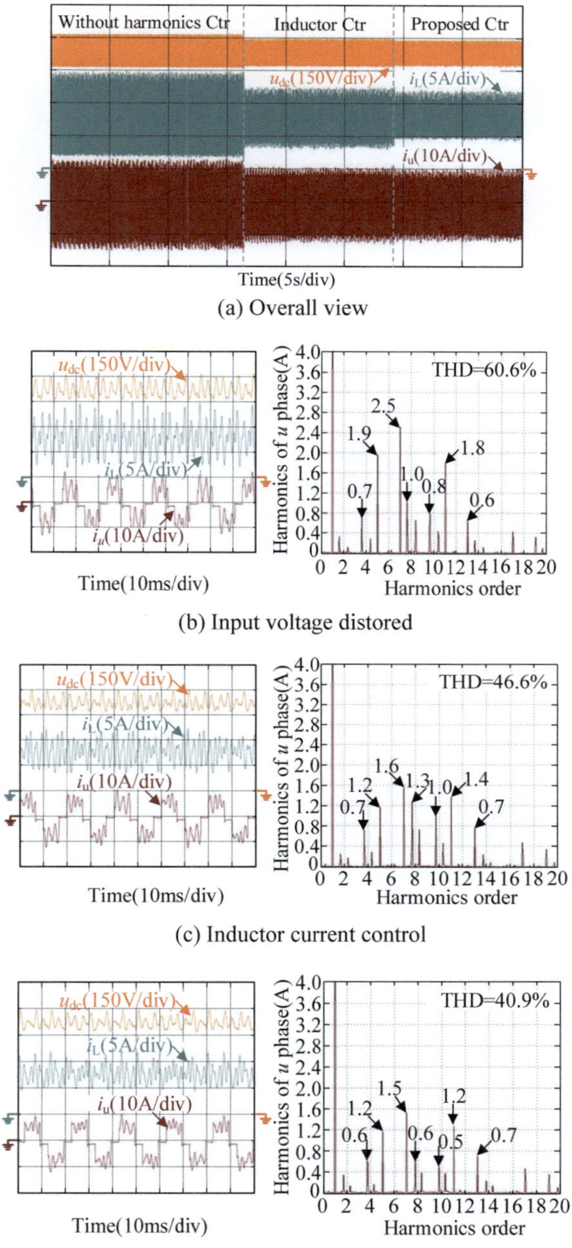

Fig. 4.33 Experimental results under grid voltage distortion condition

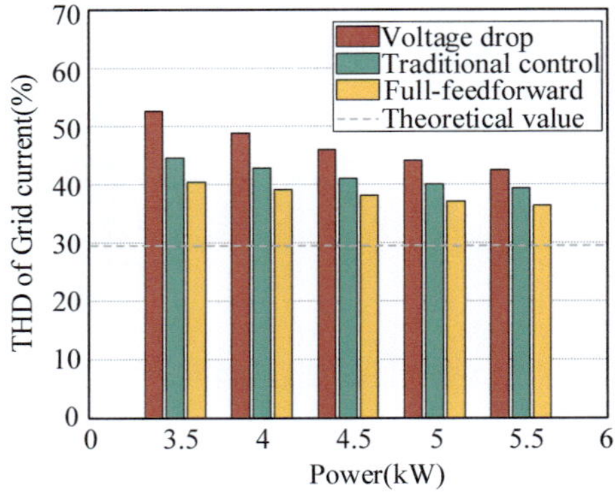

Fig. 4.34 THD of the u phase current before and after control under different operating conditions

of SVPWM is reshaped. Specifically, a phase processing unit and an amplitude processing unit based on the system model are utilized to adjust the harmonics in the motor current and DC-link voltage, reducing the harmonic energy in the grid impedance to achieve an ideal energy exchange mode, thereby enhancing the robustness of the system to grid impedance while reducing harmonics in the DC-link voltage and grid current. Simulation and experimental results demonstrate that the proposed strategy can enhance the robustness of the system to grid impedance, with the THD of the grid current reduced to the theoretical limit of 30%. Even when the grid impedance increases to 224% of its original value, the drive remains stable.

(2) Under unbalanced grid conditions, electrolytic capacitorless drive exists severe grid current harmonic issues. A grid current harmonics suppression strategy based on full-feedforward control of the rectifier voltage is proposed. By analyzing changes in the conduction angles of the rectifier bridge, the mechanism underlying changes in harmonics content in the system input voltage and inductor current is elucidated. A small-signal model of the inductance current loop is established, revealing the coupling relationship between the rectifier voltage and inductor current. Based on this, a full feedforward function of the rectifier voltage, which accounts for changes in power flow under unbalanced grid conditions, is constructed to reshape the system's grid-side output impedance, achieving decoupling of the rectifier voltage disturbance term in the inductance current control loop. An equivalent Norton circuit of the system is constructed, and system stability after applying the control strategy is verified based on cascade system criteria. The proposed strategy solves the harmonic suppression challenge faced by conventional inductance current control under unbalanced grid conditions. Finally, the effectiveness of the proposed control strategy is validated.

References

1. M. A. Awal, L. D. Flora, and I. Husain. Observer based generalized active damping for voltage source converters with LCL filters [J]. IEEE Transactions on Power Electronics, 2022, 37(1): 125–136.
2. S. Li and H. Lin. Passivity enhancement-based general design of capacitor current active damping for LCL-type grid-tied inverter [J]. IEEE Transactions on Power Electronics, 2023, 38(7): 8223–8236.
3. D. Ding, G. Zhang, G. Wang, Z. Ren, N. Mijatovic, T. Dragicevic, D. Xu. Impedance reshaping for inherent harmonics in PMSM drives with small DC-link capacitor [J]. IEEE Transactions on Power Electronics, 2022, 37(12), 14265–14279.
4. D. Wang, K. Lu, P. O. Rasmussen, L. Mathe, Y. Feng, and F. Blaabjerg. Voltage modulation using virtual positive impedance concept for active damping of small DC-link drive system [J]. IEEE Transactions on Power Electronics, 2018, 33(12), 10611–10621.
5. G. Zhang, R. Xiang, G. Wang, C. Li, G. Bi, N. Zhao, D. Xu. Hybrid pseudorandom signal injection for position sensorless SynRM drives with acoustic noise reduction [J]. IEEE Transactions on Transportation Electrification, 2022, 8(1), 1313–1325.
6. Y. Yu, Y. Shao, F. Chai, and M. Cui. Static-errorless position estimation for sensorless PMSM drives with enhanced robustness against the fullfrequency domain disturbance [J]. IEEE Transactions on Power Electronics, 2022, 37(5), 5884–5897.
7. D. Ding, S. Liu, Q. Wang, G. Wang, B. Li, G. Zhang, D. Xu, "Harmonic Energy Regulation for Grid Current Quality Improvement in Electrolytic Capacitorless PMSM Drives," IEEE Trans. Transport. Electrific., vol. 10, no. 3, pp. 6407–6417, Sept. 2024.
8. S. Shi, L. Guo, Z. Chang, C. Zhao, and P. Chen. Current controller based on active disturbance rejection control with parameter identification for PMSM servo systems [J]. IEEE Access, 2023, 11, 46882–46891.
9. Jiang Weidong, Liu Shengyu, Zhang Qingyan, et al. An improved coordination and optimization control method for swiss rectifier based on grid amplitude drop [J/OL]. Proceedings of the CSEE:1–12 [2024-05-09].
10. Tang Chengyu and Jheng Jiahe. An active power ripple mitigation strategy for three-phase grid-tied inverters under unbalanced grid voltages [J]. IEEE Transactions on Power Electronics, 2023, 38(1):27–33.
11. D. Ding, H. Xie, G. Wang, B. Li, D. Xu, Suppression Strategy of Grid-side Current Harmonics for Electrolytic Capacitor–less Permanent Magnet Motor Drives Under Unbalanced Grid Voltage, *Proceedings of the CSEE.*

Chapter 5
Motor Side Beat Phenomenon Suppression Method

5.1 Introduction

The above analysis mainly focuses on the methods for improving the power quality at grid side. In addition, the harmonic of the motor current is also an issue which cannot be ignored. In the electrolytic capacitorless drives, the significant fluctuation in the DC-link voltage results in the additional harmonics in the motor current. The interaction between the fundamental component and harmonics of the motor current produces the beat phenomenon, which increases the operating noise. This chapter studies the beat suppression method based on the delay compensation and impedance reshaping to improve the quality of motor operation.

5.2 Analysis of Current Harmonics Caused by DC-Link Voltage

5.2.1 The Influence of DC-Link Voltage Sampling Delay

The current harmonics of the motor side will be caused by the fluctuation and sampling delay of DC-link voltage, and the interaction between the harmonics and the fundamental currents will cause the beat phenomenon which increases the motor operating noise. When the impedance of the grid side is small, the DC-link voltage and the rectifier voltage are close. In this case, the equivalent sampled DC-link voltage can be expressed as

$$u_{\mathrm{dc\_e}} = u_{\mathrm{dN}}\left[1 - \sum_{n=1}^{\infty} \frac{2}{(6n)^2 - 1} cos\left(6\omega_{\mathrm{g}} nt - n\varphi_{\mathrm{dc}}\right)\right] \tag{5.1}$$

D. Ding et al., *Electrolytic Capacitorless Permanent Magnet Synchronous Motor Drives*, https://doi.org/10.1007/978-981-95-5601-4_5

where φ_{dc} is the DC-link voltage sampling delay.

In addition to the DC component of the DC-link voltage, it also contains the harmonics of different frequencies such as $6\omega_g$ and $12\omega_g$. Considering that $6\omega_g$ is the main harmonic component, the component of $6\omega_g$ is analyzed in detail.

Considering (5.1), the output voltages of inverter can be expressed as

$$\boldsymbol{u}_s \doteq \frac{\left[1 - \frac{2}{35}cos\left(6\omega_g t\right)\right]}{\left[1 - \frac{2}{35}cos\left(6\omega_g t - \varphi_d\right)\right]}\boldsymbol{u}_{sref} \tag{5.2}$$

where $\boldsymbol{u}_s$ is the motor voltage vector, $\boldsymbol{u}_{sref}$ is the reference voltage vector.

Then (5.2) can be further expressed as

$$\boldsymbol{u}_s = \frac{1 + \frac{2}{35}cos\left(6\omega_g t - \varphi_d\right) - \frac{2}{35}cos\left(6\omega_g t\right) - \left(\frac{2}{35}\right)^2 cos\left(6\omega_g t\right)cos\left(6\omega_g t - \varphi_d\right)}{1 - \left(\frac{2}{35}\right)^2 cos\left(6\omega_g t - \varphi_d\right)^2}\boldsymbol{u}_{sref}. \tag{5.3}$$

If the quadratic terms in Eq. (5.3) are ignored, (5.3) can be simplified to

$$\boldsymbol{u}_s = \left[1 + \frac{4}{35}sin(\varphi_d)\sin\left(6\omega_g t - \frac{\varphi_d}{2}\right)\right]\boldsymbol{u}_{sref}. \tag{5.4}$$

According to (5.4), if there is no DC-link voltage sampling delay, the motor voltage vector is equal to the reference voltage vector.

Using Clarke-Park transformation, the motor voltages in dq-axis can be expressed as

$$\begin{bmatrix} u_d \\ u_q \end{bmatrix} = \begin{bmatrix} |\boldsymbol{u}_{sref}|\left[1 + \frac{4}{35}\sin(\varphi_d)\sin\left(6\omega_g t - \frac{\varphi_d}{2}\right)\right]\sin(\varphi_e) \\ -|\boldsymbol{u}_{sref}|\left[1 + \frac{4}{35}\sin(\varphi_d)\sin\left(6\omega_g t - \frac{\varphi_d}{2}\right)\right]\cos(\varphi_e) \end{bmatrix}. \tag{5.5}$$

It can be seen in (5.5) that u_{dq} not only contains DC components, but also harmonic components caused by DC-link voltage fluctuations. The corresponding motor currents in dq-axis can be expressed as

$$\begin{bmatrix} i_d \\ i_q \end{bmatrix} = \begin{bmatrix} I_{d0} + I_{d1}\sin\left(6\omega_g t - \frac{\varphi_d}{2} + \varphi_{id}\right) \\ I_{q0} + I_{q1}\sin\left(6\omega_g t - \frac{\varphi_d}{2} + \varphi_{iq}\right) \end{bmatrix} \tag{5.6}$$

where φ_{id} and φ_{iq} are the phase difference between u_{dq} and i_{dq}, I_{d1} and I_{q1} are the fundamental currents of i_d and i_q caused by DC-link voltage fluctuations. Consequently, the phase current can be expressed as

$$i_a = I_{d0}\cos(\omega_e t) - I_{q0}\sin(\omega_e t) - I_{d1}\begin{bmatrix} \frac{1}{2}\sin\left(\omega_e t + 6\omega_g t - \frac{\varphi_d}{2} + \varphi_{id}\right) \\ +\frac{1}{2}\sin\left(6\omega_g t - \omega_e t - \frac{\varphi_d}{2} + \varphi_{id}\right) \end{bmatrix}$$

$$+I_{q1}\begin{bmatrix} \frac{1}{2}\cos\left(\omega_e t + 6\omega_g t - \frac{\varphi_d}{2} + \varphi_{iq}\right) \\ -\frac{1}{2}\cos\left(6\omega_g t - \omega_e t - \frac{\varphi_d}{2} + \varphi_{iq}\right) \end{bmatrix}. \tag{5.7}$$

It can be seen from (5.7) that the phase current contains three different frequency components, among which the harmonic components $6\omega_g - \omega_e$ and $6\omega_g + \omega_e$ are caused by the DC-link voltage sampling delay. The mechanism of the interaction between fundamental currents and harmonics which generates beat phenomenon will be analyzed.

5.2.2 *Beat Phenomenon Caused by the Interaction of Harmonics and the Fundamental Currents*

In order to facilitate the generation mechanism of beat phenomenon, only part of the fundamental currents and harmonics of the stator currents are analyzed, and the expression can be expressed as

$$i_{a\_p}(t) = F_1(t) + F_2(t) = I_{d0}\cos(\omega_e t) + \frac{1}{2}I_{q1}\cos\left(\omega_e t + 6\omega_g t - \frac{\varphi_d}{2} + \varphi_{iq}\right) \tag{5.8}$$

where $F_1(t)$ and $F_2(t)$ are fundamental and harmonic components of the stator currents, respectively.

In order to explore the mechanism of beat phenomenon, it is necessary to obtain the envelope of i_{a_p}. Figure 5.1 shows the beat phenomenon under different ratios of harmonic and fundamental currents $|F_2(t)|/|F_1(t)|$, where the fundamental frequency is 0.47 p.u. and the harmonic frequency is 2.38 p.u. In Fig. 5.1a, the $|F_2(t)|/|F_1(t)|$ is 0.02, and $F_1(t)$ and $F_2(t)$ are the peak and the valley values at 0.5 s, respectively. As can be seen from the figure, the envelope at this time passes through the i_{a_p} corresponding to the time of the fundamental peak, that is, the envelope is composed of points on i_{a_p} at the time of $t_0 + nT_1$. In Fig. 5.1b, the $|F_2(t)|/|F_1(t)|$ is 0.02 to 0.1, and i_{a_p} does not achieve a maximum value at the time of $t_0 + nT_1$. In this case, it is difficult to obtain the expression of the beat envelope.

Therefore, the analysis is only for i_{a_p} to obtain a maximum value at $t_0 + nT_1$, that is, the second derivative of i_{a_p} is less than 0,

$$i_{a\_p}(t_0)'' = -I_{d0}\omega_e^2\cos(\omega_e t_0) - \frac{1}{2}I_{q1}\left(\omega_e + 6\omega_g\right)^2\cos(\omega_e t_0 + 6\omega_g t_0 - \frac{\varphi_d}{2} + \varphi_{iq}) < 0. \tag{5.9}$$

In Fig. 5.1a, the fundamental currents and harmonics at time t_0 are at the maximum and minimum values, respectively. And the following expression can be obtained

$$\cos(\omega_e t_0) = 1, \cos(\omega_e t_0 + 6\omega_g t_0 - \varphi_d/2 + \varphi_{iq}) = -1. \tag{5.10}$$

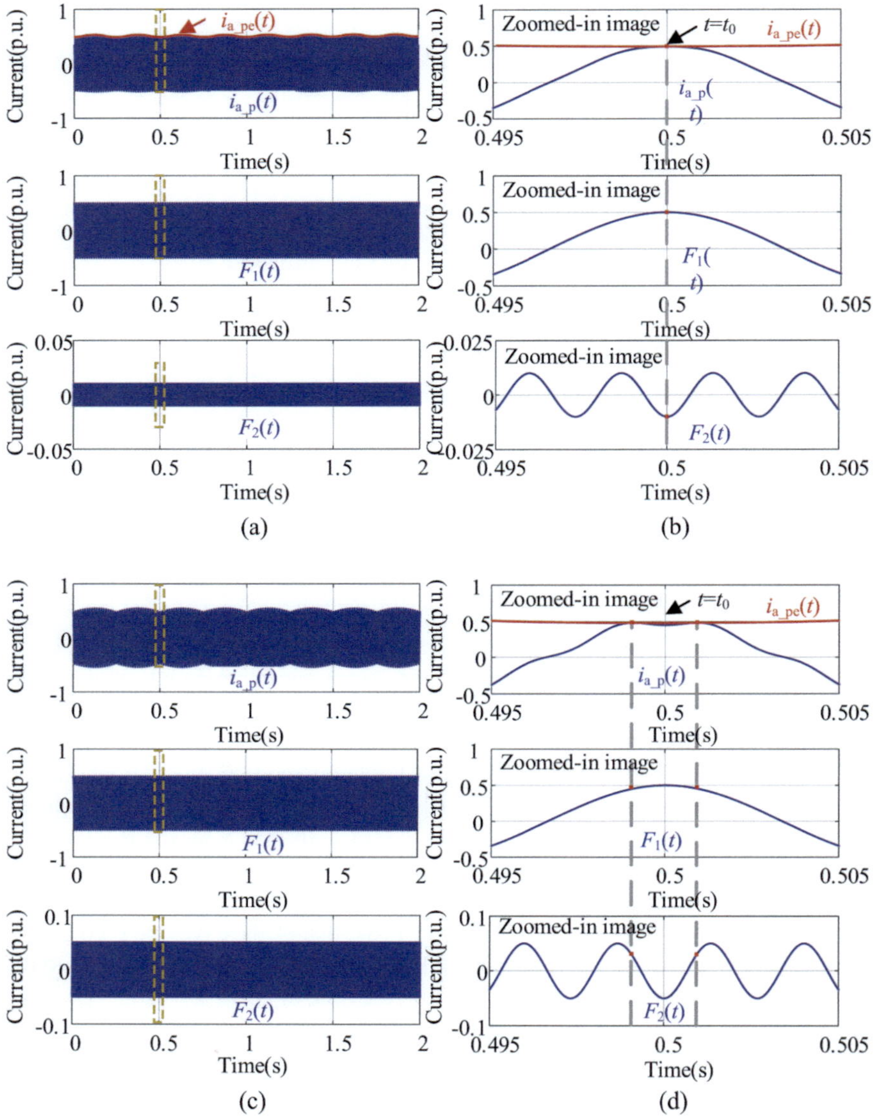

Fig. 5.1 Beat frequency phenomenon under different $|F_2(t)|/|F_1(t)|$ values, reprinted from Ref. [1], copyright 2022, with permission from Nature. **a** The value of $|F_2(t)|/|F_1(t)|$ is 0.02. **b** Zoomed-in images of (**a**). **c** The value of $|F_2(t)|/|F_1(t)|$ is 0.1. **d** Zoomed-in images of (**c**)

Therefore, the envelope consists of the maximum point of $i_{\text{a\_p}}$, when $F_1(t)$ and $F_2(t)$ meet

$$\frac{i_{q1}}{2I_{d0}} < \left(\frac{\omega_e}{\omega_e + 6\omega_g}\right)^2. \tag{5.11}$$

If the motor operating frequency is an integer, the period of $i_{\text{a\_p}}$ can be expressed as

$$T_{\text{a\_p}} = \frac{1}{\left(\frac{\omega_e}{2\pi}, \frac{(\omega_e + 6\omega_g)}{2\pi}\right)} \tag{5.12}$$

where (a, b) is the greatest common divisor operator.

In a period of $i_{\text{a\_p}}$, the number of occurrences of fundamental $F_1(t)$ can be expressed as

$$p = T_{\text{a\_p}}/T_1. \tag{5.13}$$

Thus, in a period of $i_{\text{a\_p}}$, the beat envelope consists of the following points

$$\begin{cases} i_{\text{a\_p}}(t_0 + T_1) = I_{d0} + \frac{1}{2}I_{q1}\cos\left[(t_0 + T_1)(\omega_e + 6\omega_g) - \frac{\varphi_d}{2} + \varphi_{iq}\right] \\ i_{\text{a\_p}}(t_0 + 2T_1) = I_{d0} + \frac{1}{2}I_{q1}\cos\left[(t_0 + 2T_1)(\omega_e + 6\omega_g) - \frac{\varphi_d}{2} + \varphi_{iq}\right] \\ \cdots \\ i_{\text{a\_p}}(t_0 + pT_1) = I_{d0} + \frac{1}{2}I_{q1}\cos\left[(t_0 + pT_1)(\omega_e + 6\omega_g) - \frac{\varphi_d}{2} + \varphi_{iq}\right] \end{cases}. \tag{5.14}$$

According to Eqs. (5.12), (5.13) and (5.14), it can be seen that when the fundamental frequency and harmonics present an integer multiple relationship ($p = 1$), then the envelope of $i_{\text{a\_p}}$ will be a straight line. That is, the beat phenomenon will not appear and the envelope can be approximated as

$$i_{\text{a\_pe}} = I_{d0} + \frac{1}{2}I_{q1}\cos(\omega_{\text{a\_pe}}t + \varphi_{\text{a\_pe}}) \tag{5.15}$$

where $\omega_{\text{a\_pe}}$ and $\varphi_{\text{a\_pe}}$ are frequency and initial phase of the envelope, respectively.

As can be seen from (5.15), in order to represent the envelope of beat, the frequency of beat must be obtained. In fact, the beat phenomenon is caused by the periodic change of the phase difference between fundamental currents and harmonics. And the frequency of beat is determined by the fundamental frequency and harmonic frequency. If the phase difference between fundamental and harmonic is 0 at time t_0, then the proportion of the phase shift at time $t_0 + T_1$ can be expressed as

$$b_d = \left\{\frac{\omega_e + 6\omega_g}{\omega_e}\right\} \tag{5.16}$$

where {x} is the decimal part function and b_d is the phase shift ratio at $t_0 + T_1$.

Considering that the harmonics are periodic, the phase shift ratio can be redefined as

$$b = \begin{cases} b_d & 0 \le b_d \le 0.5 \\ 1 - b_d & 0.5 < b_d < 1 \end{cases}. \tag{5.17}$$

Thus, the phase shift of harmonic at $t_0 + T_1$ is

$$\varphi_s(t_0 + nT_1) = 2\pi nb. \tag{5.18}$$

Thus, the beat envelope frequency can be expressed as

$$\omega_{a\_pe} = \frac{2\pi pb}{T_{a\_p}}. \tag{5.19}$$

From the above analysis, it can be seen that the motor current contains harmonics at the frequency of DC-link voltage fluctuation caused by the DC-link voltage sampling delay. And this part of harmonics interacts with fundamental currents to produce beat phenomenon. Because the beat frequency is determined by the periodic change of the phase difference between the fundamental frequency and the harmonic frequency, it cannot be obtained by Fourier analysis. In order to solve the beat phenomenon, it is necessary to obtain more accurate sampling value of DC-link voltage.

5.2.3 The Suppression Method of Beat Phenomenon Based on DC-Link Voltage Prediction

The suppression strategy of beat phenomenon based on vector control is shown in Fig. 5.2. By extracting the main harmonic components of the DC-link voltage and reconstructing the given voltage of the SVPWM, the suppression of beat phenomenon is achieved.

Figure 5.3 shows the prediction diagram of $6\omega_g$ component in DC-link voltage. The DC-link voltage is sampled at nT_s which is represented as $u_{dc\_s\_6\omega g}(z)$, and $u_{dc\_s\_6\omega g}(z)$ participates in the nth duty ratio signal update. At this time, the sampled value of DC-link voltage is PO and the actual value is MN. If the average DC-link voltage of $(n + 1)T_s$ and $(n + 2)T_s$ can be obtained at nT_s time, the sampling and updating error of DC-link voltage can be effectively reduced.

Based on the periodic characteristics of $6\omega_g$ component, the sample value of DC-link voltage that one step and two steps ahead can be expressed as $u_{dc\_s\_6\omega g}(z)z^{-n+1}$, $u_{dc\_s\_6\omega g}(z)z^{-n+2}$. Therefore, the average DC-link voltage at $(n + 1)T_s$ and $(n + 2)T_s$ can be approximated as

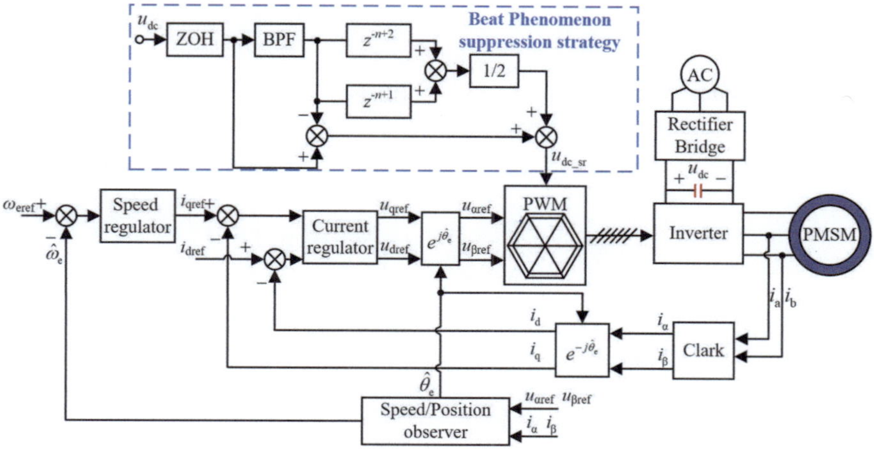

Fig. 5.2 Block diagram of vector control using beat frequency suppression strategy, reprinted from Ref. [1], copyright 2022, with permission from Nature

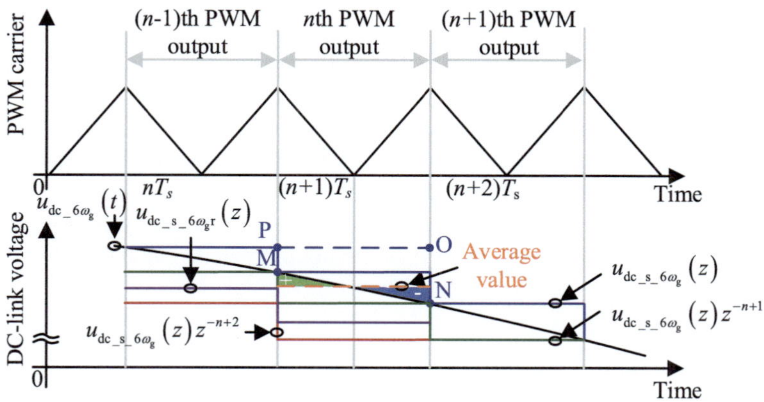

Fig. 5.3 Schematic diagram of the prediction of the $6\omega_g$ component of DC-link voltage

$$u_{dc\_s\_6\omega_g r}(z) = \frac{\left(u_{dc\_s\_6\omega_g}(z)z^{-n+2} + u_{dc\_s\_6\omega_g}(z)z^{-n+1}\right)}{2} \tag{5.20}$$

$$n = \frac{kf_s}{6f_g} \tag{5.21}$$

where $u_{dc\_s\_6\omega gr}(z)$ is the reconstructed voltage at the frequency of $6\omega_g$, f_s is the switching frequency, k is a coefficient determined by f_s and f_g.

The $6\omega_g$ component of DC-link voltage can be obtained by a bandpass filter and can be expressed as

Fig. 5.4 Relationship between motor frequency and beat frequency

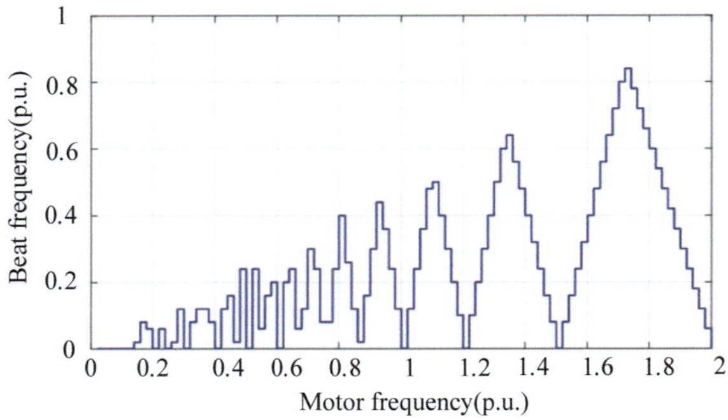

$$u_{dc\_s\_6\omega_g}(z) = G_{BPF}(z)u_{dc\_s}(z). \tag{5.22}$$

The predicted value of the $6\omega_g$ component replaces the $6\omega_g$ component in the DC-link voltage, and finally the DC-link voltage reconstruction value can be obtained

$$u_{dc\_sr}(z) = u_{dc\_s}(z) - u_{dc\_s\_6\omega_g}(z) + u_{dc\_s\_6\omega_g r}(z) \tag{5.23}$$

where u_{dc_sr} is the reconstruction value of DC-link voltage.

Figure 5.4 shows the relationship between the motor operating frequency and beat frequency. As the motor operating frequency increases, the beat frequency also increases. The beat frequency does not exceed 0.90 p.u. (282 rad/s), when the motor operating frequency changes from 0 p.u. to 2p.u. (0–628 rad/s).

According to the parameters of experiment, the center frequency of the band-pass filter is set to 6 p.u., so harmonics of beat are not extracted. The band-pass filter is designed to

$$G_{BPF}(z) = \frac{0.0078z^2 - 0.0078}{z^2 - 1.9296z + 0.9844}. \tag{5.24}$$

Because the utility grid is not completely ideal, the voltage allows a frequency range of (0.996, 1.004) p.u. and the corresponding six times the frequency of DC-link voltage is (5.975, 6.024) p.u. Figure 5.5 shows the Bode diagram of the bandpass filter with a bandwidth of 0.398 p.u. When the power grid changes within a certain frequency range, the main harmonic components of the DC-link voltage can still be effectively extracted.

The error caused by DC-link voltage sampling will produce PWM duty cycle error. Figure 5.6 shows the relationship between DC-link voltage and PWM duty cycle. According to the volt-second balance principle, the area S surrounded by the voltage instruction in a switching cycle is equal to the area surrounded by the average voltage output of the inverter. The calculation area of DC-link voltage at $(n + 1.5)T_s$ can be expressed as

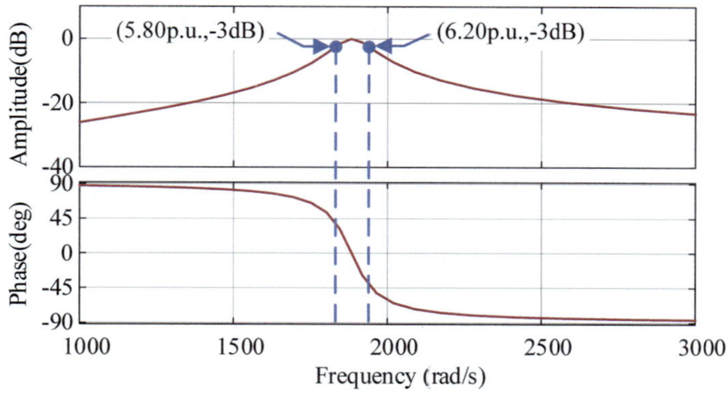

Fig. 5.5 Bode plot of the main harmonic filter in the DC-link voltage

$$
\begin{aligned}
S &= \int_{(n+1)T_s}^{(n+2)T_s} u_a^*(t)\mathrm{d}t \\
&= \int_{(n+1)T_s}^{t_1} -\frac{1}{2}u_{\mathrm{dc\_}6\omega_g}[(n+1.5)T_s]\mathrm{d}t + \int_{t_1}^{t_2} \frac{1}{2}u_{\mathrm{dc\_}6\omega_g}[(n+1.5)T_s]\mathrm{d}t \\
&\quad + \int_{t_2}^{(n+2)T_s} -\frac{1}{2}u_{\mathrm{dc\_}6\omega_g}[(n+1.5)T_s]\mathrm{d}t
\end{aligned}
\tag{5.25}
$$

Equation (5.25) can be simplified as

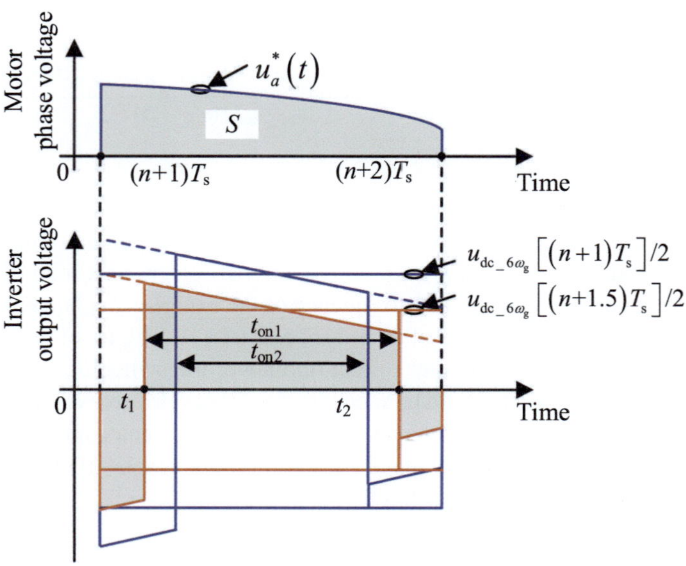

Fig. 5.6 Relationship between DC-link voltage and duty ratio

$$S = u_{\mathrm{dc\_}6\omega_g}[(n+1.5)T_s] \cdot (t_{\mathrm{on}1} - 0.5T_s). \tag{5.26}$$

The ON duration can be expressed as

$$\begin{aligned} t_{\mathrm{on}1} &= S \big/ u_{\mathrm{dc\_}6\omega_g}[(n+1.5)T_s] + T_s \big/ 2 \\ &= \frac{|\boldsymbol{u}_\mathrm{s}|\cos[\omega_\mathrm{e}(n+1)T_s + \varphi_\mathrm{e}] - |\boldsymbol{u}_\mathrm{s}|\cos[\omega_\mathrm{e}(n+2)T_s + \varphi_\mathrm{e}]}{\omega_\mathrm{e} \cdot u_{\mathrm{dc\_}6\omega_g}[(n+1.5)T_s]} + \frac{1}{2}T_s. \end{aligned} \tag{5.27}$$

By further simplifying (5.27), it can be obtained as

$$t_{\mathrm{on}1} \doteq |\boldsymbol{u}_\mathrm{s}| \cdot T_s \cdot \sin[\omega_\mathrm{e}(n+1.5)T_s + \varphi_\mathrm{e}] \big/ u_{\mathrm{dc\_}6\omega_g}[(n+1.5)T_s] + T_s \big/ 2. \tag{5.28}$$

Therefore, the duty cycle calculated by the DC-link voltage at time $(n+1.5)T_s$ can be expressed as

$$d_1 = t_{\mathrm{on}1} \big/ T_s = |\boldsymbol{u}_\mathrm{s}| \cdot \sin[\omega_\mathrm{e}(n+1.5)T_s + \varphi_\mathrm{e}] \big/ u_{\mathrm{dc\_}6\omega_g}[(n+1.5)T_s] + 1 \big/ 2. \tag{5.29}$$

Similarly, the duty cycle calculated by the DC-link voltage at nT_s can be expressed as

$$d_2 = \frac{t_{\mathrm{on}2}}{T_s} = \frac{|\boldsymbol{u}_\mathrm{s}| \cdot \sin[\omega_\mathrm{e}(n+1.5)T_s + \varphi_\mathrm{e}]}{u_{\mathrm{dc\_}6\omega_g}[(n+1)T_s]} + \frac{1}{2}. \tag{5.30}$$

According to (5.29) and (5.30), the error of duty cycle can be expressed as

$$d_{\mathrm{error}} = \frac{|\boldsymbol{u}_\mathrm{s}| \cdot \sin[\omega_\mathrm{e}(n+1.5)T_s + \varphi_\mathrm{e}] \cdot \left[u_{\mathrm{dc\_}6\omega_g}[(n+1)T_s] - u_{\mathrm{dc\_}6\omega_g}[(n+1.5)T_s]\right]}{u_{\mathrm{dc\_}6\omega_g}[(n+1)T_s] \cdot u_{\mathrm{dc\_}6\omega_g}[(n+1.5)T_s]}. \tag{5.31}$$

As can be seen from (5.31), the duty cycle error is determined by the stator voltage vector amplitude, motor operating frequency, voltage sampling time and switching period. Figure 5.7 is a quantitative diagram of the maximum duty cycle error. Considering that the switching frequency of the experiment platform is 8 kHz, the voltage difference at the time of $(n+1)T_s$ and $(n+1.5)T_s$ is set to vary within a range of (0,0.03) p.u. As a matter of fact, the lower the switching frequency, the larger the voltage difference. It can be seen that the duty cycle error will also increase with the increase of voltage difference and stator voltage vector amplitude. When the voltage difference and stator voltage vector amplitude reach 0.03 p.u. and 1 p.u., the maximum duty cycle error reaches 0.13.

Fig. 5.7 Analysis of duty ratio error

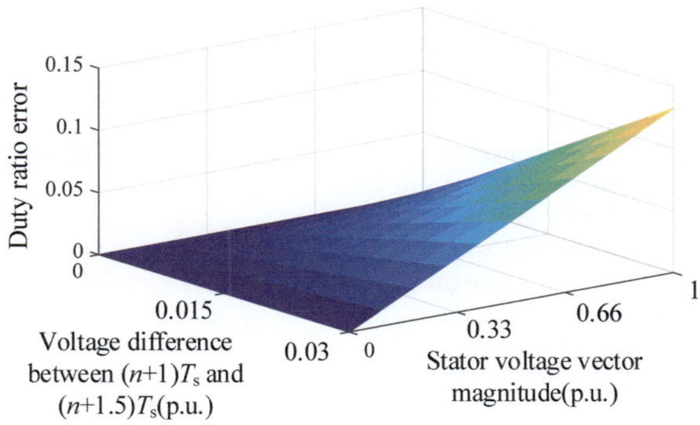

5.2.4 Experimental Result

The experimental results of suppression of beat phenomenon when the motor output power is 0.50 p.u. are shown in Fig. 5.8. According to Fig. 5.8a, before applying the proposed strategy, the phase current of motor presents an obvious beat phenomenon with a beat frequency of 0.08 p.u. When the proposed strategy is applied, the beat phenomenon is significantly suppressed and the phase current is relatively stable. Figure 5.8b and c are enlarged images before and after the application of the proposed strategy, respectively. After the application of the proposed strategy, the harmonics of the DC-link voltage fluctuation frequency in the q-axis current are significantly reduced, and the q-axis current fluctuation amplitude is from 0.17 p.u. down to 0.07 p.u. It can be seen from Fig. 5.8d and e that after applying the proposed strategy, the harmonics of phase current at the frequency in 4.52 p.u. decrease from 0.03 p.u. to 0.01 p.u., and harmonics at frequency in 7.48 p.u. reduce from 0.02 p.u. to 0.01 p.u.

The experimental results of suppression of beat phenomenon when the motor output power is 0.50 p.u. are shown in Fig. 5.9. Before applying the proposed strategy, the phase current and q-axis current of the motor show obvious beat phenomenon. After applying the proposed strategy, the beat phenomenon is effectively suppressed, and the phase current is relatively stable, as shown in Fig. 5.9a. Figure 5.9b and c are enlarged images before and after the application of the proposed strategy, respectively. It can be seen that after the application of the proposed strategy, the harmonics of the DC-link voltage fluctuation frequency in the q-axis current are significantly reduced, and the q-axis current fluctuation amplitude is from 0.09 p.u. to 0.02 p.u. It can be seen from Fig. 5.9d and e that after applying the proposed strategy, the harmonics related to the frequency of DC-link voltage fluctuation in the motor current are significantly reduced.

Figure 5.10 shows the experimental results of the motor acceleration process after applying the proposed strategy. The q-axis current changes from 0.5 p.u to 0.6 p.u. within 0.8 s, and the load torque increases with speed. It can be seen that the motor current sinusoidal degree is well in the transient process, there is no obvious beat phenomenon, and the proposed strategy will not affect the dynamic process.

Fig. 5.8 Experimental results of beat frequency suppression at the output power of 0.5 p.u., reprinted from Ref. [1], copyright 2022, with permission from Nature. **a** DC-link voltage, q-axis current and a-phase current. **b** Zoomed-in image of **a** before applying the suppression strategy. **c** Zoomed-in image of **a** after applying the suppression strategy. **d** FFT analysis of a-phase current before applying suppression strategy. **e** FFT analysis of a-phase current after applying suppression strategy

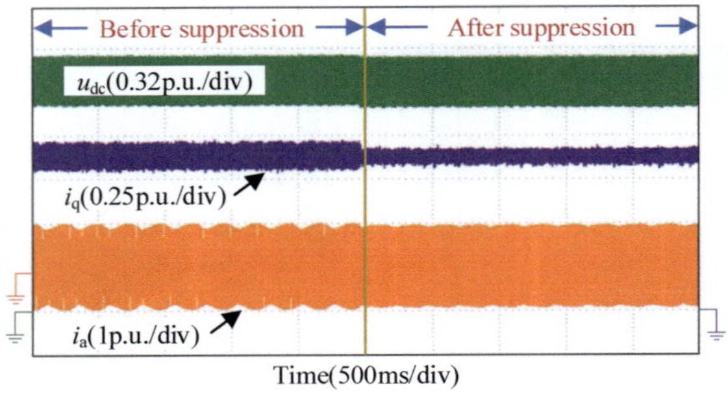

(a)

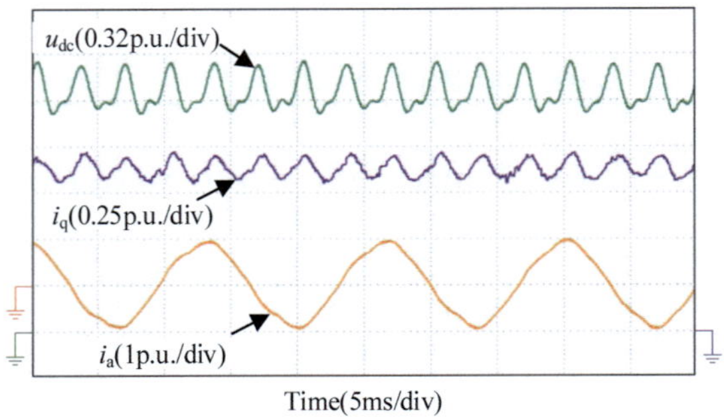

(b)

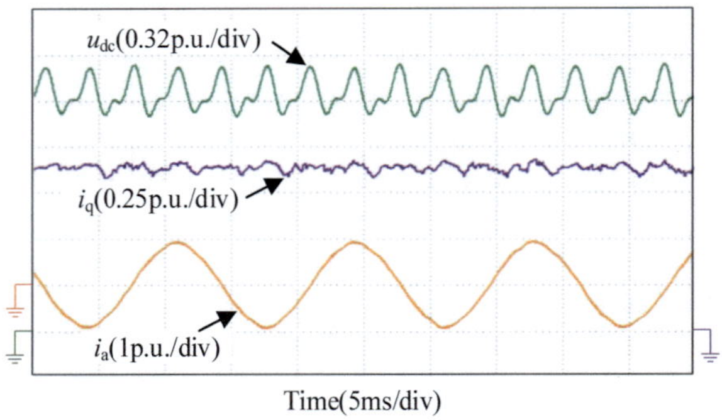

(c)

Fig. 5.8 (continued)

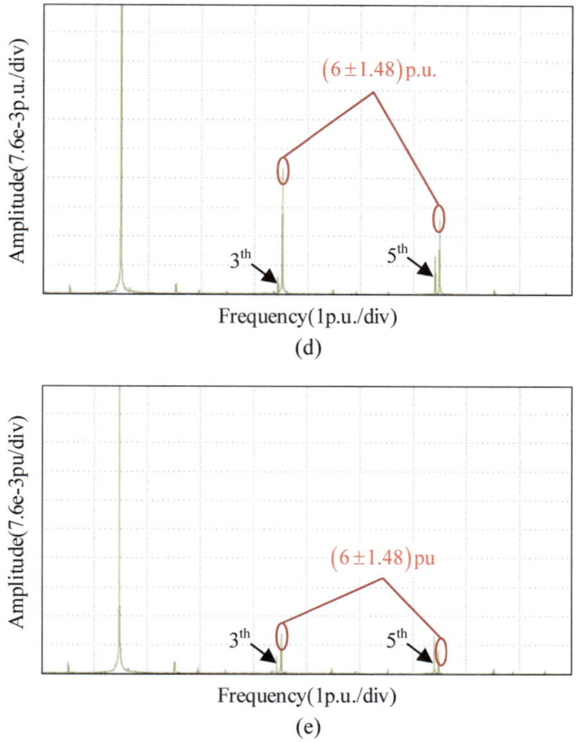

5.3 Beat Phenomenon Suppression Method Based on Impedance Remodeling

5.3.1 Formation Mechanism of Beat Phenomenon

The beat phenomenon suppression strategy based on DC-link voltage prediction is mainly realized by changing the reference voltage in the modulation process, which mainly aims at the beat phenomenon caused by the sampling and updating process in SVPWM. By adjusting the impedance characteristics of the system, the low-frequency envelope of the current in motor side can also be reduced, which is more widely used, and is also suitable for the beat phenomenon caused by the fluctuation of other parameters in vector control, and maintain the suppression effect in the full speed domain.

In vector control, in addition to the harmonics of $6\omega_g$, the d–q axis currents modulated by the fluctuating modulation have the harmonics of $6n\omega_g$, so (5.6) can be rewritten as

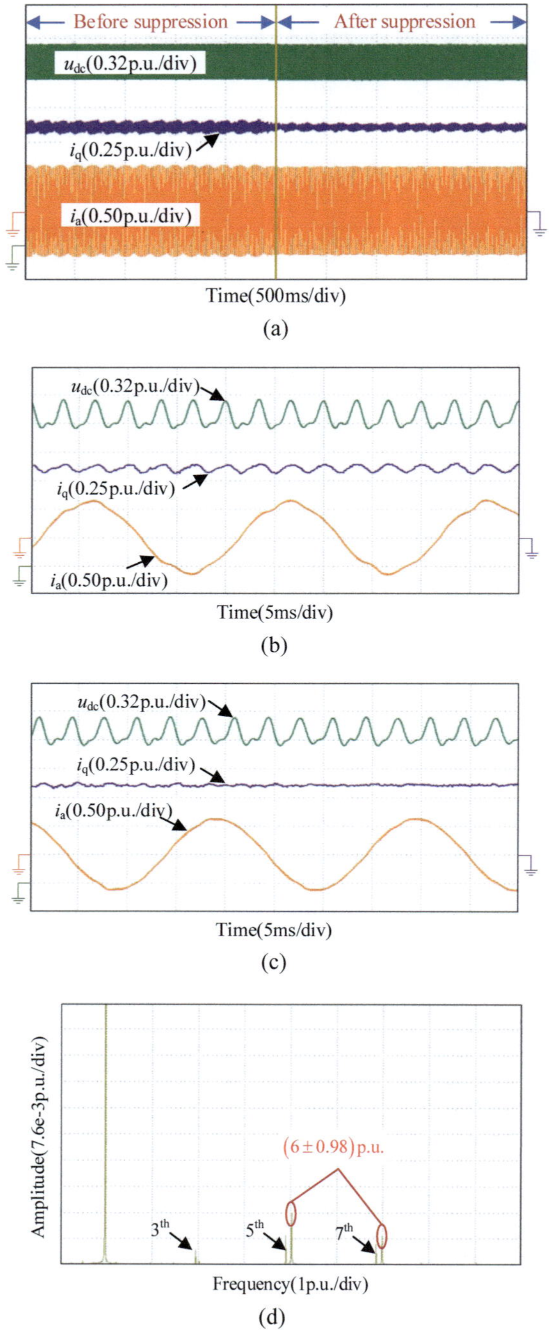

Fig. 5.9 Experimental results of beat frequency suppression at the output power of 0.23 p.u. **a** DC-link voltage, q-axis current and a-phase current. **b** Zoomed-in image of **a** before applying the suppression strategy. **c** Zoomed-in image of **a** after applying the suppression strategy. **d** FFT analysis of a-phase current before applying suppression strategy. **e** FFT analysis of a-phase current after applying suppression strategy

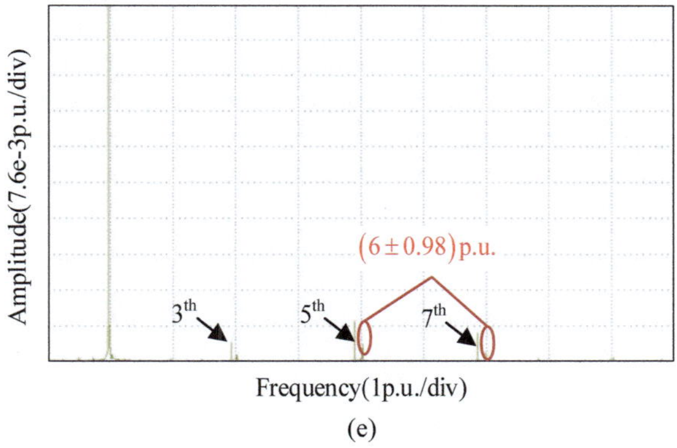

(e)

Fig. 5.9 (continued)

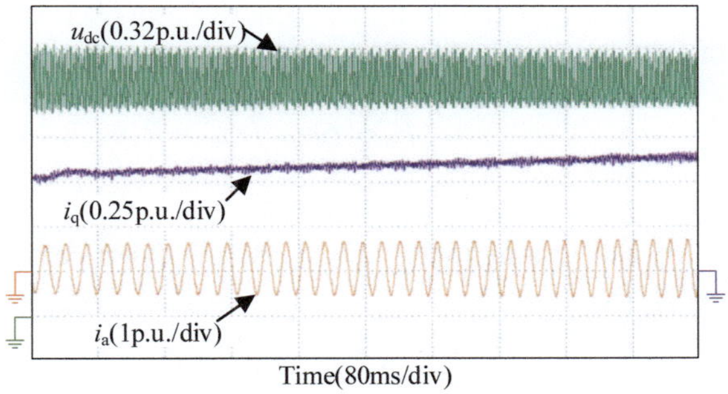

Fig. 5.10 Experimental result of the motor transient process after applying the proposed strategy

$$i_{\mathrm{d}} = I_{\mathrm{d0}} + \sum_{n=1}^{\infty} I_{\mathrm{dn}} \sin(6n\omega_{\mathrm{g}}t + \varphi_{\mathrm{dn}}),\ \ i_{\mathrm{q}} = I_{\mathrm{q0}} + \sum_{n=1}^{\infty} I_{\mathrm{qn}} \sin(6n\omega_{\mathrm{g}}t + \varphi_{\mathrm{qn}}) \tag{5.32}$$

where φ_{dn} and φ_{qn} are the phase of the nth harmonic of the d–q axis current.

Combined with Clarke and Park transformations, a-phase current i_{a} affected by the DC-link voltage fluctuations can be expressed as

$$\begin{aligned} i_{\mathrm{a}} &= \sum_{k=1}^{\infty} (i_{\mathrm{an-}} + i_{\mathrm{an+}}) + i_{\mathrm{a0}} = \sum_{k=1}^{\infty} \{I_{\mathrm{n-}} \sin[(6n\omega_{\mathrm{g}} - \omega_{\mathrm{e}})t + \varphi_{\mathrm{n-}}] \\ &\quad + I_{\mathrm{n+}} \sin[(6n\omega_{\mathrm{g}} + \omega_{\mathrm{e}})t + \varphi_{\mathrm{n+}}]\} + I_{\mathrm{a0}} \sin(\omega_{\mathrm{e}}t + \varphi_{\mathrm{e}}) \\ &= I_{\mathrm{a0}} \sin(\omega_{\mathrm{e}}t + \varphi_{\mathrm{e}}) + \sum_{n=1}^{\infty} I_{\mathrm{bn}} \sin(\omega_{\mathrm{bn}}t + \varphi_{\mathrm{bn}}) \end{aligned} \tag{5.33}$$

where i_{a0} is the fundamental wave component of the A-phase current, and $i_{\mathrm{an-,+}}$ are harmonics at $6n\omega_{\mathrm{g}}$–ω_{e} and $6n\omega_{\mathrm{g}} + \omega_{\mathrm{e}}$, respectively. $I_{\mathrm{a,0}}$ and φ_{e} are the amplitude and phase of the A-phase current fundamental wave, respectively. $i_{\mathrm{an-}}$ and $i_{\mathrm{an+}}$ are both caused by the nth harmonic of the DC-link voltage. $I_{\mathrm{n\mp}}$ and $\varphi_{\mathrm{n\mp}}$ are the amplitude and phase of the harmonics, respectively. $I_{\rm bn}$, $\omega_{\rm bn}$ and $\varphi_{\rm bn}$ are the amplitude, frequency, and phase of the phase current envelope associated with $6n\omega_{\rm g}$.

The low frequency envelope of the beat phenomenon is formed by the maximum value of the motor current. Therefore, the beat current can be derived by analyzing the harmonics at $\omega_{\rm e}t + \varphi_{\rm e} = 2\pi N + \pi/2$ ($N = 0, 1, 2, \ldots$) as

$$\begin{aligned} i_{\rm an-} + i_{\rm an+} = & -I_{\rm n-} \cos\left[12n\pi N\omega_{\rm g}/\omega_{\rm e} + \varphi_{\rm hn} + \varphi_{\rm e} + \varphi_{\rm n-}\right] \\ & + I_{\rm n+} \cos\left[12n\pi N\omega_{\rm g}/\omega_{\rm e} + \varphi_{\rm hn} - \varphi_{\rm e} + \varphi_{\rm n+}\right]\Big|_{N=0,1,2,\ldots} \end{aligned} \tag{5.34}$$

where

$$\varphi_{\rm hn} = 3n\pi\omega_{\rm g}/\omega_{\rm e} - 6n\omega_{\rm g}\varphi_{\rm e}/\omega_{\rm e}. \tag{5.35}$$

According to (5.34), the frequency of beat phenomenon is determined by the ratio of $6n\omega_{\rm g}$ to $\omega_{\rm e}$, which can be expressed as

$$R_{\rm n} = 6n\omega_{\rm g}/\omega_{\rm e} = \boldsymbol{R}\left(6n\omega_{\rm g}/\omega_{\rm e}\right) + d_{\rm n} \tag{5.36}$$

where $d_{\rm n}$ is the fractional part of $\boldsymbol{R}(\cdot)$.

By substituting (5.36) into (5.33), the motor current harmonics can be expressed as

$$\begin{aligned} i_{\rm an-} + i_{\rm an+} = & -I_{\rm n-} \cos\left[2\pi d_{\rm n}N + \varphi_{\rm bn} + \varphi_{\rm e} + \varphi_{\rm n-}\right] \\ & + I_{\rm n+} \cos\left[2\pi d_{\rm n}N + \varphi_{\rm bn} - \varphi_{\rm e} + \varphi_{\rm n+}\right]\Big|_{N=0,1,\cdots}. \end{aligned} \tag{5.37}$$

Therefore, the amplitude and frequency of the low-frequency envelope of the phase current are expressed as

$$\begin{cases} I_{\rm bn} = \sqrt{I_{\rm n-}^2 + I_{\rm n+}^2 - 2I_{\rm n-}I_{\rm n+}\cos(2\varphi_{\rm e} + \varphi_{\rm n-} - \varphi_{\rm n+})} \\ \omega_{\rm bn} = 2\pi d_{\rm n}/\left(2\pi/\omega_{\rm e}\right) = \omega_{\rm e}[R_{\rm n} - \boldsymbol{R}(R_{\rm n})] \end{cases}. \tag{5.38}$$

The phase currents under different fundamental frequencies and harmonic amplitudes are shown in Fig. 5.11a. If the grid frequency is 50 Hz and the fundamental frequency is set to 62 Hz, the frequency of beat phenomenon is inferred to be 10 Hz based on (5.38). Similarly, when the base frequency is set to 74 Hz, the frequency of beat phenomenon is 4 Hz. With the increase of harmonic amplitude, the beat phenomenon deteriorates further.

Figure 5.11b shows the beat frequency at different fundamental frequencies. When the fundamental frequency is a common factor of $6n\omega_{\rm g}$, the motor current does not appear beat phenomenon. According to Fig. 5.11, the amplitude of the beat phenomenon is proportional to the amplitude of the harmonics. Therefore, beat phenomenon suppression is achieved by reducing the motor current harmonics caused by the DC-link voltage.

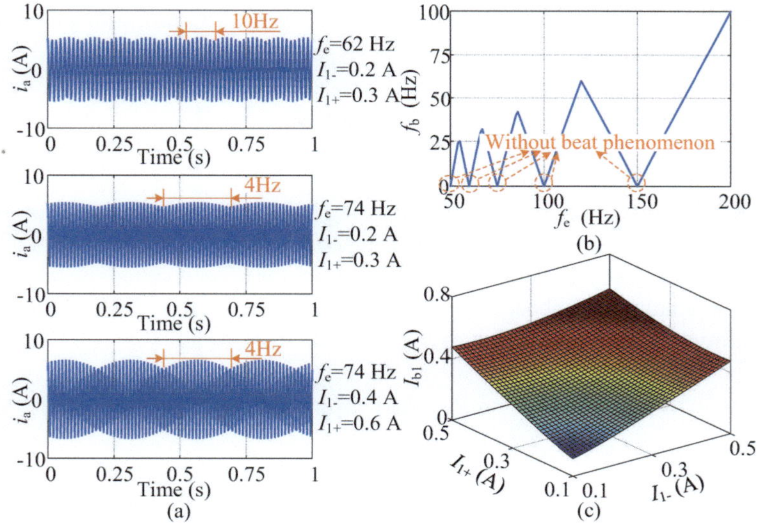

Fig. 5.11 Analysis of beat phenomenon in phase current, reprinted from Ref. [2], copyright 2022, with permission from Nature. **a** Waveforms of the phase current under different fundamental frequencies and harmonic amplitudes. **b** Beat frequency under different fundamental frequencies. **c** Beat amplitude under different amplitudes of harmonics

The electrolytic capacitorless drive system can be regarded as a cascade system composed of LC link and PMSM. As shown in Fig. 5.12, the source of the system is the rectification voltage u_g, and the response is the beat current. Therefore, it is significant to establish the impedance model of beat phenomenon suppression.

According to the inverter power balance and (2.28), the transfer function $\boldsymbol{K}_m$ between the DC-link voltage and the rectified voltage can be expressed as

$$\boldsymbol{K}_m = \Delta u_{dc} / \Delta u_g = 1 / \left[1 + (L_g s + R_g)(\boldsymbol{Y}_m + C_{dc} s)\right]. \tag{5.39}$$

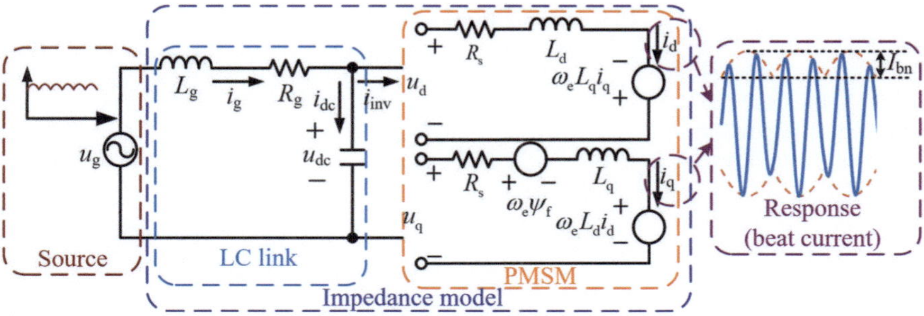

Fig. 5.12 Model of the electrolytic capacitor-less drive

Finally, the d–q axis current harmonics caused by the DC-link voltage fluctuation can be expressed as

$$\begin{cases} I_{\text{dn}} = U_{\text{gn}} |\boldsymbol{K}_{\text{mn}} \boldsymbol{W}_{\text{dn}}|, I_{\text{qn}} = U_{\text{gn}} |\boldsymbol{K}_{\text{mn}} \boldsymbol{W}_{\text{qn}}| \\ \varphi_{\text{dn}} = \varphi_{\text{n}} + \angle \boldsymbol{W}_{\text{dn}}, \varphi_{\text{qn}} = \varphi_{\text{n}} + \angle \boldsymbol{W}_{\text{qn}} \end{cases}. \tag{5.40}$$

where $\boldsymbol{K}_{\text{mn}}$ represents $\boldsymbol{K}_{\text{m}}(j6n\omega_{\text{g}})$. According to (2.23), the d–q axis currents are affected by system parameters and operating conditions.

As shown in (5.40), the impedance model from the source voltage to the d–q axis currents is obtained. The beat phenomenon is analyzed by the relationship between phase current harmonics and d–q axis current fluctuations as

$$i_{\text{a\_n}\mp} = 0.5\{I_{\text{dn}} \sin[(6n\omega_{\text{g}} \mp \omega_{\text{e}})t + \varphi_{\text{dn}}] + I_{\text{qn}} \sin[(6n\omega_{\text{g}} \mp \omega_{\text{e}})t + \varphi_{\text{qn}} \mp \pi/2]\}. \tag{5.41}$$

Therefore, the amplitude and phase of the above harmonics in A-phase current can be expressed as

$$\begin{cases} I_{\text{n}\mp} = \sqrt{I_{\text{dn}}^2 + I_{\text{qn}}^2 \mp 2I_{\text{dn}}I_{\text{qn}} \sin(\varphi_{\text{dn}} - \varphi_{\text{qn}})} \Big/ 2 \\ \varphi_{\text{n}-} - \varphi_{\text{n}+} = \arctan 2I_{\text{qn}}I_{\text{dn}} \cos(\varphi_{\text{dn}} - \varphi_{\text{qn}}) \Big/ \left(I_{\text{qn}}^2 - I_{\text{dn}}^2\right) \end{cases}. \tag{5.42}$$

By substituting (5.40) and (5.42) into (5.38), the influence of rectified voltage harmonics $U_{\text{grec,k}}$ on beat phenomenon is expressed as

$$\frac{I_{\text{bn}}}{U_{\text{gn}}} = \frac{|\boldsymbol{K}_{\text{mn}}|}{\sqrt{2}} \cdot \sqrt{\begin{aligned} &|\boldsymbol{W}_{\text{dn}}|^2 + |\boldsymbol{W}_{\text{qn}}|^2 - 2\cos(2\varphi_{\text{e}} + \varphi_{\text{n}-} - \varphi_{\text{n}+})[|\boldsymbol{W}_{\text{dn}}|^4 \\ &+|\boldsymbol{W}_{\text{qn}}|^4 + 2|\boldsymbol{W}_{\text{dn}}\boldsymbol{W}_{\text{qn}}|^2 \cos 2\{\angle \boldsymbol{W}_{\text{dn}} - \angle \boldsymbol{W}_{\text{qn}}\}]^{0.5} \end{aligned}}. \tag{5.43}$$

Figure 5.13 shows $I_{\text{bn}}/U_{\text{gn}}$ with different bandwidths of the current controller. When n is less than 3, the value of I_{bn} decreases with the increase of bandwidth. So the increase of bandwidth can reduce the beat current at $6\omega_{\text{g}}$ and $12\omega_{\text{g}}$.

According to the FFT of the rectifier voltage in (3.36), the amplitude of the beat current caused by the DC-link voltage fluctuation can be expressed as

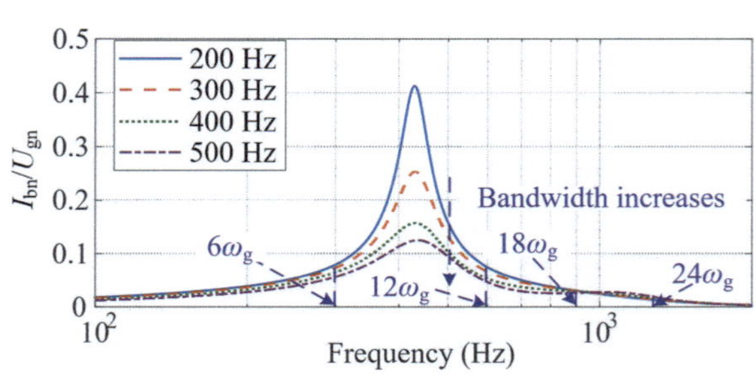

Fig. 5.13 Ratios of $I_{\text{bn}}/U_{\text{gn}}$ with different bandwidths of the current controllers

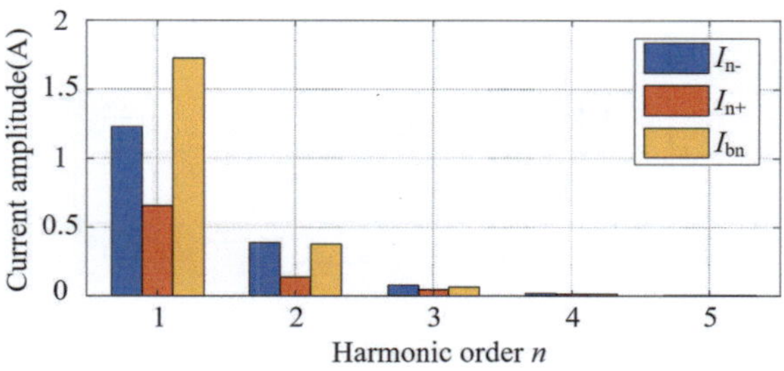

Fig. 5.14 Amplitude of harmonics in the motor current

$$I_{\text{bn}} = \frac{6U_{\text{L}}|\boldsymbol{K}_{\text{mn}}|}{\pi(36k^2-1)}\sqrt{\begin{array}{l}|\boldsymbol{W}_{\text{dn}}|^2 + \left|\boldsymbol{W}_{\text{qn}}\right|^2 - 2\cos(2\varphi_{\text{e}} + \varphi_{\text{n}-} - \varphi_{\text{n}+})[|\boldsymbol{W}_{\text{dn}}|^4 \\ +\left|\boldsymbol{W}_{\text{qn}}\right|^4 + 2\left|\boldsymbol{W}_{\text{dn}}\boldsymbol{W}_{\text{qn}}\right|^2 \cos 2\{\angle\boldsymbol{W}_{\text{dn}} - \angle\boldsymbol{W}_{\text{qn}}\}]^{0.5}\end{array}} \tag{5.44}$$

where U_{L} is the effective value of the line voltage.

Figure 5.14 shows the amplitudes of $I_{\text{n}-}$, $I_{\text{n}+}$ and I_{bn} at different harmonic frequencies. It can be seen that the main component is the harmonics of the beat current related to $6\omega_{\text{g}}(n = 1)$ and $12\omega_{\text{g}}(n = 1)$. These dominant harmonics are the main reason for the beat phenomenon and need to be suppressed.

5.3.2 *Impedance Remodeling Method*

According to the impedance analysis, the amplitude of the beat current can be derived from the fluctuation of the d–q axis current, and the frequencies of the main components are $6\omega_{\text{g}}$ and $12\omega_{\text{g}}$. Therefore, the beat phenomenon can be suppressed by remodeling the impedance between the d–q axis current and the DC-link voltage at the same frequency. In the proposed method, impedance remodeling is achieved by modifying the angle of the current vector. Figure 5.15 shows the beat phenomenon suppression strategy based on impedance remodeling at the dominant frequency. The motor speed and the rotor position are observed by the position and speed observer. Based on vector control, the dominant harmonics ($u_{\text{dc\_1}}$ and $u_{\text{dc\_2}}$) in the DC-link voltage are extracted by a bandpass filter, and then the regulated current is generated and applied to the drive system by changing the angle of the current vector with the regulated angle $\Delta\theta_{\text{e}}$. The regulated current angle is $\hat{\theta}_{\text{en}}$, which is applied to the Park transform of the current feedback loop. The advantage of impedance remodeling is that the influence of steady-state value can be eliminated in the process of small-signal modeling, indicating that the impedance model is not affected by changes in motor speed and q-axis current after applying the proposed method.

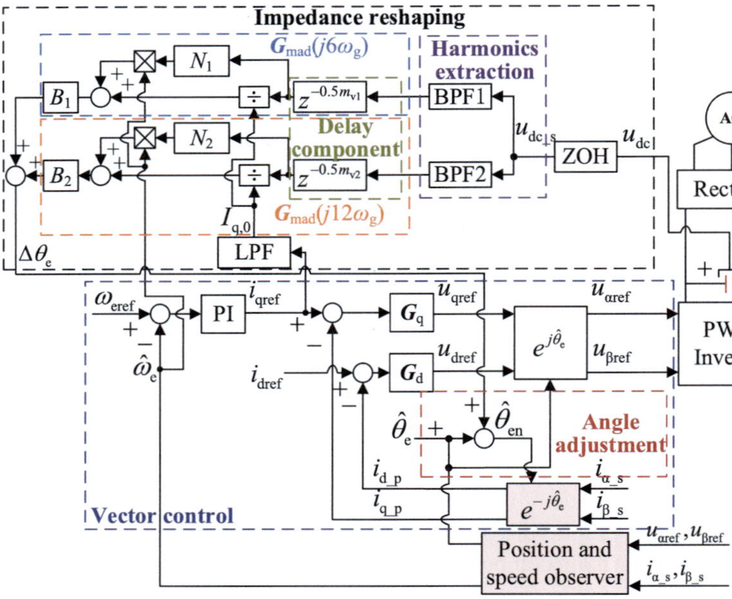

Fig. 5.15 Block diagram of beatless control strategy based on impedance reshaping at dominant harmonics, reprinted from Ref. [2], copyright 2022, with permission from Nature

The equivalent feedback current can be expressed as

$$\begin{bmatrix} i_{\text{d\_p}} \\ i_{\text{q\_p}} \end{bmatrix} = \begin{bmatrix} \cos(\hat{\theta}_e + \Delta\theta_e) & \sin(\hat{\theta}_e + \Delta\theta_e) \\ -\sin(\hat{\theta}_e + \Delta\theta_e) & \cos(\hat{\theta}_e + \Delta\theta_e) \end{bmatrix} \begin{bmatrix} i_{\alpha\_s} \\ i_{\beta\_s} \end{bmatrix} \tag{5.45}$$

where $i_{\text{d\_p}}$ and $i_{\text{q\_p}}$ are d–q axis currents after applying the regulated angle, i_{α_s} and i_{β_s} are sampling currents in α–β axis.

As the position and speed observer converges, the estimated rotor position is equal to the equivalent sampling position. Considering the sampling delay of the stator current, Eq. (5.45) can be rewritten as

$$\begin{bmatrix} i_{\text{d\_p}} \\ i_{\text{q\_p}} \end{bmatrix} = e^{-0.5sT_s} \begin{bmatrix} \cos\Delta\theta_e & \sin\Delta\theta_e \\ -\sin\Delta\theta_e & \cos\Delta\theta_e \end{bmatrix} \begin{bmatrix} i_d \\ i_q \end{bmatrix} \tag{5.46}$$

where $\Delta\theta_e$ is the small disturbance. According to the Bessel function, Eq. (5.46) can be simplified to

$$\begin{bmatrix} I_{\text{d\_p0}} + \Delta i_{\text{d\_p}} \\ I_{\text{q\_p0}} + \Delta i_{\text{q\_p}} \end{bmatrix} = \begin{bmatrix} I_{d0} + e^{-0.5sT_s}\Delta i_d + I_{q0}\Delta\theta_e \\ I_{q0} + e^{-0.5sT_s}\Delta i_q - I_{d0}\Delta\theta_e \end{bmatrix} \tag{5.47}$$

where $I_{\text{d\_p,0}}$, $I_{\text{q\_p,0}}$, $\Delta i_{\text{d\_p}}$ and $\Delta i_{\text{q\_p}}$ are DC components and small signal of d–q axis currents after regulation.

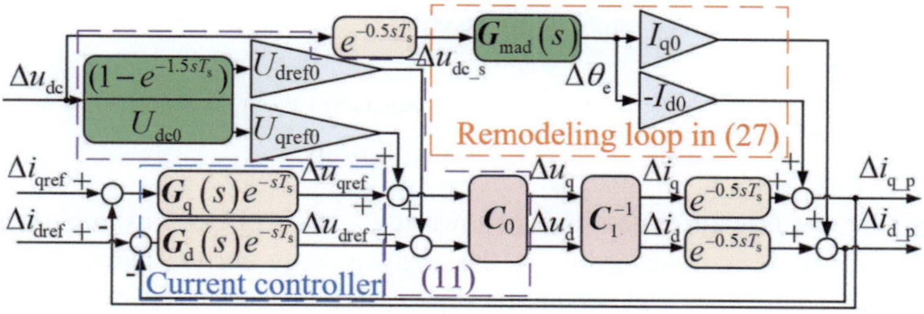

Fig. 5.16 Small-signal model with the impedance remodeling strategy, reprinted from Ref. [2], copyright 2022, with permission from Nature

As shown in (5.47), the regulation of the current vector angle is equivalent to introducing additional regulating currents $I_{q0}\Delta\theta_e$ and $-I_{d0}\Delta\theta_e$ in the d–q axis. The amplitude of the regulating current introduced in the d–q axis can change adaptively to the operating conditions. Figure 5.16 shows the small signal model after applying the proposed beat phenomenon suppression strategy. It can be seen that the feedback control loop from the DC-link voltage to the motor current is constructed to remodel the impedance of the driver, thus suppressing current harmonics.

From (2.14) and (5.47), the state equation in Fig. 5.16 can be derived, as shown in (5.48), which has an additional regulating current component compared to (2.22).

$$\begin{bmatrix} \boldsymbol{A}_{1d} & \boldsymbol{A}_{1q} \\ \boldsymbol{A}_{2d} & \boldsymbol{A}_{2q} \end{bmatrix}\begin{bmatrix} \Delta i_d \\ \Delta i_q \end{bmatrix} = \frac{\left(1-\mathrm{e}^{-1.5sT_s}\right)\Delta u_{dc}}{U_{dc,0}}\begin{bmatrix} U_{dref0} \\ U_{qref0} \end{bmatrix} + \mathrm{e}^{-sT_s}\begin{bmatrix} -I_{q0}\boldsymbol{G}_d \\ I_{d0}\boldsymbol{G}_q \end{bmatrix}\Delta\theta_e. \quad (5.48)$$

Then, transfer functions $\boldsymbol{W}_{dbt}$ and $\boldsymbol{W}_{qbt}$ between the DC-link voltage and the motor current can then be remodeled as

$$\begin{bmatrix} \boldsymbol{W}_{dbt} \\ \boldsymbol{W}_{qbt} \end{bmatrix} = \begin{bmatrix} \boldsymbol{W}_d \\ \boldsymbol{W}_q \end{bmatrix} + \frac{\mathrm{e}^{-1.5sT_s}\boldsymbol{G}_{bt}}{\boldsymbol{A}_{2q}\boldsymbol{A}_{1d} - \boldsymbol{A}_{2d}\boldsymbol{A}_{1q}}\begin{bmatrix} -I_{q0}\boldsymbol{G}_d\boldsymbol{A}_{2q} - I_{d0}\boldsymbol{G}_q\boldsymbol{A}_{1q} \\ I_{d0}\boldsymbol{G}_q\boldsymbol{A}_{1d} + I_{q0}\boldsymbol{G}_d\boldsymbol{A}_{2d} \end{bmatrix}. \quad (5.49)$$

In (5.49), $\boldsymbol{G}_{bt}$ is the transfer function from $\Delta\theta_e$ to Δu_{dc}, which is critical to reduce the amplitude of the beat current caused by the DC-link voltage.

Since the system operates without beat phenomenon, the harmonic amplitude of the stator current vector is 0. When field weakening control is not performed, I_{d0} is usually set to 0 A. Therefore, the small signal of the stator current vector is approximately equal to the small signal of the q-axis current, and $\boldsymbol{W}_{qbt}$ is equal to 0. Then,

$$\frac{\left(1-\mathrm{e}^{-1.5sT_s}\right)\left(U_{qref0}\boldsymbol{A}_{1d} - U_{dref0}\boldsymbol{A}_{2d}\right)}{U_{dc,0}} + \mathrm{e}^{-1.5sT_s}\boldsymbol{G}_{bt}I_{q,0}\boldsymbol{G}_d\boldsymbol{A}_{2d} = 0. \quad (5.50)$$

According to (5.50), $\boldsymbol{G}_{\mathrm{bt}}$ can be expressed as

$$\boldsymbol{G}_{\mathrm{bt}}=\frac{\Delta\theta_{\mathrm{e}}}{\Delta u_{\mathrm{dc}}}=\frac{\left(\mathrm{e}^{1.5sT_{\mathrm{s}}}-1\right)\left(U_{\mathrm{dref0}}\boldsymbol{A}_{2\mathrm{d}}-U_{\mathrm{qref0}}\boldsymbol{A}_{1\mathrm{d}}\right)}{U_{\mathrm{dc0}}I_{\mathrm{q0}}\boldsymbol{G}_{\mathrm{d}}\boldsymbol{A}_{2\mathrm{d}}}. \tag{5.51}$$

However, the transfer function $\boldsymbol{G}_{\mathrm{bt}}$ shown in (5.51) is difficult to implement and requires further design. According to the analysis, only current harmonics with a frequency of $6n\omega_{\mathrm{g}}$ ($n=1, 2$) need to be suppressed. Instead of designing $\boldsymbol{G}_{\mathrm{bt}}$ in the full frequency band, amplitude and phase characteristics can be designed at specific frequencies, which can lead to better harmonic suppression. By substituting (2.23) and PMSM voltage equation into (5.51), $\boldsymbol{G}_{\mathrm{bt}}$ can be expressed as

$$\boldsymbol{G}_{\mathrm{bt}}\approx\frac{\left(R_{\mathrm{s}}I_{\mathrm{q0}}+\omega_{\mathrm{e}}\psi_{\mathrm{f}}\right)\left(R_{\mathrm{s}}+L_{\mathrm{d}}s+\mathrm{e}^{-1.5sT_{\mathrm{s}}}\boldsymbol{G}_{\mathrm{d}}\right)+\omega_{\mathrm{e}}^{2}L_{\mathrm{d}}L_{\mathrm{q}}I_{\mathrm{q0}}}{U_{\mathrm{dc0}}I_{\mathrm{q0}}\omega_{\mathrm{e}}L_{\mathrm{d}}\boldsymbol{G}_{\mathrm{d}}\big/\left(1-\mathrm{e}^{1.5sT_{\mathrm{s}}}\right)} \tag{5.52}$$

where ψ_{f} is the stator flux.

Ignoring the resistance loss, $\boldsymbol{G}_{\mathrm{bt}}$ at a frequency of $6n\omega_{\mathrm{g}}$ can be expressed as

$$\begin{aligned}\boldsymbol{G}_{\mathrm{bt}}(j6n\omega_{\mathrm{g}})\approx{}&\frac{1-\mathrm{e}^{j9n\omega_{\mathrm{g}}T_{\mathrm{s}}}}{L_{\mathrm{d}}U_{\mathrm{dc,0}}\left(K_{\mathrm{id}}+j6nK_{\mathrm{pd}}\omega_{\mathrm{g}}\right)}\Big\{j6n\omega_{\mathrm{e}}L_{\mathrm{d}}L_{\mathrm{q}}\omega_{\mathrm{g}}\\&+\left[\mathrm{e}^{-j9n\omega_{\mathrm{g}}T_{\mathrm{s}}}\left(K_{\mathrm{id}}+j6nK_{\mathrm{pd}}\omega_{\mathrm{g}}\right)+j6nR_{\mathrm{s}}\omega_{\mathrm{g}}-36n^{2}L_{\mathrm{d}}\omega_{\mathrm{g}}^{2}\right]\psi_{\mathrm{f}}\big/I_{\mathrm{q,0}}\Big\}.\end{aligned} \tag{5.53}$$

It can be seen from Eq. (5.53) that the steady-state values in $\boldsymbol{G}_{\mathrm{bt}}$ can be separated into ω_{e} and I_{q0}. Although $\boldsymbol{G}_{\mathrm{bt}}$ is complex in (5.53), it can be expressed as (5.54) because only the amplitude and phase need to be designed at a specific frequency. Therefore, the frequency characteristic at $6n\omega_{\mathrm{g}}$ can be expressed as

$$\boldsymbol{G}_{\mathrm{bt}}(j6n\omega_{\mathrm{g}})=B_{\mathrm{n}}(1\big/I_{\mathrm{q,0}}+N_{\mathrm{n}}\omega_{\mathrm{e}})\mathrm{e}^{-0.5m_{\mathrm{vn}}\cdot j6k\omega_{\mathrm{g}}T_{\mathrm{s}}} \tag{5.54}$$

where B_{k} and m_{vk} are the amplitudes and delays at $6n\omega_{\mathrm{g}}$ ($n=1, 2$).

The amplitude characteristics of $\boldsymbol{G}_{\mathrm{bt}}$ at $6\omega_{\mathrm{g}}$ and $12\omega_{\mathrm{g}}$ should be the same before and after design. Therefore, the proportionality coefficient B_{n} and N_{n} are expressed as

$$\begin{cases}B_{\mathrm{n}}=\dfrac{\left|1-\mathrm{e}^{j9n\omega_{\mathrm{g}}T_{\mathrm{s}}}\right|\left|\mathrm{e}^{-j9n\omega_{\mathrm{g}}T_{\mathrm{s}}}\left(K_{\mathrm{id}}+j6nK_{\mathrm{pd}}\omega_{\mathrm{g}}\right)-36L_{\mathrm{d}}n^{2}\omega_{\mathrm{g}}^{2}+j6nR_{\mathrm{s}}\omega_{\mathrm{g}}\right|\psi_{\mathrm{f}}}{L_{\mathrm{d}}U_{\mathrm{dc,0}}\left|K_{\mathrm{id}}+j6nK_{\mathrm{pd}}\omega_{\mathrm{g}}\right|}\\N_{\mathrm{n}}=\dfrac{6nL_{\mathrm{d}}L_{\mathrm{q}}\omega_{\mathrm{g}}}{\left|\mathrm{e}^{-j9n\omega_{\mathrm{g}}T_{\mathrm{s}}}\left(K_{\mathrm{id}}+j6nK_{\mathrm{pd}}\omega_{\mathrm{g}}\right)-36L_{\mathrm{d}}n^{2}\omega_{\mathrm{g}}^{2}+j6nR_{\mathrm{s}}\omega_{\mathrm{g}}\right|\psi_{\mathrm{f}}}\end{cases}. \tag{5.55}$$

The delay link shown in Fig. 5.15 is designed to regulate the phase of the extracted DC-link voltage harmonics. The purpose of setting the equivalent delay is to improve

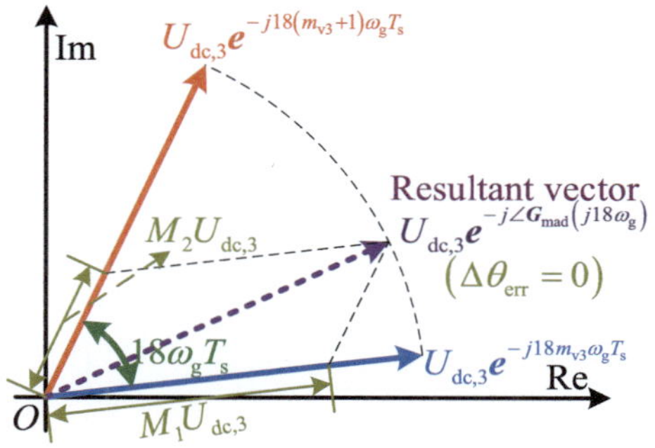

Fig. 5.17 Vector schematic diagram of the phase regulation for the drives with high resonant frequency

the accuracy of the design delay by obtaining the sampling voltage that differs by one switching cycle or the average of the same sampling voltage. To ensure consistent phase characteristics before and after design, the equivalent delay design of $\boldsymbol{G}_{\text{bt}}$ is

$$m_{\text{v1}} = \boldsymbol{R}\left[-\frac{2\angle \boldsymbol{G}_{\text{bt}}(j6\omega_{\text{g}})}{6\omega_{\text{g}}T_{\text{s}}}\right], m_{\text{v2}} = \boldsymbol{R}\left[-\frac{2\angle \boldsymbol{G}_{\text{bt}}(j12\omega_{\text{g}})}{12\omega_{\text{g}}T_{\text{s}}}\right]. \tag{5.56}$$

The application of $\boldsymbol{R}[\cdot]$ introduces the phase error $\Delta\theta_{\text{err}}$ between the theoretical regulation angle and the realizable regulation angle. The equivalent delay in (5.56) reduces the maximum phase regulation error to $1.5n\omega_{\text{g}}T_{\text{s}}$ compared to the direct delay DC-link voltage harmonics.

However, as the resonant frequency increases, the beat current associated with the high frequency needs to be suppressed, and the phase regulation error will increase. In order to further improve the regulation accuracy, the improved phase regulation vector related to $18\omega_{\text{g}}$ is shown in Fig. 5.17. The phase of the obtained vector can be remodeled into the phase of $\boldsymbol{G}_{\text{bt}}(j18n\omega_{\text{g}})$ by regulating the sampling voltage amplitude, which differs by one switching cycle, to $M_1U_{\text{dc,3}}$ and $M_2U_{\text{dc,3}}$ respectively. The phase regulation error $\Delta\theta_{\text{err}}$ can then be reduced to 0.

At the same time, two bandpass filters BPF1 and BPF2 are used to extract the DC-link voltage harmonics with a frequency of $6n\omega_{\text{g}}$ ($k = 1, 2$). According to (5.54), $\boldsymbol{G}_{\text{bt}}$ can be designed as

$$\boldsymbol{G}_{\text{bt}} = B_1\left(1/I_{\text{q,0}} + N_1\omega_{\text{e}}\right)\boldsymbol{G}_{\text{BPF1}}\text{e}^{-0.5m_{\text{v1}}T_{\text{s}}s} + B_2\left(1/I_{\text{q,0}} + N_2\omega_{\text{e}}\right)\boldsymbol{G}_{\text{BPF2}}\text{e}^{-0.5m_{\text{v2}}T_{\text{s}}s}. \tag{5.57}$$

The Bode diagram of $\boldsymbol{G}_{\text{bt}}$ before and after design is shown in Fig. 5.18. The frequency response characteristics of the simplified before and after reconstruction are very close at the frequencies of $6\omega_{\text{g}}$ and $12\omega_{\text{g}}$. At the same time, the amplitude at other frequencies is much lower than the amplitude at $6\omega_{\text{g}}$ and $12\omega_{\text{g}}$. DC-link voltage harmonics of other frequencies do not negatively affect the beat phenomenon.

According to (5.53) and (5.57), $\boldsymbol{G}_{\text{bt}}$ is mainly determined by the parameters of the motor and the bandwidth of the control loop by isolating the steady-state values of the

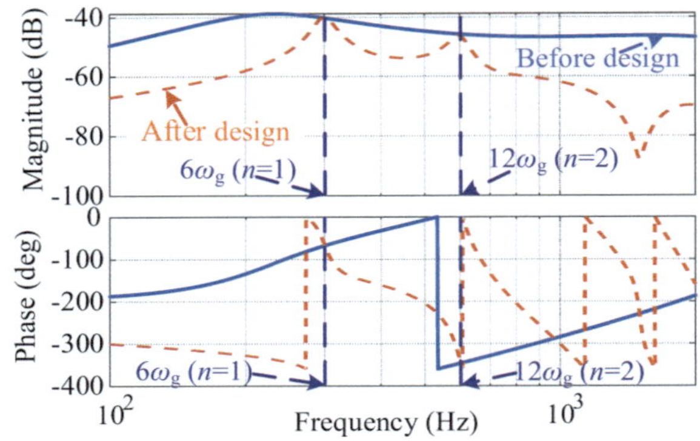

Fig. 5.18 Bode diagrams of G_{mad} before and after the design

PMSM. Therefore, $\boldsymbol{G}_{\mathrm{bt}}$ can be designed by selecting the appropriate amplitude and equivalent delay at a frequency of $6n\omega_{\mathrm{g}}$, and the proposed impedance remodeling method is not affected by motor speed and q-axis current.

5.3.3 Analysis of Beat Phenomenon Suppression Ability

The design of B_{n} and m_{vn} is essential for achieving satisfactory harmonic suppression effect. It is necessary to derive the influence of DC-link voltage fluctuation on the beat current after the control is applied. After impedance remodeling is applied, the transfer function $\boldsymbol{Y}_{\mathrm{mbt}}$ of the input admittance of the inverter can be expressed as

$$\begin{aligned}\boldsymbol{Y}_{\mathrm{mbt}} &= 3\boldsymbol{W}_{\mathrm{dbt}}\left(U_{\mathrm{d,0}} + \omega_{\mathrm{e}}L_{\mathrm{d}}I_{\mathrm{q0}} + L_{\mathrm{d}}I_{\mathrm{d0}}s + I_{\mathrm{d0}}R_{s}\right)/\,2U_{\mathrm{dc0}} \\ &\quad + 3\boldsymbol{W}_{\mathrm{qbt}}\left(U_{\mathrm{q0}} - \omega_{\mathrm{e}}L_{\mathrm{q}}I_{\mathrm{d0}} + L_{\mathrm{q}}I_{\mathrm{q0}}s + I_{\mathrm{q0}}R_{\mathrm{s}}\right)/\,2U_{\mathrm{dc0}} - I_{\mathrm{inv0}}/\,U_{\mathrm{dc0}}. \end{aligned} \tag{5.58}$$

Therefore, the transfer function $\boldsymbol{K}_{\mathrm{mbt}}$ between the rectifier voltage and the DC-link voltage after the control is applied can be derived as

$$\boldsymbol{K}_{\mathrm{mbt}} = 1/\left[1 + \left(L_{\mathrm{g}}s + R_{\mathrm{g}}\right)(\boldsymbol{Y}_{\mathrm{mbt}} + C_{\mathrm{dc}}s)\right]. \tag{5.59}$$

According to (5.44), the ratio between the beat current and the rectifier voltage fluctuation is

$$\frac{I_{\mathrm{bn,bt}}}{U_{\mathrm{gn}}} = \frac{|\boldsymbol{K}_{\mathrm{mbtn}}|}{\sqrt{2}} \cdot \sqrt{\begin{array}{c}|\boldsymbol{W}_{\mathrm{dbtn}}|^{2} + |\boldsymbol{W}_{\mathrm{qbtn}}|^{2} - 2\cos(2\varphi_{\mathrm{e}} + \varphi_{\mathrm{n-}} - \varphi_{\mathrm{n+}}) \\ \cdot\sqrt{|\boldsymbol{W}_{\mathrm{dbtn}}|^{4} + |\boldsymbol{W}_{\mathrm{qbtn}}|^{4} + 2|\boldsymbol{W}_{\mathrm{dbtn}}\boldsymbol{W}_{\mathrm{qbtn}}|^{2}\cos 2\{\angle\boldsymbol{W}_{\mathrm{dbtn}} - \angle\boldsymbol{W}_{\mathrm{qbtn}}\}}\end{array}} \tag{5.60}$$

where $I_{\mathrm{bn,bt}}$ are envelope amplitudes related to $6n\omega_{\mathrm{g}}$ after control is applied. $\boldsymbol{W}_{\mathrm{dbtn}}$, $\boldsymbol{W}_{\mathrm{qbtn}}$ and $\boldsymbol{K}_{\mathrm{mbtn}}$ represent $\boldsymbol{W}_{\mathrm{dbt}}\,(j6n\omega_{\mathrm{g}})$, $\boldsymbol{W}_{\mathrm{qbt}}\,(j6n\omega_{\mathrm{g}})$ and $\boldsymbol{K}_{\mathrm{mbt}}\,(j6n\omega_{\mathrm{g}})$, respectively.

In Fig. 5.19, the plane shows the ratios between $I_{bn,bt}$ and I_{bn} at different impedance values. The *dashed plane shows the amplitude ratio without control. As shown in* Fig. 5.19a, *when* B_1 *and* m_{v1} are selected as 3.0e^{-2} and 7, harmonics caused by DC-link voltages with frequencies of $6\omega_g$ can be reduced to 12%. According to Fig. 5.19b, the values of B_2 and m_{v2} can be selected as 1.5e^{-1} and 2.

Figure 5.20 shows the amplitude of the beat envelope caused by DC-link voltage harmonics with frequencies of $6\omega_g$ and $12\omega_g$ under different operating conditions. In the case of design coefficients ($B_1 = 3.0$e^{-2}, $m_{v1} = 7$), I_{b1} can be effectively reduced by comparing the two surfaces shown in Fig. 5.20a. The proposed beat phenomenon suppression method based on impedance remodeling is effective over a wide range of speeds and torques when using the designed coefficients.

In order to further study the robustness of the control strategy to PMSM parameters, the ratio between the beat current and the rectifier voltage under different values

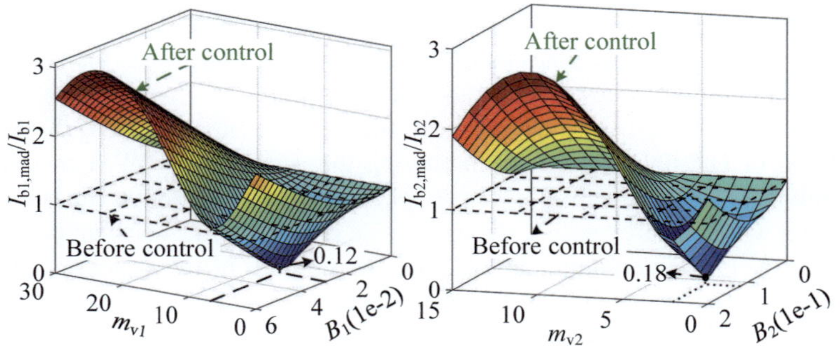

(a) At $6\omega_g$ (b) At $12\omega_g$

Fig. 5.19 Ratio between the amplitudes of beat envelope before and after control under different impedances, reprinted from Ref. [2], copyright 2022, with permission from Nature

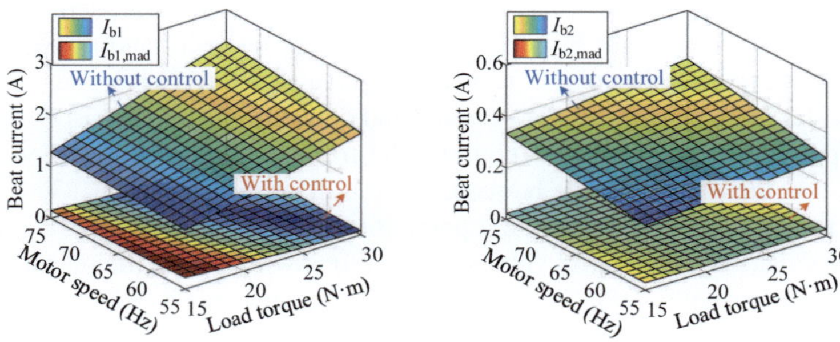

(a) Beat envelop caused by the harmonics of 6ωg (b) Beat envelop caused by the harmonics of 12ωg

Fig. 5.20 Theoretical analysis of beat harmonics under different operating conditions with and without the proposed impedance reshaping strategy, reprinted from Ref. [2], copyright 2022, with permission from Nature

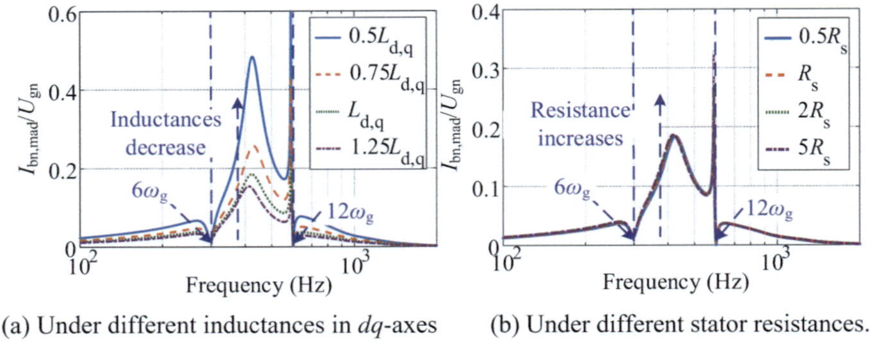

Fig. 5.21 Ratios between the beat current and rectified voltage under different PMSM parameters, reprinted from Ref. [2], copyright 2022, with permission from Nature

of L_d, L_q and R_s is shown in Fig. 5.21. The impedance remodeling method can keep the ratio constant at $6\omega_g$, and the change of $I_{bn,mad}/U_{gn}$ at $12\omega_g$ is less than 0.01 when $L_{d,q}$ is reduced by 0.5 times or R_s is increased by 5 times.

The traditional control strategy solves the beat phenomenon caused by the periodic fluctuation of DC-link voltage by improving the accuracy of PWM modulation. Although these methods can achieve beat phenomenon suppression, the stability of the system will decrease because the motor side is closer to the constant power load after duty cycle compensation. In this part, the stability of the system is analyzed after applying the proposed impedance remodeling strategy.

The transfer function $\boldsymbol{K}_{gbt}$ between DC current Δi_L and rectifier voltage Δu_g after control can be derived as

$$\boldsymbol{K}_{gbt} = \frac{\Delta i_L}{\Delta u_g} = \frac{\boldsymbol{Y}_{mbt} + C_{dc}s}{1 + \left(L_g s + R_g\right)\left(\boldsymbol{Y}_{mbt} + C_{dc}s\right)}. \tag{5.61}$$

The Bode diagrams of $\boldsymbol{K}_{mbt}$ and $\boldsymbol{K}_{gbt}$ with and without beat phenomenon suppression methods are shown in Fig. 5.22. By comparing the solid and dotted lines in Fig. 5.22, the peak value of LC resonance increases from 13 dB and −8 dB to 22 dB and 0.9 dB respectively. The dashed and dotted lines show the Bode diagram when applying the proposed method and the traditional damping method. It can be seen that the amplitude can be reduced by 4 dB at $6\omega_g$ and $12\omega_g$. This means that the proposed beat phenomenon suppression strategy can reduce the beat current harmonics caused by DC-link voltage fluctuations without sacrificing the system stability. The harmonics of DC-link voltage and grid current related to $6\omega_g$ and $12\omega_g$ can also be suppressed, and the damping performance is close to that of traditional damping methods.

In addition, there is a limit to the amplitude of the regulation angle to ensure the stability improvement of the beat phenomenon suppression strategy. According to the system model in Fig. 5.23, according to the ratio of LC link impedance to PMSM

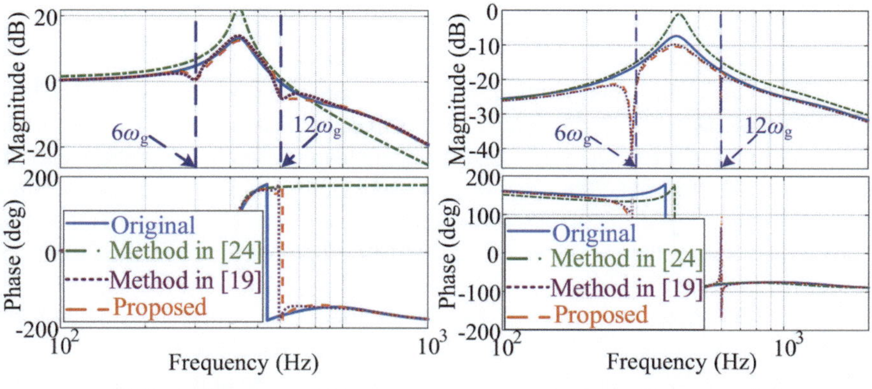

(a) Bode diagrams of $\boldsymbol{K}_{\mathrm{mmad}}$ (b) Bode diagrams of $\boldsymbol{K}_{\mathrm{gmad}}$

Fig. 5.22 Harmonic evaluation with and without the control methods, reprinted from Ref. [2], copyright 2022, with permission from Nature

impedance, the stability of the drive system can be evaluated by Nyquist stability criterion. The admittance $\boldsymbol{Y}_{\mathrm{f}}$ of LC link can be expressed as

$$\boldsymbol{Y}_{\mathrm{f}} = \left(L_{\mathrm{g}} C_{\mathrm{dc}} s^2 + R_{\mathrm{g}} C_{\mathrm{dc}} s + 1\right) / \left(L_{\mathrm{g}} s + R_{\mathrm{g}}\right). \tag{5.62}$$

The Nyquist diagram of $\boldsymbol{Y}_{\mathrm{mmad}}/\boldsymbol{Y}_{\mathrm{f}}$ with different amplitudes of $\Delta\theta_{\mathrm{e}}$ amplitude is shown in Fig. 5.23. When the amplitude is less than 20°, the proposed method can improve the stability of the system. However, when $(\Delta\theta_{\mathrm{e}})_{\max}$ is too high, the stability deteriorates. According to Fig. 5.23, the limit of the current vector regulation angle can be set to 20°.

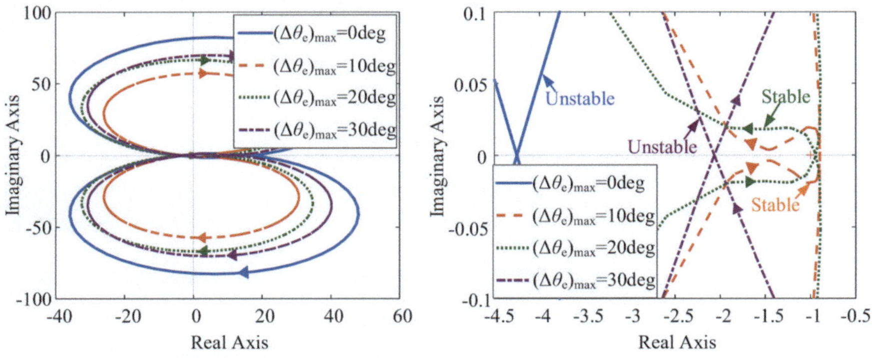

(a) Overall Nyquist plots (b) Detailed view in (a)

Fig. 5.23 Nyquist plots of $\boldsymbol{Y}_{\mathrm{mmad}}/\boldsymbol{Y}_{\mathrm{f}}$ with different amplitudes of $\Delta\theta_{\mathrm{e}}$

5.3.4 Experimental Results

Figure 5.24 shows the experimental results at the motor frequency of 74 Hz. The beat frequency is 4 Hz, which is the same as the result calculated in (5.38). The amplitude of the adjusting angle is set as 10°, and the beat phenomenon in the a-phase current can be reduced effectively. The peak-to-peak value of the DC-link voltage ripple falls from 120 to 85 V. The fluctuation of the current vector can be reduced from 2.1 to 1.2 A, and the amplitude of the low-frequency envelope can be suppressed from 1.2 to 0.2 A.

Figure 5.25 shows the effect of the conventional beat suppression method by PWM modulation proposed in [1]. As can be seen, the amplitude of the envelope is reduced from 1.8 to 0.4 A. However, the peak-to-peak value of the DC-link voltage ripple increases from 120 to 153 V, which means the *LC* resonance is aggravated by the conventional method.

Figure 5.26 shows the FFT analysis of the *a*-phase current and DC-link voltage at the motor frequency of 74 Hz. By comparing Fig. 5.26a and b, the harmonics caused by the fluctuated DC-link voltage can be reduced effectively. Specifically, the harmonic at the frequency of 226 Hz can be reduced from 0.46 to 0.19 A, and the

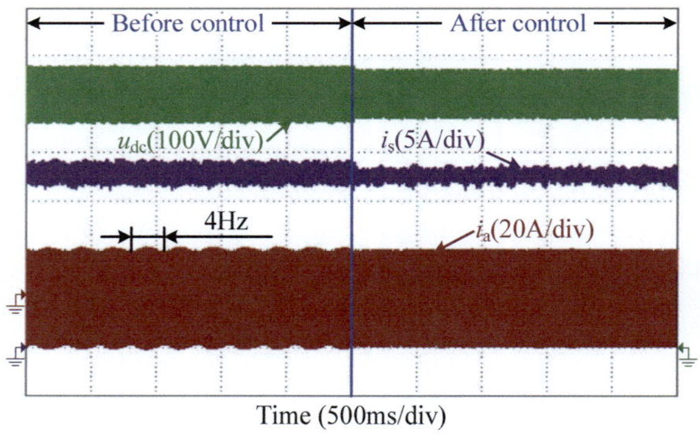

Fig. 5.24 Experimental results with the impedance reshaping at the motor frequency of 74 Hz

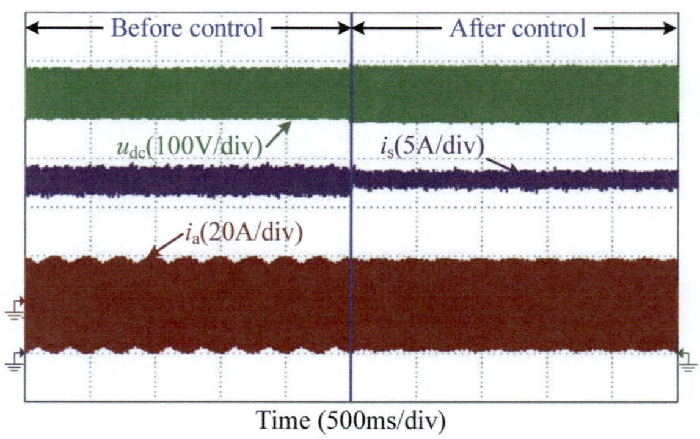

Fig. 5.25 Experimental results with the conventional method at the motor frequency of 74 Hz

harmonic at the frequency of 674 Hz can be reduced from 0.08 to 0.01 A. Figure 5.26c and d shows the FFT results of the DC-link voltage without and with the proposed method. The harmonics at the frequency of $6\omega_g$ can be reduced from 44.8 V to 31.2 V, and the harmonic at $12\omega_g$ can be reduced from 16.8 V to 11.3 V. According to the results shown in Fig. 5.26e, the conventional control will increase the harmonic at $6\omega_g$ by 23.1 V.

In order to further verify the damping performance, the FFT analysis of the grid current are shown in Fig. 5.27. With the impedance reshaping, the harmonics related to $6\omega_g$ and $12\omega_g$ can be reduced by 55% and 48% respectively.

Figure 5.28 shows the experimental results at the motor frequency of 62 Hz. As can be seen in Fig. 5.28, the amplitude of the low-frequency envelope can be suppressed from 1.25 to 0.15 A, and the spikes are the amplitude of the beat current after control, which are less obvious than those before control. The beat frequency is 10 Hz, which is determined by the frequency of fundamental current and harmonics.

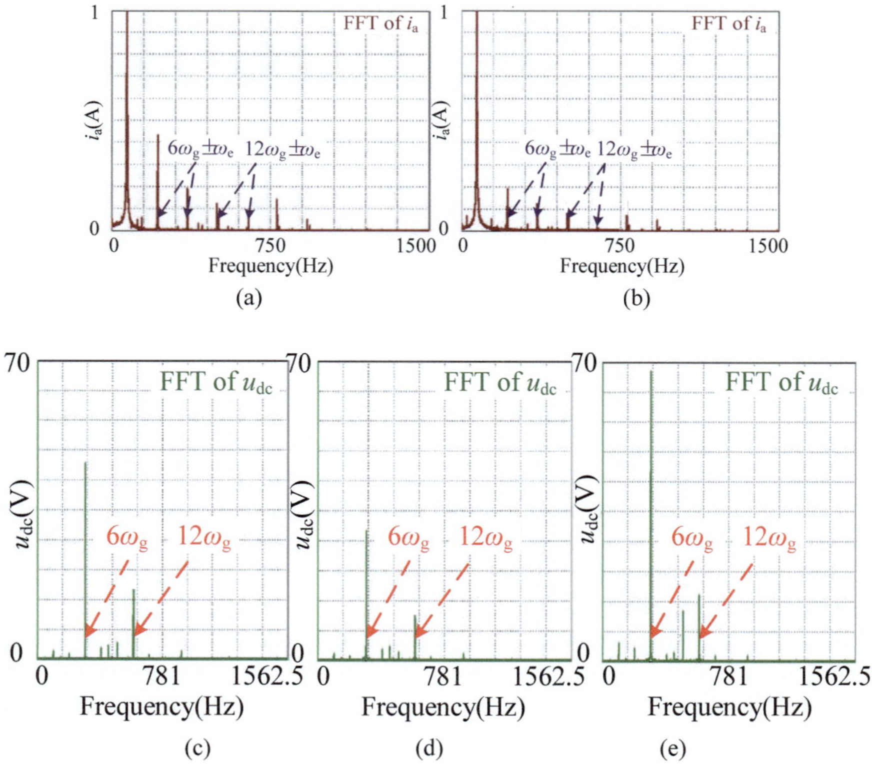

Fig. 5.26 Experimental results of harmonic analysis, reprinted from Ref. [2], copyright 2022, with permission from Nature. **a** Phase current before control. **b** Phase current after the impedance reshaping. **c** DC-link voltage before impedance reshaping. **d** DC-link voltage after impedance reshaping. **e** DC-link voltage after the conventional control

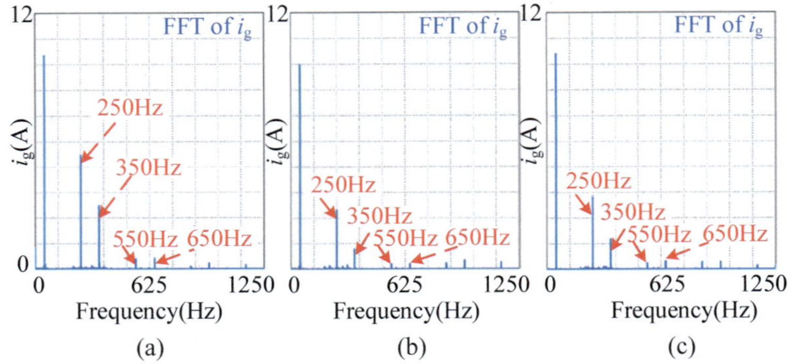

Fig. 5.27 FFT of the grid current, reprinted from Ref. [2], copyright 2022, with permission from Nature. **a** Without the control method. **b** With the proposed method. **c** With the conventional method

Fig. 5.28 Experimental results at the motor frequency of 62 Hz

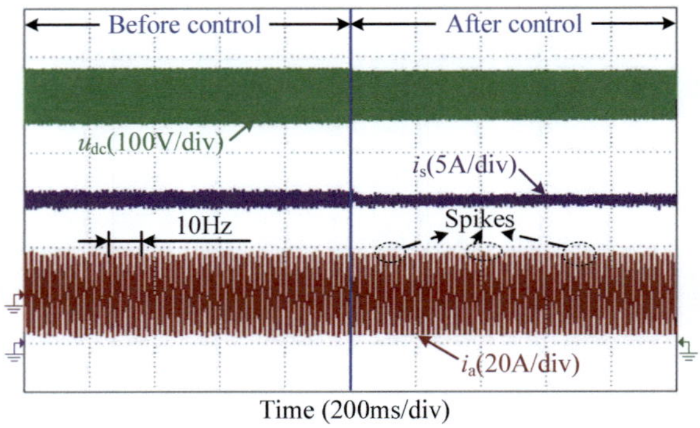

Figure 5.29 shows the amplitude of beat current under different output power conditions. The amplitude of the beat envelope can be reduced to approximate 20% after using the proposed method, and the beat phenomenon can be suppressed effectively accordingly. The current error between the theoretical results and experimental results remains lower than 0.05 A. Meanwhile, compared with the results of reduced DC-link capacitance drives with proposed method, the amplitude of the beat envelope with electrolytic capacitor is lower by 0.02 A.

Figure 5.30a shows the THD of the phase current under different output power conditions. The THD can be reduced by 1.5% after using the conventional method. And the THD of phase current with the proposed method can be further reduced by 0.5%, which improves the suppression effect to the beat phenomenon. Figure 5.30b shows the results of system efficiencies under different values of output power. The system efficiencies can be increased by over 1.5% with the proposed method in the output power between 1.7 kW and 5.2 kW.

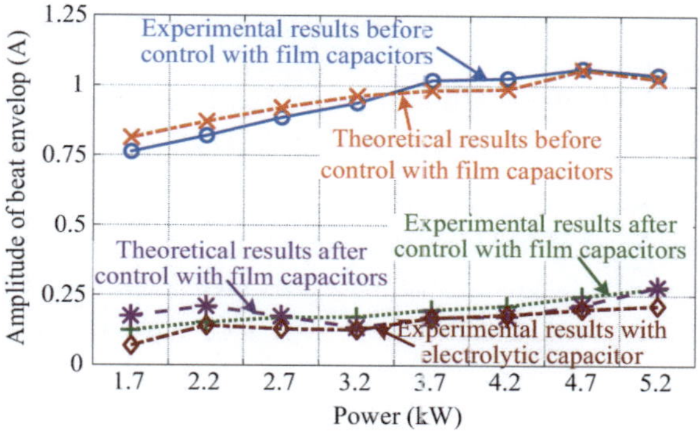

Fig. 5.29 Amplitude of beat envelope under different output power conditions

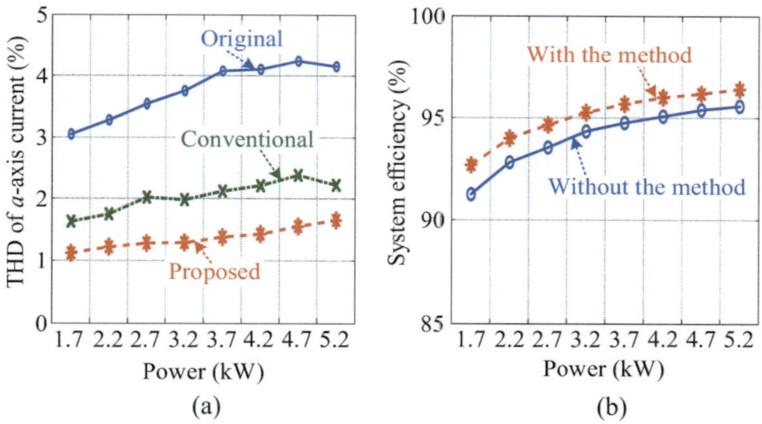

Fig. 5.30 Experimental results under different output power conditions. **a** THD of the motor currents. **b** System efficiencies

5.4 Closed-Loop Beat Suppression Method Based on Feature Signal Extraction

5.4.1 Analysis of Beat Currents Under the Coupling of d–q Axis Current Harmonics

To improve efficiency or expand the operating range, the drive system needs to apply flux-weakening (FW) control or maximum torque per ampere (MTPA) control. At this time, the d-axis current reference is negative, resulting in the mutual coupling of d–q axis current information in the beat current. The DC-link voltage usually participates in the generation of the d-axis current reference, which aggravates the distortion of the motor current and the beat phenomenon. To further enhance the effect of beat suppression, it is necessary to analyze the beat current under the coupling of d–q axis current harmonics.

According to (5.37), the constituent points I_{ap} (N) of the beat envelope of the machine-side current can be expressed as ($N = 0, 1, 2,\ldots$)

$$\begin{aligned} I_{ap}(N) = \sum_{k=1}^{\infty} \{ & I_{n+} \cos\left[12n\pi N\omega_g / \omega_e + \varphi_{hn} - \varphi_e + \varphi_{k+}\right] \\ & - I_{n-} \cos\left[12n\pi N\omega_g / \omega_e + \varphi_{hn} + \varphi_e + \varphi_{k-}\right]\} + I_{a,0}. \end{aligned} \quad (5.63)$$

Therefore, the variables of the beat envelope can be expressed as

$$I_{bn} = |I_{n+}e^{j(\varphi_{n+} - \varphi_e)} - I_{n-}e^{j(\varphi_{n-} + \varphi_e)}|, \ \omega_{bn} = 6n\omega_g - \omega_e \cdot \boldsymbol{R}(6n\omega_g / \omega_e). \quad (5.64)$$

According to Eq. (5.64), if I_{bn} can be reduced, the beat phenomenon will be effectively suppressed. However, I_{bn} is related not only to $I_{n\mp}$ and $\varphi_{n\mp}$, but also to the fundamental phase φ_e. Therefore, I_{bn} cannot be obtained by directly extracting harmonics from the phase current. Meanwhile, the frequency of $i_{an-,+}$ changes with the variation of the motor speed, further increasing the difficulty of extracting the amplitude of the beat envelope. The d–q axis currents $i_{d,q}$ can be regarded as a model composed of harmonics with a frequency of $6n\omega_g$ and average components, and can be extracted by a band-pass filter (BPF) and a low-pass filter (LPF). Therefore, I_{bn} can be characterized by the d–q axis currents

$$\begin{cases} i_{a0} = I_{d0} \cos \omega_e t - I_{q,0} \sin \omega_e t \\ i_{an\pm} = 0.5\{I_{dn} \sin[(6n\omega_g \pm \omega_e)t + \varphi_{dn}] \mp I_{qn} \cos[(6n\omega_g \pm \omega_e)t + \varphi_{qn}]\} \end{cases}. \quad (5.65)$$

According to (5.65), the harmonics of phase current can be expressed as

$$I_{n\mp}e^{j\varphi_{n\mp}} = 0.5(I_{dn}e^{j\varphi_{dn}} \mp jI_{qn}e^{j\varphi_{qn}}). \quad (5.66)$$

According to (5.64) and (5.66), the amplitude of the beat current can be expressed as

$$I_{bn} = 0.5|I_{dn}e^{j(\varphi_{dn} - \varphi_{qn})}(e^{-j\varphi_e} - e^{j\varphi_e}) + jI_{qn}(e^{-j\varphi_e} + e^{j\varphi_e})|. \quad (5.67)$$

According to (5.67), I_{bn} is determined by the amplitudes and phases of the harmonics of the d–q axis currents. If only the amplitudes of the current harmonics are suppressed, such as the solution proposed in [1], I_{bn} may not be suppressed due to the change in the phase difference φ_{dn}–φ_{qn}. With the change of φ_{dn}–φ_{qn}, the result of I_{bn} is shown in Fig. 5.31. By comparing points A and B under the condition of the same I_{q1} and I_{d1}, the change in φ_{dn}–φ_{qn} increases the amplitude of the beat current by 4.8 A. Suppose that points B and C are the points without and with the strategy in [1] applied respectively, and the dotted line shows the control trajectory of I_{b1}. I_{d1} and

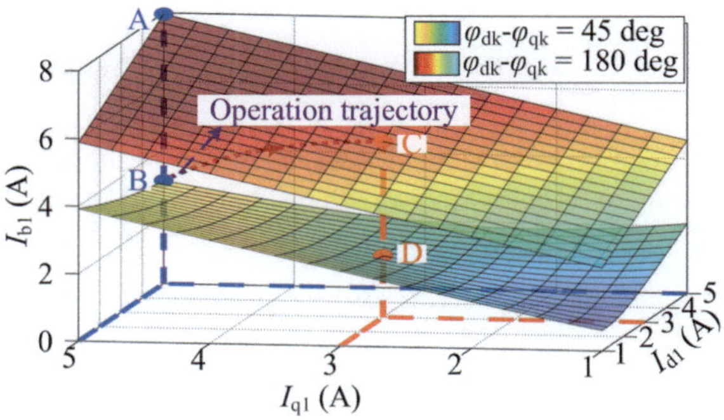

Fig. 5.31 Impact of amplitude of current harmonic on the amplitude of beat current under the condition of d–q axis phase difference variation, reprinted from Ref. [3], copyright 2022, with permission from Nature

I_{q1} are suppressed from 5 to 3 A. However, the control strategy changes φ_{dn}–φ_{qn} from 45° to 180°, increasing I_{b1} from 3.8 to 5.2 A and worsening the beat phenomenon. Therefore, suppressing the beat phenomenon by only controlling the harmonics of the d-axis or q-axis current cannot maintain the effect under the condition of current coupling.

In order to further analyze the relationship between the amplitude of the beat current and the harmonics of the d–q axis currents, the vector diagram of the relationship between the beat current and the motor current harmonics is shown in Fig. 5.32. It can be seen that the harmonics of d–q axis current vectors are $I_{dn}e^{j\varphi_{dn}}$ and $I_{dn}e^{j\varphi_{qn}}$. According to Eqs. (5.33) and (5.65), I_{d0}, I_{q0} and the fundamental wave of the motor current can be expressed as

$$e^{j\varphi_e} = (-I_{q0} + jI_{d0})/I_{a0},\ e^{-j\varphi_e} = (-I_{q0} - jI_{d0})/I_{a0}. \tag{5.68}$$

According to (5.64), (5.67) and (5.68), the beat current vector and the d–q axis current harmonic vectors can be expressed as

$$I_{bn}e^{j(\varphi_{bn}-\varphi_{hn}-\varphi_{qn})} = \left[I_{d0}I_{dn}e^{j(\varphi_{dn}-\varphi_{qn})} + I_{q0}I_{qn}\right] \Big/ \sqrt{I_{d0}^2 + I_{q0}^2}. \tag{5.69}$$

According to Eq. (5.69), I_{bn} and the harmonics of d–q axis currents can be expressed in the time domain as

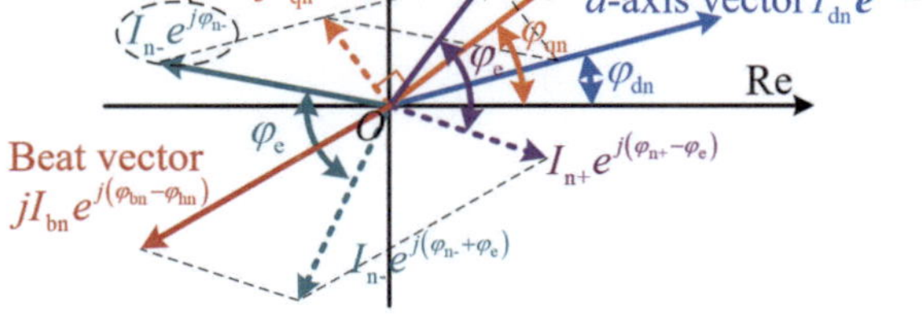

Fig. 5.32 Vector diagram of the relationship between the beat current and the motor current harmonics, reprinted from Ref. [3], copyright 2022, with permission from Nature

$$I_{\mathrm{bn}}\sin(6n\omega_{\mathrm{g}}t+\varphi_{\mathrm{bn}}-\varphi_{\mathrm{hn}})=\left[\begin{array}{l}I_{\mathrm{d0}}I_{\mathrm{dn}}\sin(6n\omega_{\mathrm{g}}t+\varphi_{\mathrm{dn}})\\+I_{\mathrm{q0}}I_{\mathrm{qn}}\sin(6n\omega_{\mathrm{g}}t+\varphi_{\mathrm{qn}})\end{array}\right]\Big/I_{\mathrm{a0}}. \tag{5.70}$$

Therefore, the characteristic signal i_{btn} of the amplitude of the beat current can be extracted through the average and harmonics of the d–q axis currents

$$i_{\mathrm{btn}}=\left(I_{\mathrm{d,}}i_{\mathrm{dn}}+I_{\mathrm{q0}}i_{\mathrm{qn}}\right)\Big/\sqrt{I_{\mathrm{d0}}^2+I_{\mathrm{q0}}^2} \tag{5.71}$$

where i_{dn} and i_{qn} are the harmonics of the d–q axis current at the frequency of $6n\omega_{\mathrm{g}}$, respectively.

Finally, the amplitude of the beat current envelope is

$$I_{\mathrm{bn}}=\sqrt{I_{\mathrm{d0}}^2I_{\mathrm{dn}}^2+I_{\mathrm{q0}}^2I_{\mathrm{qn}}^2+2I_{\mathrm{d0}}I_{\mathrm{dn}}I_{\mathrm{q0}}I_{\mathrm{qn}}\cos\left(\varphi_{\mathrm{qn}}-\varphi_{\mathrm{dn}}\right)}\Big/I_{\mathrm{a0}}. \tag{5.72}$$

Under the condition of $i_{\mathrm{d}}=0$ control, there is no coupling between the harmonics on the d–q axis, and the beat current is determined only by the harmonics of the q-axis current. Therefore, the beat phenomenon can be suppressed by the strategy proposed in [2]. However, in FW control or MTPA control, I_{d0} is negative. As shown in Eq. (5.72), at this time, there is a coupling phenomenon between the d-axis current harmonics and the q-axis current in the beat current, which leads to the deterioration of the beat phenomenon. The comparison block diagram between the beat current and the characteristic signal is shown in Fig. 5.33. The frequency of the characteristic signal i_{bt1} is equal to the fundamental frequency $6\omega_{\mathrm{g}}$ of the DC-link voltage, and its amplitude is equal to I_{b1}. Therefore, even if there is coupling among the harmonics in the d–q axis, the characteristic signal can still represent the amplitude of the beat current.

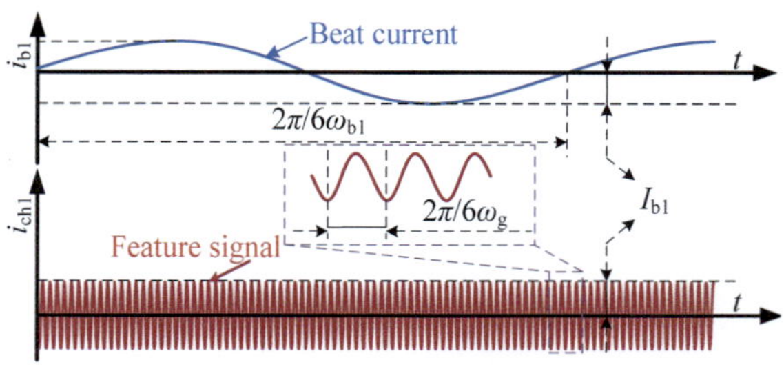

Fig. 5.33 Comparison between the beat current and the characteristic signal

5.4.2 *Method Based on Harmonic Decoupling and Beat Characteristic Signal Extraction*

The control block diagram of the beat suppression strategy based on the closed-loop of characteristic signals is shown in Fig. 5.34. The dominant harmonics and averages ($i_{d,q1}$, $i_{d,q2}$, I_{d0} and I_{q0}) of the d–q axis currents are extracted by the Band Pass Filter (BPF) and the Low Pass Filter (LPF) respectively, and the characteristic signals i_{bt1} and i_{bt2} with frequencies of $6\omega_g$ and $12\omega_g$ are generated based on Eq. (5.71). The output of the control loop is the voltage regulation angle $\Delta\theta_u$, which is superimposed on the estimated position of the motor. The adjusted voltage angle is θ_{un}, and the control is carried out through the iPark transformation of the voltage reference. The advantage of the closed-loop of the characteristic signal of the amplitude of beat envelope is that it can achieve direct control over the beat phenomenon by eliminating the coupling of the d–q axis current harmonics in the beat envelope.

The feedback Δi_{bt} of the control loop includes i_{bt1} and i_{bt2}, which can be expressed as

$$\Delta i_{ch} = i_{ch1} + i_{ch2} = \left(I_{d0}\Delta i_d + I_{q0}\Delta i_q\right)\left(\boldsymbol{G}_{BPF1} + \boldsymbol{G}_{BPF2}\right)/I_{a0}. \tag{5.73}$$

According to Fig. 5.33, the characteristic signal is a periodic sinusoidal signal. Therefore, in the closed loop of Δi_{bt}, better performance can be obtained by using

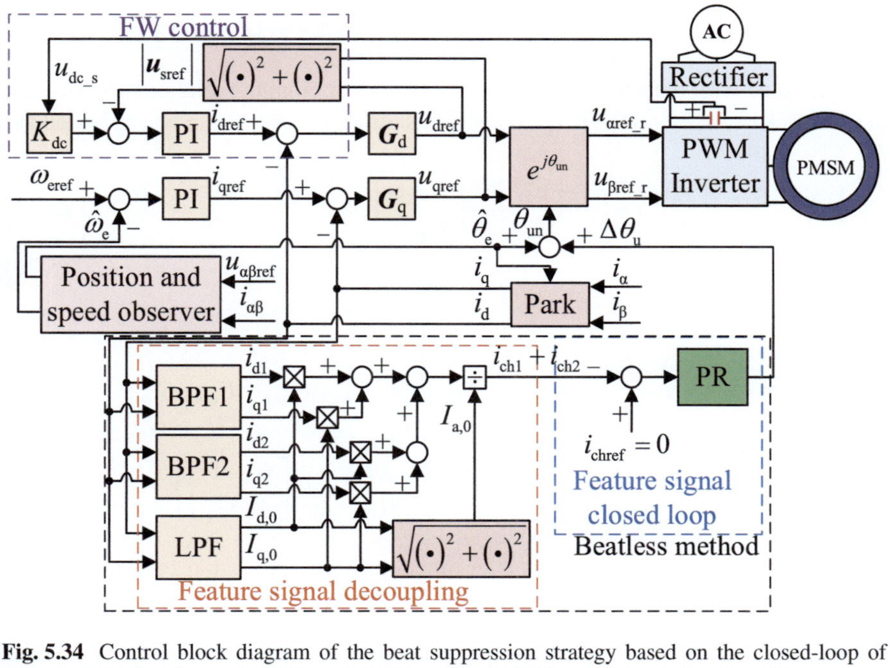

Fig. 5.34 Control block diagram of the beat suppression strategy based on the closed-loop of characteristic signals, reprinted from Ref. [3], copyright 2022, with permission from Nature

proportional-resonant (PR) control than by using proportional integral (PI) control. The resonant frequency is determined by the internal model of the sinusoidal signal and is set to $6\omega_g$ and $12\omega_g$. Hence, its transfer function is the same as that in Eq. (3.60).

According to Eq. (3.60), the controller transfer function $\boldsymbol{G}_{PR}$ contains the internal models of the characteristic signals i_{bt1} and i_{bt2}. According to the internal model theory, the control loop can perform high-precision control on the beat amplitude. The relationship between $\Delta\theta_u$ and the d–q axis currents is as follows

$$\begin{cases} \Delta\theta_u = \boldsymbol{G}_{md} e^{-0.5sT_s} \Delta i_d + \boldsymbol{G}_{mq} e^{-0.5sT_s} \Delta i_q \\ \boldsymbol{G}_{mdq} = -I_{dq,0} \boldsymbol{G}_{PR} (\boldsymbol{G}_{BPF1} + \boldsymbol{G}_{BPF2}) / I_{a0} \end{cases}. \tag{5.74}$$

As shown in Eq. (5.74), the output of the PR controller depends on the difference between i_{btref} and $i_{bt1} + i_{bt2}$. According to (3.63), regulating the voltage angle is equivalent to injecting additional small signals $-U_{qref0}\Delta\theta_u$ and $U_{dref0}\Delta\theta_u$ into the voltage angle. The injected additional voltage signals are used to reshape the impedance relationship between the amplitude of beat current and the rectified voltage.

After the angle compensation strategy is applied according to Eq. (2.14), the small signals of actual voltages of the d–q axis can be expressed as

$$\begin{aligned} \begin{bmatrix} \Delta u_d \\ \Delta u_q \end{bmatrix} = \boldsymbol{C}_0 \Bigg\{ & \frac{\left(1 - e^{-1.5sT_s}\right)\Delta u_{dc}}{U_{dc0}} \begin{bmatrix} U_{dref0} \\ U_{qref0} \end{bmatrix} \\ & + e^{-sT_s} \begin{bmatrix} \Delta u_{dref} - U_{qref0} e^{-0.5sT_s} (\boldsymbol{G}_{md}\Delta i_d + \boldsymbol{G}_{mq}\Delta i_q) \\ \Delta u_{qref} + U_{dref0} e^{-0.5sT_s} (\boldsymbol{G}_{md}\Delta i_d + \boldsymbol{G}_{mq}\Delta i_q) \end{bmatrix} \Bigg\}. \end{aligned} \tag{5.75}$$

Combined with the impedance model in the FW region, the relationship between the DC-link voltage and the d–q axis currents after the control can be expressed as

$$\begin{bmatrix} \boldsymbol{A}_{1dfw} & \boldsymbol{A}_{1qfw} \\ \boldsymbol{A}_{2dfw} & \boldsymbol{A}_{2qfw} \end{bmatrix} \begin{bmatrix} \Delta i_d \\ \Delta i_q \end{bmatrix} = \left\{ \frac{1 - e^{-1.5sT_s}}{U_{dc,0}} \begin{bmatrix} U_{dref0} \\ U_{qref0} \end{bmatrix} + \begin{bmatrix} K_{dc} U_{sref0} e^{-1.5sT_s} \boldsymbol{G}_{dc} \\ 0 \end{bmatrix} \right\} \Delta u_{dc} \tag{5.76}$$

where

$$\begin{aligned} \begin{bmatrix} \boldsymbol{A}_{1dfw} & \boldsymbol{A}_{1qfw} \\ \boldsymbol{A}_{2dfw} & \boldsymbol{A}_{2qfw} \end{bmatrix} = \boldsymbol{C}_0^{-1}\boldsymbol{C}_1 + e^{-1.5sT_s} \Bigg\{ & \begin{bmatrix} U_{qref0}\boldsymbol{G}_{md} & U_{qref0}\boldsymbol{G}_{mq} \\ -U_{dref0}\boldsymbol{G}_{md} & -U_{dref0}\boldsymbol{G}_{mq} \end{bmatrix} \\ & -\boldsymbol{G}_{dc} \begin{bmatrix} U_{dref0}\boldsymbol{G}_d & U_{qref0}\boldsymbol{G}_q \\ 0 & 0 \end{bmatrix} + \begin{bmatrix} \boldsymbol{G}_d & 0 \\ 0 & \boldsymbol{G}_q \end{bmatrix} \Bigg\}. \end{aligned} \tag{5.77}$$

Define $\boldsymbol{W}_{dfw}$ and $\boldsymbol{W}_{qfw}$ as the transfer functions between the d–q axis currents and the DC-link voltage after control

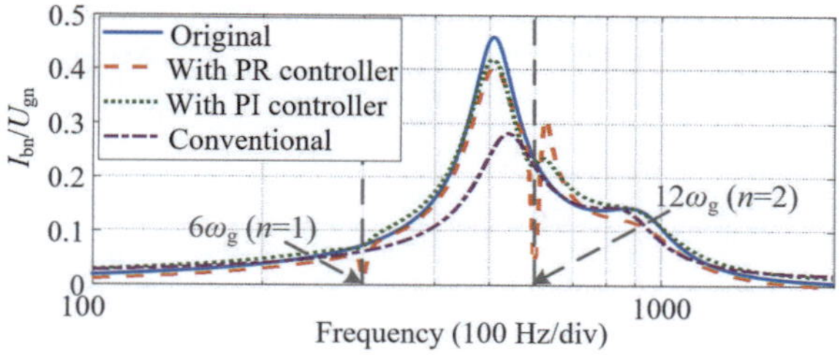

Fig. 5.35 $I_{bk}/U_{grec,k}$ before and after applying the beat suppression strategy

$$\begin{bmatrix} \boldsymbol{W}_{dfw} \\ \boldsymbol{W}_{qfw} \end{bmatrix} = \frac{1}{\Delta u_{dc}} \begin{bmatrix} \Delta i_d \\ \Delta i_q \end{bmatrix} = \frac{1}{\boldsymbol{A}_{2qfw}\boldsymbol{A}_{1dfw} - \boldsymbol{A}_{2dfw}\boldsymbol{A}_{1qfw}} \left\{ K_{dc} U_{sref0} e^{-1.5sT_s} \cdot \begin{bmatrix} \boldsymbol{G}_{dc}\boldsymbol{A}_{2qfw} \\ -\boldsymbol{G}_{dc}\boldsymbol{A}_{2dfw} \end{bmatrix} + \frac{1 - e^{-1.5sT_s}}{U_{dc,0}} \begin{bmatrix} U_{dref0}\boldsymbol{A}_{2qfw} - U_{qref0}\boldsymbol{A}_{1qfw} \\ U_{qref0}\boldsymbol{A}_{1dfw} - U_{dref0}\boldsymbol{A}_{2dfw} \end{bmatrix} \right\}. \tag{5.78}$$

In Eq. (5.78), the transfer functions $\boldsymbol{G}_{md}$ and $\boldsymbol{G}_{mq}$ can reshape the impedance relationships of Δi_d, Δi_q and Δu_{dc}, and can affect the ratio of I_{bk} and Δu_{grec}.

$\boldsymbol{K}_{mfw}$ is defined as the transfer function of the rectified voltage and the DC-link voltage after control is applied. Therefore, the ratio of the beat current amplitude I_{bk} and the harmonic of rectified voltage after control can be expressed as

$$I_{bn}/U_{gn} = |I_{d0}\boldsymbol{K}_{mfwn}\boldsymbol{W}_{dfwn} + I_{q0}\boldsymbol{K}_{mfwn}\boldsymbol{W}_{qfwn}|/I_{a0} \tag{5.79}$$

where $\boldsymbol{W}_{d,qfwk}$ and $\boldsymbol{K}_{mfwk}$ represent $\boldsymbol{W}_{d,qfw}(j6n\omega_g)$ and $\boldsymbol{K}_{mfw}(j6n\omega_g)$.

The ratio of I_{bn}/U_{gn} before and after applying the beat suppression strategy is shown in Fig. 5.35. By comparing the solid line and the dotted line, the change in the ratio at $6\omega_g$ and $12\omega_g$ is less than 0.01, indicating that the traditional method cannot eliminate the influence of the d-axis current harmonic on the beat phenomenon. After using PI control, the ratio of I_{bn}/U_{gn} at $6\omega_g$ and $12\omega_g$ can only be reduced by no more than 0.01, which shows that the PI controller cannot effectively suppress the beat phenomenon. After using PR control, the ratio of I_{bn}/U_{gn} at the same frequencies can be reduced by 0.075 and 0.22, respectively. Therefore, under the condition that there is d–q axis currents coupling in the beat current, the suppression strategy based on characteristic signal extraction has better performance than traditional strategies.

5.4.3 Analysis on the Adaptability of the Strategy to Changes in Working Conditions

Closed-loop control has the ability of feedback adjustment, which improves the adaptability of the drive system. According to the power balance of the inverter, the transfer function $\boldsymbol{Y}_{mfw}$ of the input admittance of the inverter after the control is applied can be expressed as

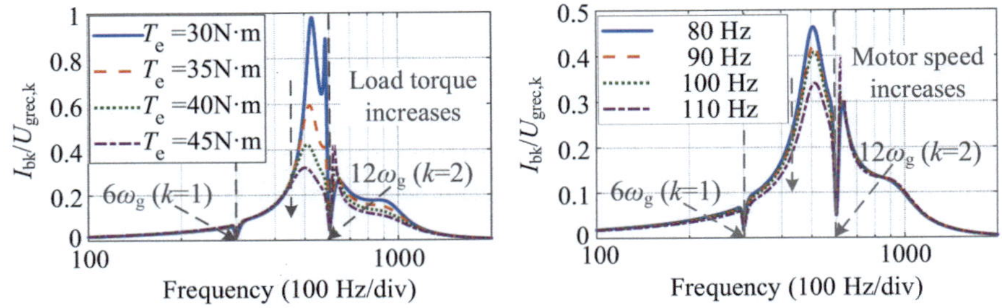

(a) Different load torques (b) Different motor frequencies

Fig. 5.36 Ratio between the beat current and the rectified voltage under different operating conditions, reprinted from Ref. [3], copyright 2022, with permission from Nature

$$\begin{aligned}\boldsymbol{Y}_{\text{mfw}} = \Delta i_{\text{inv}} / \Delta u_{\text{dc}} = 3\boldsymbol{W}_{\text{mfw}}\left(U_{\text{d0}} + \omega_{\text{e}} L_{\text{d}} I_{\text{q0}} + L_{\text{d}} I_{\text{d0}} s + I_{\text{d0}} R_s\right) / 2U_{\text{dc0}} \\ + 3\boldsymbol{W}_{\text{mfw}}\left(U_{\text{q0}} - \omega_{\text{e}} L_{\text{q}} I_{\text{d0}} + L_{\text{q}} I_{\text{q0}} s + I_{\text{q0}} R_{\text{s}}\right) / 2U_{\text{dc0}} - I_{\text{inv0}} / U_{\text{dc0}}.\end{aligned} \quad (5.80)$$

Therefore, $\boldsymbol{K}_{\text{mfw}}$ can be expressed as

$$\boldsymbol{K}_{\text{mfw}} = \Delta u_{\text{dc}} / \Delta u_{\text{g}} = 1/[1 + (L_{\text{g}} s + R_{\text{g}})(\boldsymbol{Y}_{\text{mfw}} + C_{\text{dc}} s)]. \quad (5.81)$$

$\boldsymbol{W}_{\text{dk}}$, $\boldsymbol{W}_{\text{qk}}$ and $\boldsymbol{K}_{\text{mk}}$ contain system parameters and operating variables. The influence of operating conditions on the control strategy can be verified according to (5.79). The ratio of $I_{\text{bn}}/U_{\text{gn}}$ under different load torques and motor frequencies is shown in Fig. 5.36. When T_{e} and ω_{e} vary within the ranges of 30–45 N m and 80–110 Hz, respectively, the control strategy based on the closed loop of characteristic signals can keep the ratio of $I_{\text{bn}}/U_{\text{gn}}$ unchanged at $6\omega_{\text{g}}$, and the variation of $I_{\text{bn}}/U_{\text{gn}}$ at $12\omega_{\text{g}}$ is less than 0.01. The proposed strategy can maintain its effectiveness under different load torques and operating frequencies.

According to the FFT of u_{grec}, the beat current caused by the DC-link voltage fluctuation can be expressed as

$$I_{\text{bk}} = 6U_{\text{g}} |I_{\text{d,0}} \boldsymbol{K}_{\text{mfwk}} \boldsymbol{W}_{\text{dfwk}} + I_{\text{q,0}} \boldsymbol{K}_{\text{mfwk}} \boldsymbol{W}_{\text{qfwk}}| / \pi (36k^2 - 1) I_{\text{a,0}}. \quad (5.82)$$

Figure 5.37 shows the amplitudes of the beat envelope under different operating conditions caused by the DC-link voltage harmonics. By comparing the two surfaces shown in Fig. 5.37a, the control strategy can effectively reduce I_{b1}. When the motor frequency and load torque vary within the ranges of 85–100 Hz and 30–45 N m, respectively, the change in the amplitude of the beat envelope is less than 0.05 A. The control strategy can also effectively reduce the amplitude of the beat envelope related to $12\omega_{\text{g}}$, as shown in Fig. 5.37b. Therefore, the proposed control strategy can maintain its effectiveness within a wide range of speeds and torques.

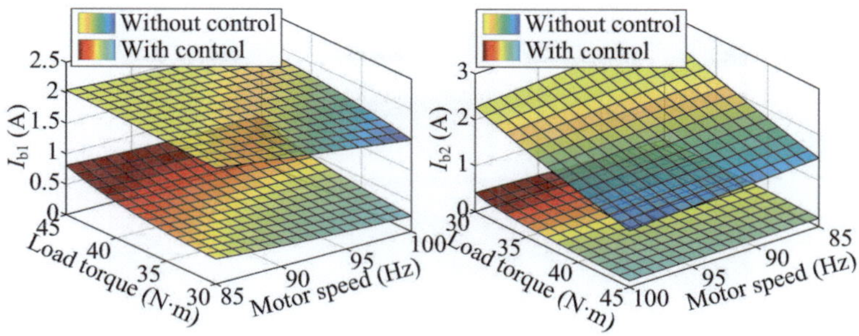

(a) Amplitude of beat current I_{b1}. (b) Amplitude of beat current I_{b2}

Fig. 5.37 Harmonics of beat current under different operating conditions, reprinted from Ref. [3], copyright 2022, with permission from Nature

5.4.4 Experimental Verification

Figure 5.38 shows the experimental results when the motor frequency is 99 Hz. The waveforms from top to bottom are the DC-link voltage, the beat characteristic signal, the d-axis current reference and the a-phase current. The beat frequency is 3 Hz, and the amplitude of the low-frequency envelope of the current can be suppressed from 2.00 to 0.32 A. Figure 5.38b and c are enlarged views of the experimental results before and after applying impedance reshaping. Among them, the average of the d-axis current reference i_{dref} is −3 A, indicating that the system is operating in the FW region. The frequency of the characteristic signal is the same as that of the DC-link pulsation, and the amplitude is equal to the amplitude of the beat envelope.

Figure 5.39 shows the FFT analysis of the a-phase current and the characteristic signal when the motor frequency is 99 Hz. As shown in Fig. 5.39a, the a-phase current contains harmonics caused by the DC-link voltage ripple with frequencies of 201, 399, 501 and 699 Hz, as well as harmonics generated by the inverter switching process with frequencies of 297, 495 and 697 Hz. It can be seen from Fig. 5.39b that frequency coupling generates harmonics at 195, 397, 597 and 795 Hz. By comparing Fig. 5.39a and c, the control strategy can effectively suppress the above three types of harmonics. The harmonics at 201 Hz and 399 Hz are reduced from 0.62 A and 0.42 A to 0.18 A and 0.15 A, respectively. The harmonics at 501 Hz and 699 Hz are reduced from 0.38 A and 0.18 A to 0.12 A and 0.11 A, respectively. The harmonics at 495 Hz and 697 Hz are reduced by 0.28 A and 0.15 A, respectively. The harmonics at 195 Hz and 795 Hz are reduced by 0.21 A and 0.09 A, respectively. Figure 5.39e and f show the FFT of the characteristic signal before and after applying the control strategy, where the harmonics at $6\omega_g$ and $12\omega_g$ can be reduced from 1.0 A and 0.50 A to 0.16 A and 0.20 A, respectively.

To verify the effectiveness of the strategy under the $i_d = 0$ control, Fig. 5.40 shows the experimental results when the motor frequency is 85 Hz. The frequencies of the beat currents caused by the DC-link voltage harmonics at $6\omega_g$ and $12\omega_g$ are 40 Hz

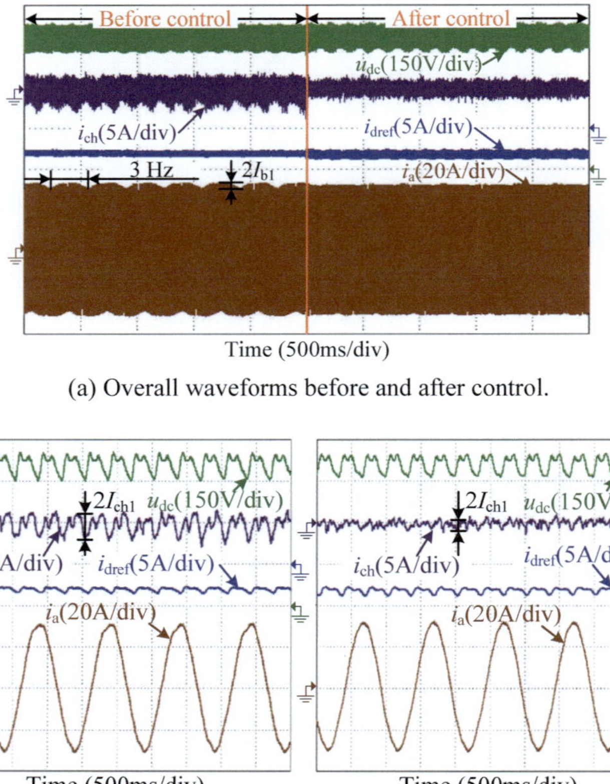

(a) Overall waveforms before and after control.

(b) Enlarged view before control. (c) Enlarged view after control.

Fig. 5.38 Experimental results of impedance reshaping when the motor frequency is 99 Hz, reprinted from Ref. [3], copyright 2022, with permission from Nature

and 5 Hz respectively. The low-frequency envelope is mainly generated by the $12\omega_g$ harmonic, and its amplitude can be suppressed from 1.25 A to 0.15 A.

Figure 5.41a and c show the effectiveness of the traditional method proposed in Ref. [1]. As shown in Fig. 5.41a, under the id = 0 control, the amplitude of the beat characteristic signal is equal to the amplitude of the q-axis current harmonic. By suppressing the q-axis current harmonic, the amplitude of the beat frequency current can be reduced from 1.25 to 0.46 A. However, in Fig. 5.41c, the traditional strategy increases the amplitude of the d-axis current fluctuation from 1.50 to 2.70 A, causing the beat envelope to increase to 2.40 A. Therefore, in the case of d–q axis current coupling, suppressing the current harmonics in the q-axis or d-axis alone cannot achieve effective beat suppression. Figure 5.41b and d show the influence of the strategy based on feature extraction on current harmonics. By directly controlling the beat characteristic signal, under the id = 0 control and FW control, the amplitude of the beat frequency envelope can be suppressed to 0.15 A and 0.32 A, respectively. This

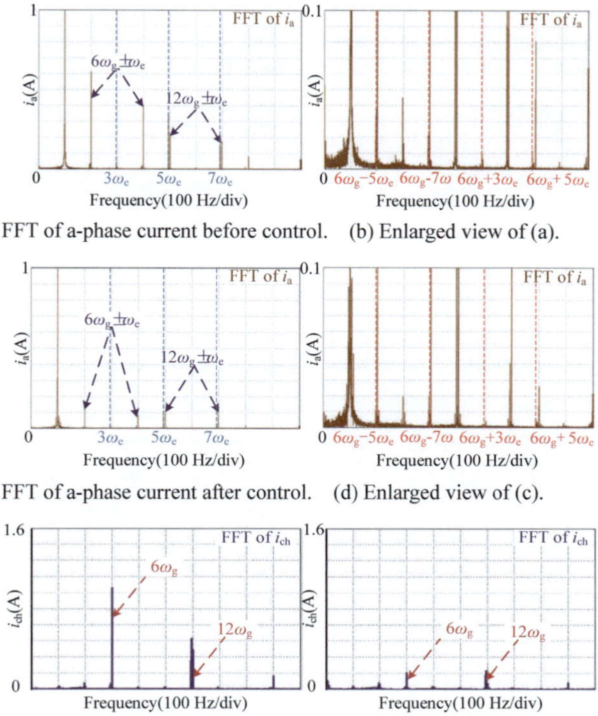

(a) FFT of a-phase current before control. (b) Enlarged view of (a).

(c) FFT of a-phase current after control. (d) Enlarged view of (c).

(e) FFT of the characteristic signal before control. (f) FFT of the characteristic signal after impedance reshaping.

Fig. 5.39 Experimental results of harmonic analysis, reprinted from Ref. [3], copyright 2022, with permission from Nature

Fig. 5.40 Experimental results of adopting the traditional strategy when the motor frequency is 85 Hz

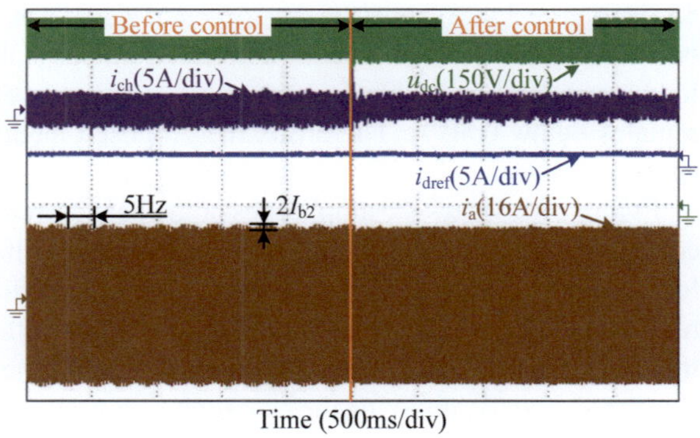

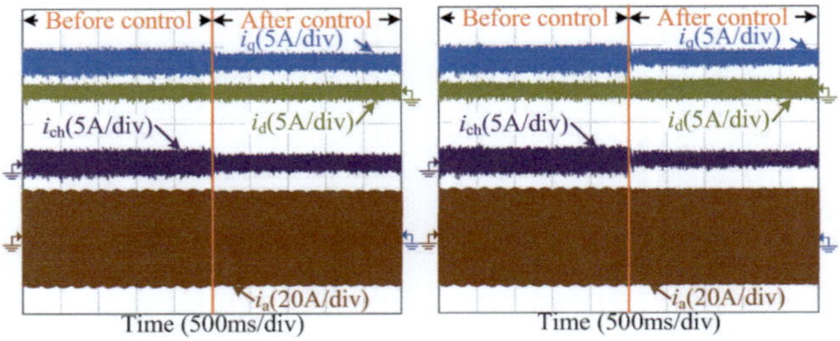

(a) Applying the traditional control strategy under the id = 0 control. (b) Applying the impedance reshaping strategy under the id = 0 control.

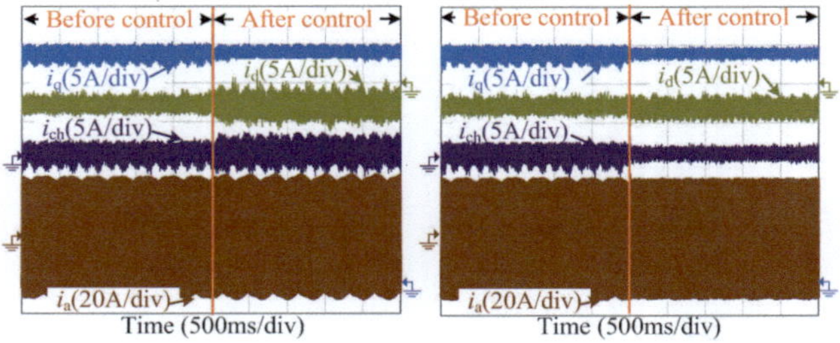

(c) Applying the traditional control strategy under the FW control. (d) Applying the impedance reshaping strategy under the FW control.

Fig. 5.41 Experimental results of applying different control strategies

indicates that even when the d–q axis current harmonics are coupled, the closed-loop strategy for the beat characteristic signal is still effective.

The experimental waveforms with the change of rotational speed are shown in Fig. 5.42a. The operating frequency changes from 85 to 99 Hz and then back to 85 Hz. Figure 5.42b shows the experimental waveforms when the load torque changes between 38 and 45 N m. The motor speed remains stable, and during the changes of load torque and operating frequency, the amplitude of the low-frequency envelope remains below 0.40 A. Therefore, the proposed method is adaptable to torque and speed pulsations.

Figure 5.43 shows the amplitudes of the beat current under different output power conditions. After applying the improved topology, the amplitude of the beat envelope can be suppressed by 50%. Outside the FW region, the traditional method can reduce the amplitude by about 60%. However, when the direct current component of the d-axis current is negative, the control effect of the traditional method is significantly reduced. By applying the beat suppression strategy based on feature extraction, regardless of whether the dq-axis current harmonics are coupled or not, the amplitude of the beat envelope is reduced to below 20%. Meanwhile, compared with the experimental results of the electrolytic capacitorless drive system with the proposed method applied, the amplitude of the beat envelope of the electrolytic capacitor drive system is decreased by 0.08 A.

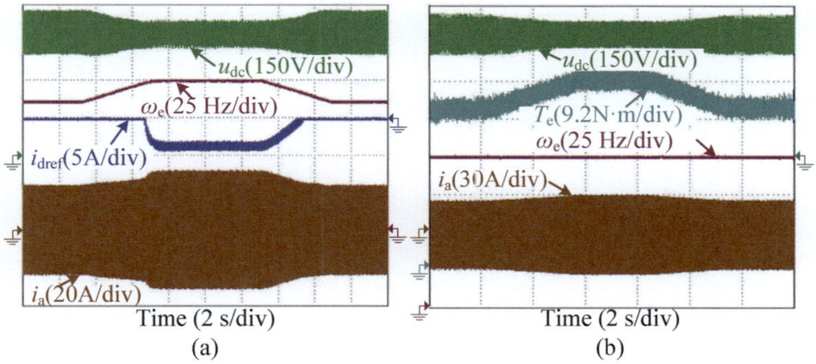

(a) Speed change. (b) Torque change.

Fig. 5.42 Dynamic experimental waveforms of the control strategy

Fig. 5.43 Amplitude of the beat envelope under different output powers

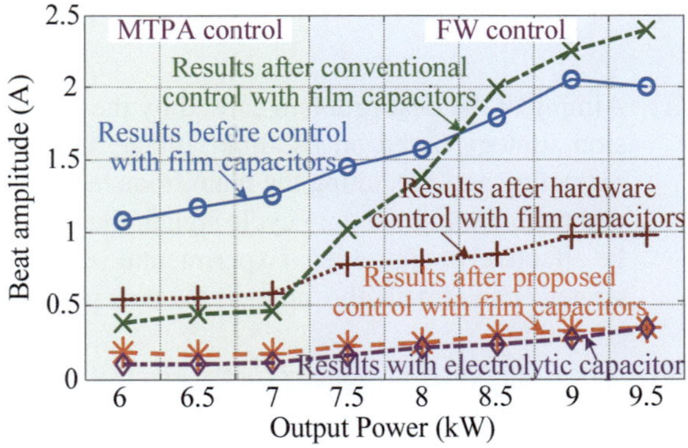

Figure 5.44 shows the copper loss and system efficiency under different output power conditions. Outside the FW region, both the traditional control strategy and the proposed control strategy can reduce the copper loss by more than 8 W and increase the efficiency by more than 1.5%. However, when the average of the d-axis current is negative, the effect of the traditional control strategy is significantly reduced. By adopting the proposed control strategy, under the coupling of d–q axis current harmonics, the system efficiency is increased by 1.8%.

5.5 Summary

Based on revealing the generation mechanism of the beat phenomenon in the electrolytic capacitorless PMSM drive system, this paper has studied a full-speed range suppression method for various beats to reduce the current harmonics on the motor side. The main work is summarized as follows

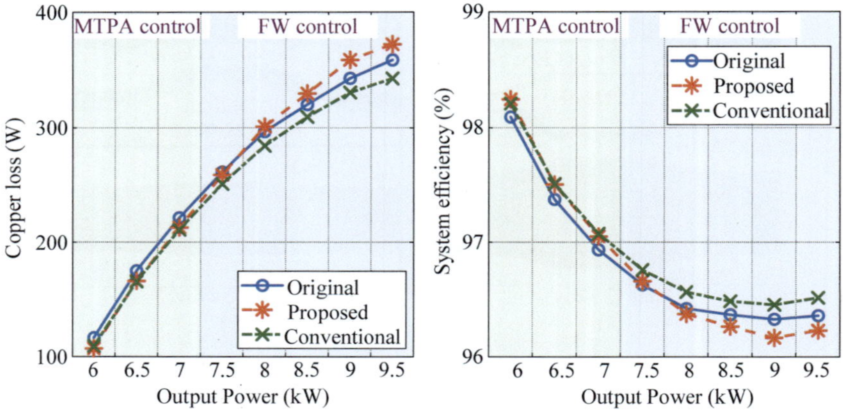

(a) Copper loss before and after control. (b) System efficiency before and after control.

Fig. 5.44 Experimental results under different output power conditions

(1) Aiming at the beat problem caused by the modulation process, a beat suppression strategy based on DC-link voltage harmonic prediction is proposed. By extracting and predicting the main frequency harmonics in the DC-link voltage and regenerating the duty cycle signal, the beat phenomenon of the system can be effectively suppressed. Experimental results show that the proposed control strategy can effectively suppress the beat phenomenon under different working conditions.
(2) A beat suppression strategy for the electrolytic capacitorless drive system based on impedance reshaping is proposed. The harmonics in the beat current are analyzed from the perspective of impedance. On the basis of the impedance model, the adjustment current is generated by modifying the current angle, and the beat phenomenon is suppressed by reshaping the impedance of the PMSM drive system under the dominant harmonics. By generating the adjustment current at the dominant harmonics of the motor current vector, the control strategy is not affected by the changes in the steady-state values of torque and speed, thus ensuring the effectiveness of beat suppression within a wider operating range. Experimental results show that the proposed control strategy can suppress the harmonics of the grid-side current and DC-link voltage while improving the power quality on the motor side.
(3) A strategy for extracting and suppressing the beat characteristics on the motor side based on the decoupling of d–q axis current harmonics is proposed to suppress the beat phenomenon under the condition of harmonic coupling of the d–q axis currents. A beat suppression method with a closed-loop of the beat amplitude is adopted to improve the adaptability to the motor speed and torque. The characteristic signals of the average of the d–q axis currents and the harmonics are used to reconstruct the beat current amplitude, which is then

used to reshape the impedance relationship between the beat current amplitude and the input side, realizing the direct suppression of the amplitude of the beat envelope. Experimental results show that the proposed control strategy can effectively suppress the beat phenomenon both inside and outside the FW region.

References

1. D. Ding, N. Zhao, G. Wang, G. Zhang, X. Zhang, N. Mijatovic and D. Xu. Suppression of beat phenomenon for electrolytic capacitorless motor drives accounting for sampling delay of DC-link voltage [J]. IEEE Transactions on Industrial Electronics, 2022, 69(2): 1167–1176.
2. D. Ding, R. Gao, Q. Wang, B. Li, G. Zhang, G. Wang, L. Wu, and D. Xu. Beatless Control Strategy Based on Impedance Reshaping for PMSM Drives With Small DC-Link Capacitors [J]. IEEE Transactions on Industrial Electronics, 2024, 41(7): 6829–6840.
3. R. Gao, Z. Fu, D. Ding, G. Wang, G. Zhang and D. Xu, Beat Phenomenon Feature Extraction Based Harmonic Decoupling for Electrolytic Capacitorless Motor Drives[J]. IEEE Transactions on Industrial Electronics, 2025, 72(2): 1193–1204.

Chapter 6
Overmodulation Methods with Low Distortion

6.1 Introduction

Influenced by the periodic fluctuation of DC-link voltage, the stator voltage distortion phenomenon occurs in the electrolytic capacitor-less motor drive system when using the traditional overmodulation strategy. Therefore, the THD of motor current is increased and the utilization of DC-link voltage is reduced. In this chapter, the switching aliasing caused by the fluctuation of DC-link voltage and reference voltage vector is solved by optimizing the overmodulation voltage boundary, and the operation effect of the drive system in the overmodulation region is improved.

6.2 Analysis of Voltage Distortion Caused by Conventional Dual-Mode Overmodulation

6.2.1 Analysis of the Fluctuation of Voltage Reference

The influence of DC-link voltage fluctuation on voltage boundary hexagon is shown as Fig. 6.1. As can be seen, the size of the hexagon voltage boundary changes with the fluctuation of the DC-link voltage [1, 2]. And when the motor operates in the overmodulation region, additional harmonics will be generated, affecting the operation performance of the motor drive [3].

In the overmodulation regions, limited by voltage output, the modulation index fluctuation cannot compensate for the DC-link voltage fluctuation, resulting in the output voltage vector not being equal to the reference voltage vector. Thus, the amplitude of the reference of voltage vector $|\boldsymbol{u}_{sref}|$ fluctuates at the same frequency, which can be expressed as [4]

D. Ding et al., *Electrolytic Capacitorless Permanent Magnet Synchronous Motor Drives*, https://doi.org/10.1007/978-981-95-5601-4_6

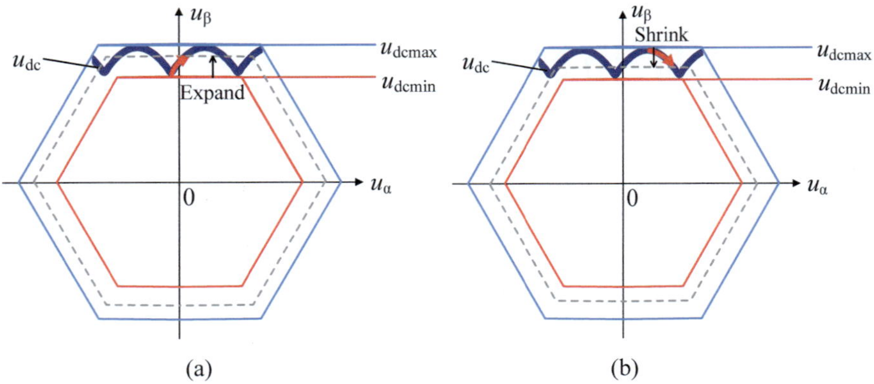

Fig. 6.1 Influence of DC-link voltage fluctuation on voltage boundary hexagon. **a** When DC-link voltage increases. **b** When DC-link voltage decreases

$$|\boldsymbol{u}_{\text{sref}}| = U_{\text{sref0}} + \sum_{n=1}^{\infty} U_{\text{sref\_n}} \sin(6n\omega_g t + \theta_{\text{un}}) \tag{6.1}$$

where $U_{\text{sref\_0}}$, $U_{\text{sref\_n}}$, and θ_{un} are the DC component of reference voltage vector, the amplitude and the phase of the nth reference voltage vector harmonic, respectively.

Aiming at the limitations of the conventional dual-mode overmodulation control strategy in the electrolytic capacitorless drive system, the proposed overmodulation strategy based on fixed voltage boundary sets the switching boundary between the overmodulation region I and the overmodulation region II with the maximum value of the DC-link voltage u_{dcmax}. After setting the fixed voltage boundary, the trajectories of the voltage reference and output voltage vectors when $\boldsymbol{u}_{\text{sref}}$ has periodic fluctuations are shown in Fig. 6.2. Among them, the changing region constantly changes between the overmodulation region I and the linear region due to the fluctuation of the DC-link voltage; $\boldsymbol{u}_{\text{sref}}(t_1) - \boldsymbol{u}_{\text{sref}}(t_7)$ are the voltage vectors reference at different moments, and $\boldsymbol{u}_{\text{s}}(t_1) - \boldsymbol{u}_{\text{s}}(t_7)$ are the output voltage vectors at different moments.

In Fig. 6.2a, $\boldsymbol{u}_{\text{sref}}(t)$ is located on the boundary of overmodulation region I. At the time of t_1, t_2, t_4, t_6, and t_7, $\boldsymbol{u}_{\text{sref1}}$ is in overmodulation region I, and the output voltage vector is limited to the boundary of the voltage hexagon. At the time of t_3 and t_5, $\boldsymbol{u}_{\text{sref1}}$ is in overmodulation region II, and the output voltage vector is limited to the vertices of the voltage hexagon. From the trajectory of $\boldsymbol{u}_{\text{s1}}$, the periodic fluctuation of the voltage vector reference leads to switching aliasing. For example, $\boldsymbol{u}_{\text{sref}}$ locates in OVM region I at the time of t_2, but it jumps into OVM region II at the time of t_3 directly, which does not make full use of the output voltage capacity in OVM region I. In Fig. 6.2b, $\boldsymbol{u}_{\text{sref2}}(t)$ locates in the boundary of overmodulation region II. At the time of t_1, t_2, t_3, t_5, t_6, and t_7, $\boldsymbol{u}_{\text{sref2}}$ is in overmodulation region II. At the time of t_4, $\boldsymbol{u}_{\text{sref2}}$ is in overmodulation region I. $\boldsymbol{u}_{\text{sref}}$ locates in OVM region II at the time of t_3, but it jumps back to OVM region I at the time of t_4, which does not make full use of the output voltage capacity in OVM region II to enter the six-step operation.

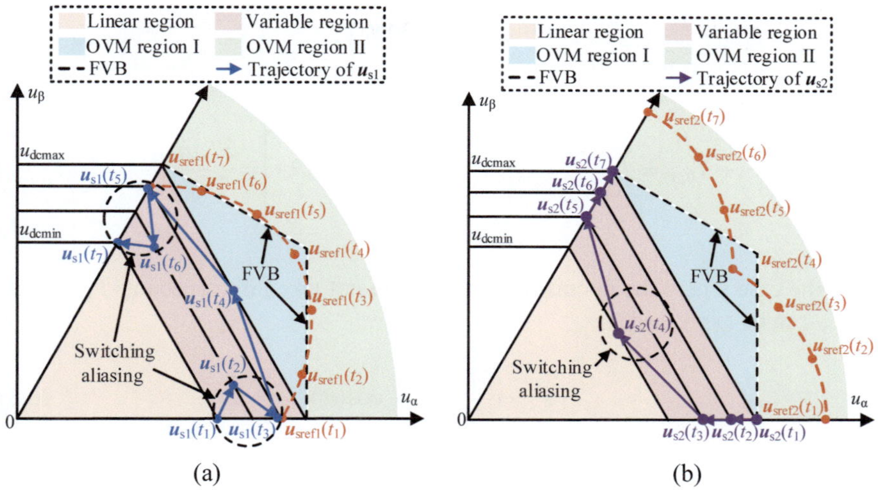

Fig. 6.2 Trajectories of reference and output voltage vectors with switching aliasing, reprinted from Ref. [7], copyright 2022, with permission from Nature. **a** Boundary of OVM region I. **b** Boundary of OVM region II

6.2.2 Analysis of the Aliasing Process of Modulation Region Switching

The fluctuation of DC-link voltage causes the space vector hexagon to contract and expand, which will lead to the further distortion if the motor operates in the overmodulation regions. In addition, the DC-link voltage is used to calculate the modulation formula of SVPWM, so the calculation of the action time of fundamental vectors (T_i and T_{i+1}) will be affected.

After the fixed voltage boundary (FVB) is established, the determination of the modulation region in Sector I shown in Fig. 6.2, as well as that of the output voltage vector, depends on the calculation of the action time of fundamental vectors, specifically T_{4_fv} and T_{6_fv}, which are derived from u_{dcmax}. And these can be expressed as [5]

$$\begin{cases} T_{4\_fv} = \frac{\sqrt{3}|u_{sref}|}{u_{dcmax}} T_s \sin\left(\frac{\pi}{3} - \theta_t\right) \\ T_{6\_fv} = \frac{\sqrt{3}|u_{sref}|}{u_{dcmax}} T_s \sin(\theta_t) \end{cases} \tag{6.2}$$

where T_s is the PWM period, and θ is the angle between $\boldsymbol{u}_{sref}$ and the α-axis.

The determination time of modulation region T_{Ux_fv} ($x = 4$, $0 \le \theta \le \pi/6$; x = 6, $\pi/6 \le \theta \le \pi/3$) is calculated by the action time of fundamental vectors as shown in (6.2). In OVM region I, $T_{U4\_fv} < T_s$ and $T_{U6\_fv} < T_s$. In OVM region II, $T_{U4\_fv} > T_s$ or $T_{U6\_fv} > T_s$. T_{U4_fv} and T_{U6_fv} can be expressed as [6]

$$T_{\mathrm{U4\_fv}}=T_{4\_\mathrm{fv}}+\frac{T_{6\_\mathrm{fv}}}{2}=\frac{\sqrt{3}|u_{\mathrm{sref}}|}{u_{\mathrm{dcmax}}}T_{\mathrm{s}}\left[\sin\left(\frac{\pi}{3}-\theta_{\mathrm{t}}\right)+\frac{\sin(\theta)}{2}\right]=\frac{3T_{\mathrm{s}}|u_{\mathrm{sref}}|}{2u_{\mathrm{dcmax}}}\cos(\theta_{\mathrm{t}}), \tag{6.3}$$

$$T_{\mathrm{U6\_fv}}=\frac{T_{4\_\mathrm{fv}}}{2}+T_{6\_\mathrm{fv}}=\frac{\sqrt{3}|u_{\mathrm{sref}}|}{u_{\mathrm{dcmax}}}T_{\mathrm{s}}\left[\frac{\sin\left(\frac{\pi}{3}-\theta_{\mathrm{t}}\right)}{2}+\sin(\theta_{\mathrm{t}})\right]=\frac{3T_{\mathrm{s}}|u_{\mathrm{sref}}|}{2u_{\mathrm{dcmax}}}\cos\left(\theta_{\mathrm{t}}-\frac{\pi}{3}\right). \tag{6.4}$$

In the case where the fluctuations of $\boldsymbol{u}_{\mathrm{sref}}$ are not taken into account, $T_{\mathrm{U4\_fv}}$ and $T_{\mathrm{U6\_fv}}$ can be expressed as

$$T_{\mathrm{U4\_fv}}=\frac{3T_{\mathrm{s}}U_{\mathrm{sref\_0}}}{2u_{\mathrm{dcmax}}}\cos(\theta_{\mathrm{t}}), \tag{6.5}$$

$$T_{\mathrm{U6\_fv}}=\frac{3T_{\mathrm{s}}U_{\mathrm{sref\_0}}}{2u_{\mathrm{dcmax}}}\cos\left(\theta_{\mathrm{t}}-\frac{\pi}{3}\right). \tag{6.6}$$

Based on Eqs. (6.5) and (6.6), it is evident that the changes in $T_{\mathrm{U4\_fv}}$ and $T_{\mathrm{U6\_fv}}$ are solely related to θ_{t}. After setting the FVB in the voltage hexagon, if the fluctuation of the designated voltage vector is ignored, a continuous and smooth transition from overmodulation region I to overmodulation region II can be achieved as the designated voltage vector increases. Taking into account the fluctuation of u_{sref}, $T_{\mathrm{U4\_fv}}$ and $T_{\mathrm{U6\_fv}}$ can be formulated as

$$T_{\mathrm{U4\_fv}}=\frac{3T_{\mathrm{s}}\left[U_{\mathrm{sref0}}+\sum_{n=1}^{\infty}U_{\mathrm{sref\_n}}\sin\left(6n\omega_{\mathrm{g}}t+\varphi_{\mathrm{sref\_n}}\right)\right]}{2u_{\mathrm{dcmax}}}\cos(\theta_{\mathrm{t}}), \tag{6.7}$$

$$T_{\mathrm{U6\_fv}}=\frac{3T_{\mathrm{s}}\left[U_{\mathrm{sref0}}+\sum_{n=1}^{\infty}U_{\mathrm{sref\_n}}\sin\left(6n\omega_{\mathrm{g}}t+\varphi_{\mathrm{sref\_n}}\right)\right]}{2u_{\mathrm{dcmax}}}\cos\left(\theta_{\mathrm{t}}-\frac{\pi}{3}\right). \tag{6.8}$$

According to Eqs. (6.7) and (6.8), the variations of $T_{\mathrm{U4\_fv}}$ and $T_{\mathrm{U6\_fv}}$ are related to θ as well as the alternating quantity $u_{\mathrm{sref\_ac}}$ of the reference voltage vector u_{sref}. Figure 6.3 shows the modulation area determination time when the reference voltage vector u_{sref} is located at the boundary of overmodulation region I after the establishment of fixed voltage boundary. As can be seen from the figure, once the fixed voltage boundary is set, there are parts of $T_{\mathrm{U4\_fv}}$ and $T_{\mathrm{U6\_fv}}$ that exceed T_{s}, which indicates an early transition into overmodulation region II. Due to the fluctuation of the reference voltage vector, when approaching the boundary of overmodulation region I, the system experiences random hopping back and forth between overmodulation region I and overmodulation region II.

Figure 6.4a displays the trajectories of the reference and output voltage vectors during the switching aliasing in the boundary of overmodulation region I. Because of

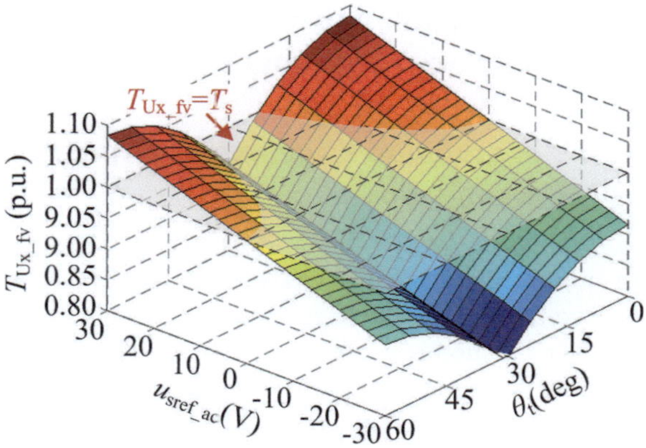

Fig. 6.3 Determination time of modulation region T_{Ux_fv} in the boundary of OVM I

the fluctuation of the reference voltage vector, the system stays within overmodulation region I along trajectory 1, yet it enters overmodulation region II along trajectory 2. It shifts into overmodulation region II prematurely without making full use of overmodulation region I, and this causes an increase in the Total Harmonic Distortion (THD) of the system. Figure 6.4b presents the trajectories of the reference voltage and output voltage vectors during the switching aliasing at the boundary of overmodulation region II. The system remains in overmodulation region II along trajectory 3, but then jumps back to overmodulation Region I along trajectory 4. It jumps back to overmodulation region I without fully entering the six-step operation state, which leads to a reduction in the output voltage capacity of the system.

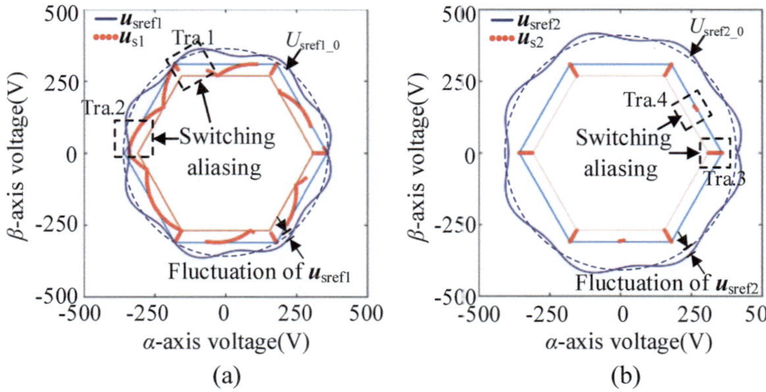

Fig. 6.4 Phenomenon of modulation with switching aliasing, reprinted from Ref. [7], copyright 2022, with permission from Nature. **a** Boundary of OVM region I. **b** Boundary of OVM region II

6.3 Overmodulation Strategy Based on Dynamic Voltage Boundary Adjustment

6.3.1 Modulation Switching Algorithm Based on Dynamic Voltage Boundaries

To prevent the switching aliasing brought about by the fluctuation of the reference voltage vector, an overmodulation strategy based on dynamic voltage boundary (DVB) regulation is proposed, as illustrated in Fig. 6.5. In space vector pulse width modulation, the action time of fundamental vectors T_{i_av} and T_{i+1_av} is computed using the actual DC-link voltage u_{dc} so as to determine the modulation regions. In the linear region, the actual DC-link voltage u_{dc} is chosen to take part in the modulation. In the overmodulation regions, the dynamic voltage u_{dc_bv} is utilized to separate overmodulation region I and II. Consequently, the voltage vectors are output in different modulation modes in overmodulation region I and II respectively.

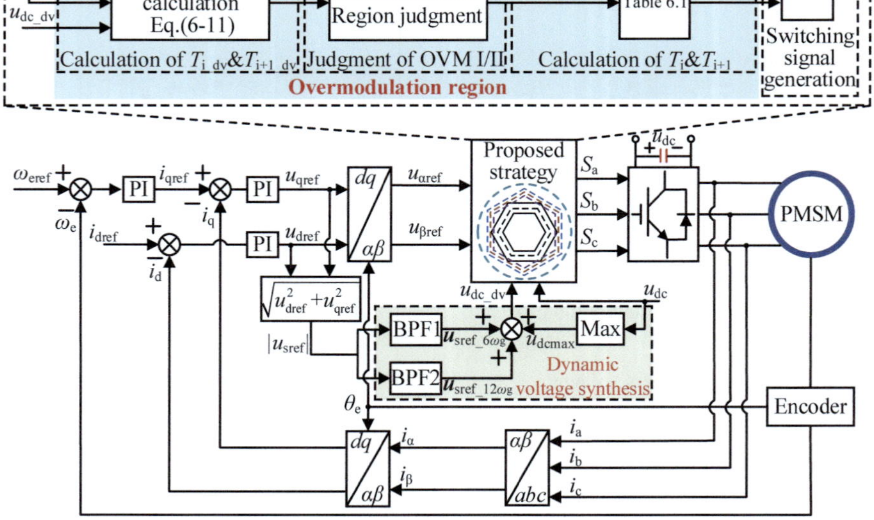

Fig. 6.5 Block diagram of the dynamic voltage boundary regulation method, reprinted from Ref. [7], copyright 2022, with permission from Nature

The dq-axis reference voltage is employed to generate the reference voltage vector u_{sref}. In the dynamic voltage generation module, the harmonics with frequencies of $6\omega_{\text{g}}$ and $12\omega_{\text{g}}$ of the reference voltage vector are extracted by two bandpass filters (BPF1 and BPF2). Then, these extracted harmonics are combined with the maximum value of the DC-link voltage u_{dcmax} to form the dynamic voltage $u_{\text{dc\_dv}}$ which can be expressed as

$$u_{\text{dc\_dv}} = u_{\text{dc}max} + u_{\text{sref\_}6\omega_{\text{g}}} + u_{\text{sref\_}12\omega_{\text{g}}} \tag{6.9}$$

where u_{dcmax} is the maximum value of the DC-link voltage, $\boldsymbol{u}_{\text{sref\_}6\omega\text{g}}$ and $\boldsymbol{u}_{\text{sref\_}12\omega\text{g}}$ are the harmonics of reference voltage vector at frequencies of $6\omega_{\text{g}}$ and $12\omega_{\text{g}}$, respectively.

In sector I, the action time of fundamental vectors is calculated respectively by means of the actual DC-link voltage u_{dc} and the dynamic voltage $u_{\text{dc\_dv}}$. It can be expressed as

$$\begin{cases} T_{4\_\text{av}} = \frac{\sqrt{3}T_{\text{s}}}{u_{\text{dc}}}\left(\frac{\sqrt{3}}{2}u_{\alpha\text{ref}} - \frac{1}{2}u_{\beta\text{ref}}\right) \\ T_{6\_\text{av}} = \frac{\sqrt{3}T_{\text{s}}}{u_{\text{dc}}}u_{\beta\text{ref}} \end{cases}, \tag{6.10}$$

$$\begin{cases} T_{4\_\text{dv}} = \frac{\sqrt{3}T_{\text{s}}}{u_{\text{dc\_dv}}}\left(\frac{\sqrt{3}}{2}u_{\alpha\text{ref}} - \frac{1}{2}u_{\beta\text{ref}}\right) \\ T_{6\_\text{dv}} = \frac{\sqrt{3}T_{\text{s}}}{u_{\text{dc\_dv}}}u_{\beta\text{ref}} \end{cases} \tag{6.11}$$

where T_{4_av} and T_{6_av} are the action time of fundamental vectors calculated by u_{dc}, T_{4_dv} and T_{6_dv} are the action time of fundamental vectors calculated by $u_{\text{dc\_dv}}$, and $u_{\alpha\text{ref}}$ and $u_{\beta\text{ref}}$ are the voltages of $\alpha\beta$-axis.

Based on (6.10), the sum of the action times of fundamental vectors can be obtained. When the reference voltage vector $\boldsymbol{u}_{\text{sref}}$ is located in the linear region, the following condition is satisfied

$$T_{4\_\text{av}} + T_{6\_\text{av}} = \frac{\sqrt{3}|u_{\text{sref}}|}{u_{\text{dc}}}T_{\text{s}}\cos\left(\theta - \frac{\pi}{3}\right) \le T_{\text{s}}. \tag{6.12}$$

Then, the actual action time of fundamental vectors T_4 and T_6 can be represented as

$$\begin{cases} T_4 = T_{4\_\text{av}} \\ T_6 = T_{6\_\text{av}} \end{cases}. \tag{6.13}$$

When the reference voltage vector $\boldsymbol{u}_{\text{sref}}$ is located in the overmodulation region, the action time of fundamental vectors meets the following condition

$$T_{4\_\text{av}} + T_{6\_\text{av}} = \frac{\sqrt{3}|u_{\text{sref}}|}{u_{\text{dc}}}T_{\text{s}}\cos\left(\theta - \frac{\pi}{3}\right) > T_{\text{s}}. \tag{6.14}$$

When the reference voltage vector $\boldsymbol{u}_{\text{sref}}$ is located in overmodulation region I, the determination time of the modulation region, namely $T_{\text{U4\_dv}}$ and $T_{\text{U6\_dv}}$, can be calculated in accordance with (6.11), and it satisfies the following condition

$$T_{\text{U4\_dv}} = T_{\text{4\_dv}} + \frac{T_{\text{6\_dv}}}{2} = \frac{3T_{\text{s}}|u_{\text{sref}}|}{2u_{\text{dc\_dv}}}\cos(\theta) \le T_{\text{s}}, \tag{6.15}$$

$$T_{\text{U6\_dv}} = \frac{T_{\text{4\_dv}}}{2} + T_{\text{6\_dv}} = \frac{3T_{\text{s}}|u_{\text{sref}}|}{2u_{\text{dc\_dv}}}\cos\left(\theta - \frac{\pi}{3}\right) \le T_{\text{s}} \tag{6.16}$$

Then, the actual action time of fundamental vectors T_4 and T_6 can be represented as

$$\begin{cases} T_4 = \frac{T_{\text{4\_dv}}}{T_{\text{4\_dv}}+T_{\text{6\_dv}}} \\ T_6 = \frac{T_{\text{6\_dv}}}{T_{\text{4\_dv}}+T_{\text{6\_dv}}} \end{cases}. \tag{6.17}$$

When the reference voltage vector $\boldsymbol{u}_{\text{sref}}$ is located in overmodulation region II, the determination time of modulation regions $T_{\text{U4\_dv}}$ and $T_{\text{U6\_dv}}$ meets the following condition

$$T_{\text{U4\_ dv}} = T_{\text{4\_dv}} + T_{\text{6\_dv}}/2 > T_{\text{s}}, \tag{6.18}$$

$$T_{\text{U6\_ dv}} = T_{\text{4\_dv}}/2 + T_{\text{6\_dv}} > T_{\text{s}}. \tag{6.19}$$

Then, the actual action time of fundamental vectors T_4 and T_6 can be formulated as

$$\begin{cases} T_4 = T_{\text{s}} \\ T_6 = 0 \end{cases}, \ 0 \le \theta \le \frac{\pi}{6}, \tag{6.20}$$

$$\begin{cases} T_4 = 0 \\ T_6 = T_{\text{s}} \end{cases}, \ \frac{\pi}{6} \le \theta \le \frac{\pi}{3}. \tag{6.21}$$

The modulation region, decision condition, and the actual action time of fundamental vectors are presented in Table 6.1.

The trajectories of the reference and output voltage vectors after the proposed strategy is adopted at the boundary of overmodulation region I are depicted in Fig. 6.6a. As can be observed, due to the periodic fluctuation of the reference voltage vector, the dynamic voltage boundary is formed by the dynamic voltages at different times (t_1 ~ t_7). At time t_2, if the strategy based on the fixed voltage boundary is adopted, the reference voltage vector $\boldsymbol{u}_{\text{sref1}}$ is located in overmodulation region I. However, after adopting the strategy based on the dynamic voltage boundary, $\boldsymbol{u}_{\text{sref1}}$ is located in overmodulation region I and outputs the voltage vector $\boldsymbol{u}_{\text{s1}}(t_2)$. At time t_3,

Table 6.1 Actual action time of fundamental vectors

Modulation region	Decision condition	Action time
Linear region	$T_{\mathrm{i\_av}} + T_{\mathrm{i+1\_av}} \leq T_{\mathrm{s}}$	$T_{\mathrm{i}} = T_{\mathrm{i\_av}}$ $T_{\mathrm{i+1}} = T_{\mathrm{i+1\_av}}$
OVM region I	$T_{\mathrm{i\_av}} + T_{\mathrm{i+1\_av}} > T_{\mathrm{s}}$ and $T_{\mathrm{i\_dv}} + T_{\mathrm{i+1\_dv}}/2 \leq T_{\mathrm{s}}$ and $T_{\mathrm{i\_dv}}/2 + T_{\mathrm{i+1\_dv}} \leq T_{\mathrm{s}}$	$T_{\mathrm{i}} = T_{\mathrm{i\_dv}}/(T_{\mathrm{i\_dv}}+T_{\mathrm{i+1\_dv}})$ $T_{\mathrm{i+1}} = T_{\mathrm{i+1\_dv}}/(T_{\mathrm{i\_dv}}+T_{\mathrm{i+1\_dv}})$
OVM region II	$T_{\mathrm{i\_dv}} + T_{\mathrm{i+1\_dv}}/2 > T_{\mathrm{s}}$ or $T_{\mathrm{i\_dv}}/2 + T_{\mathrm{i+1\_dv}} > T_{\mathrm{s}}$	$T_{\mathrm{i}} = T_{\mathrm{s}}$ and $T_{\mathrm{i+1}} = 0$ or $T_{\mathrm{i}} = 0$ and $T_{\mathrm{i+1}} = T_{\mathrm{s}}$

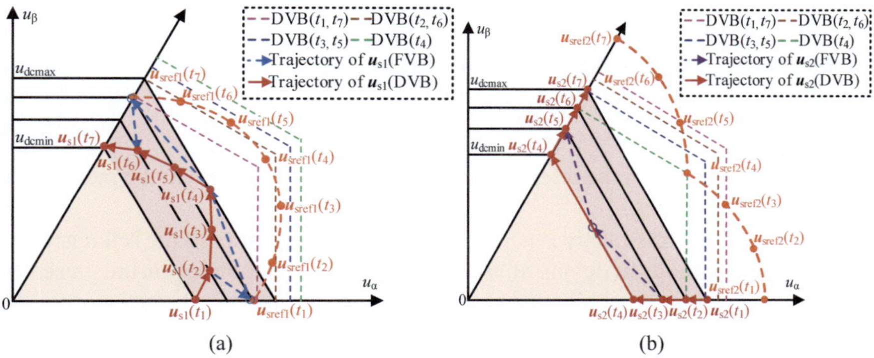

Fig. 6.6 Trajectories of reference and output voltage vectors without switching aliasing after adopting the proposed strategy, reprinted from Ref. [7], copyright 2022, with permission from Nature. **a** Boundary of OVM region I. **b** Boundary of OVM region II

if the strategy based on the fixed voltage boundary is adopted, $\boldsymbol{u}_{\mathrm{sref1}}$ is located in overmodulation region II. But after adopting the strategy based on the dynamic voltage boundary, $\boldsymbol{u}_{\mathrm{sref1}}$ is located in overmodulation region I and outputs the voltage vector $\boldsymbol{u}_{\mathrm{s1}}(t_3)$. The remaining moments in the analysis process follow the same pattern.

The trajectories of the reference and output voltage vectors after the proposed strategy is adopted at the boundary of overmodulation region II are shown in Fig. 6.6b. As can be seen, the dynamic voltage boundary are formed by the dynamic voltages at different times ($t_1 \sim t_7$). At time t_3, if the strategy based on the fixed voltage boundary is adopted, the reference voltage vector $\boldsymbol{u}_{\mathrm{sref2}}$ is located in overmodulation region II. After adopting the strategy based on the dynamic voltage boundary, $\boldsymbol{u}_{\mathrm{sref2}}$ remains in overmodulation region II and outputs the voltage vector $\boldsymbol{u}_{\mathrm{s2}}(t_3)$. At time t_4, if the strategy based on the fixed voltage boundary is adopted, $\boldsymbol{u}_{\mathrm{sref2}}$ is located in overmodulation region I. However, after adopting the strategy based on the dynamic voltage boundary, $\boldsymbol{u}_{\mathrm{sref2}}$ is located in overmodulation region II and outputs the voltage vector $\boldsymbol{u}_{\mathrm{s2}}(t_4)$. The remaining moments in the analysis process follow the same principle.

The dynamic voltage boundary is set in real time in line with the fluctuation of reference voltage vectors. As a result, the overmodulation region is determined based

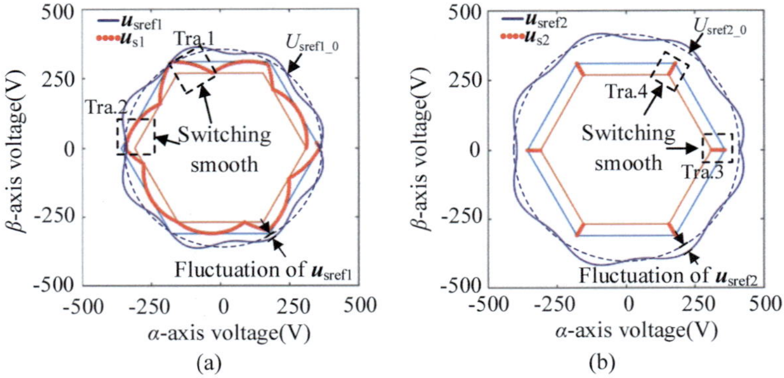

Fig. 6.7 Phenomenon of modulation without switching aliasing after adopting the proposed method. **a** Boundary of OVM region I. **b** Boundary of OVM region II

on the DC component of the reference voltage vector so as to prevent the modulation from switching aliasing.

After the proposed strategy is adopted, the vector trajectories of the reference and output voltage without switching aliasing in the overmodulation region are presented in Fig. 6.7. In order to achieve smooth switching, the proposed strategy sets the dynamic voltage boundary in real time according to the fluctuation of the reference voltage vector. Consequently, the overmodulation regions can be divided by the DC component of the amplitude of the reference voltage vector. At the boundary of overmodulation region I, the proposed strategy stops the reference voltage vector from prematurely jumping into overmodulation region II. As depicted in Fig. 6.7a, the output voltage capability in overmodulation region I can be fully utilized. At the boundary of overmodulation region II, the proposed strategy prevents the reference voltage vector from jumping back to overmodulation region I, as shown in Fig. 6.7b.

6.3.2 Harmonic Characteristics of Inverter Output Voltage

In the SVPWM, the reference modulation index (RMI) is defined to calculate the action time of fundamental vectors, which can be expressed as [8]

$$\mathrm{RMI_{av}} = \frac{\sqrt{3}|u_{\mathrm{sref}}|}{u_{\mathrm{dc}}} = \frac{\sqrt{3}\left|U_{\mathrm{sref0}} + \sum\limits_{n=1}^{\infty} U_{\mathrm{sref\_n}} \sin\left(6n\omega_{\mathrm{g}}t + \varphi_{\mathrm{sref\_n}}\right)\right|}{U_{\mathrm{dc0}} + \sum\limits_{n=1}^{\infty} U_{\mathrm{dc\_n}} \sin\left(6n\omega_{\mathrm{g}}t + \varphi_{\mathrm{dc\_n}}\right)} \tag{6.22}$$

where $\mathrm{RMI_{av}}$ is calculated using the DC-link voltage (u_{dc}).

Both the DC-link voltage and the reference voltage vector have fluctuations at six times the grid frequency. As can be seen from (6.22), the different phase relationships between the DC-link voltage and the reference voltage vector result in the introduction of harmonics in the reference modulation index. The overmodulation control strategy based on fixed voltage boundary selects the maximum value of the DC-link voltage (u_{dcmax}) to eliminate the harmonics of the RMI, and this can be expressed as

$$\begin{aligned}\mathrm{RMI}_{\mathrm{fv}} &= \frac{\sqrt{3}|u_{\mathrm{sref}}|}{u_{\mathrm{dcmax}}} = \frac{\sqrt{3}\left|U_{\mathrm{sref0}}+\sum\limits_{n=1}^{\infty} U_{\mathrm{sref\_k}}\sin\left(6n\omega_{\mathrm{g}}t+\varphi_{\mathrm{sref\_n}}\right)\right|}{u_{\mathrm{dcmax}}} \\ &= \mathrm{RMI}_{\mathrm{fv\_0}} + \sum_{n=1}^{\infty}\mathrm{RMI}_{\mathrm{fv\_}}\ 6n\omega_{\mathrm{g}}\sin\left(6n\omega_{\mathrm{g}}t+\varphi_{\mathrm{fb\_n}}\right)\end{aligned} \quad (6.23)$$

where $\mathrm{RMI}_{\mathrm{fv\_0}}$ is the DC component of the $\mathrm{RMI}_{\mathrm{fv}}$, $\mathrm{RMI}_{\mathrm{fv\_6n\omega g}}$ and $\varphi_{\mathrm{fb\_n}}$ are the amplitude and phase of the 6nth $\mathrm{RMI}_{\mathrm{fv}}$, respectively.

According to (6.23), there are harmonics at six times the grid frequency in the $\mathrm{RMI}_{\mathrm{fv}}$, which mainly consist of $6\omega_{\mathrm{g}}$ and $12\omega_{\mathrm{g}}$ harmonics. In the proposed strategy, the dynamic voltage is generated by the $6\omega_{\mathrm{g}}$ and $12\omega_{\mathrm{g}}$ harmonics of the reference voltage vector. Moreover, the RMI can be expressed when primarily taking into account the $6\omega_{\mathrm{g}}$ and $12\omega_{\mathrm{g}}$ harmonics of the DC-link voltage.

$$\begin{aligned}\mathrm{RMI}_{\mathrm{dv}} &= \frac{\sqrt{3}|u_{\mathrm{sref}}|}{u_{\mathrm{dc\_dv}}} = \frac{\sqrt{3}\left|U_{\mathrm{sref0}}+U_{\mathrm{sref\_6\omega g}}\sin\left(6\omega_{\mathrm{g}}t+\varphi_{\mathrm{sref\_6\omega g}}\right)+U_{\mathrm{sref\_12\omega g}}\sin\left(12\omega_{\mathrm{g}}t+\varphi_{\mathrm{sref\_12\omega g}}\right)\right|}{u_{\mathrm{dcmax}}+U_{\mathrm{sref\_6\omega g}}\sin\left(6\omega_{\mathrm{g}}t+\varphi_{\mathrm{sref\_6\omega g}}\right)+U_{\mathrm{sref\_12\omega g}}\sin\left(12\omega_{\mathrm{g}}t+\varphi_{\mathrm{sref\_12\omega g}}\right)} \\ &= \mathrm{RMI}_{\mathrm{dv\_0}} + \mathrm{RMI}_{\mathrm{dv\_6\omega g}}\sin\left(6\omega_{\mathrm{g}}t+\varphi_{\mathrm{dv\_6\omega g}}\right) + \mathrm{RMI}_{\mathrm{dv\_12\omega g}}\sin\left(12\omega_{\mathrm{g}}t+\varphi_{\mathrm{dv\_12\omega g}}\right)\end{aligned} \quad (6.24)$$

where $\mathrm{RMI}_{\mathrm{dv\_0}}$ is the DC component of the $\mathrm{RMI}_{\mathrm{dv}}$, $\mathrm{RMI}_{\mathrm{dv\_n}}$ and $\varphi_{\mathrm{dv\_n}}$ are the amplitude and phase of the kth $\mathrm{RMI}_{\mathrm{dv}}$, respectively.

Since RMI fluctuates around the DC component, the overmodulation regions can be determined by $\mathrm{RMI}_{\mathrm{dv\_0}}$. In the linear region, the range of RMI is 0–1. In the OVM region I, the range of RMI is 1–1.155. In the OVM region II, the range of RMI is 1.155–1.133.

Since both the DC-link voltage u_{dc} and the reference voltage vector u_{sref} fluctuate at six times the grid frequency, a period of 1/6 cycle is selected, which is denoted as $\omega_{\mathrm{g}}t$ for the convenience of description. The phase difference between the harmonics of the voltage vector and the harmonics of the DC-link voltage at the same frequency is derived and expressed as $\varphi_{\mathrm{dc\_sref}}$. The theoretical analysis results of the RMI under different phase relationships between u_{dc} and u_{sref} are presented in Fig. 6.8. Compared with the strategy based on fixed voltage boundary, the proposed strategy based on dynamic voltage boundary is capable of reducing the fluctuation of the RMI.

For the convenience of analysis, the higher harmonics can be neglected and the impacts of the $6\omega_{\mathrm{g}}$, $12\omega_{\mathrm{g}}$, $6\omega_{\mathrm{e}}$, and $12\omega_{\mathrm{e}}$ harmonics of the output voltage vector $\boldsymbol{u}_{\mathrm{s}}$ are mainly taken into account. From the relationship between the stator voltages and $\boldsymbol{u}_{\mathrm{s}}$, it can be understood that the harmonics of the stator voltages are mainly composed of $5\omega_{\mathrm{e}}$, $7\omega_{\mathrm{e}}$, $6\omega_{\mathrm{g}}$–ω_{e}, and $6\omega_{\mathrm{g}}+\omega_{\mathrm{e}}$ frequency harmonics. When the dynamic voltage boundary is employed, the proposed strategy decreases the harmonic content of the

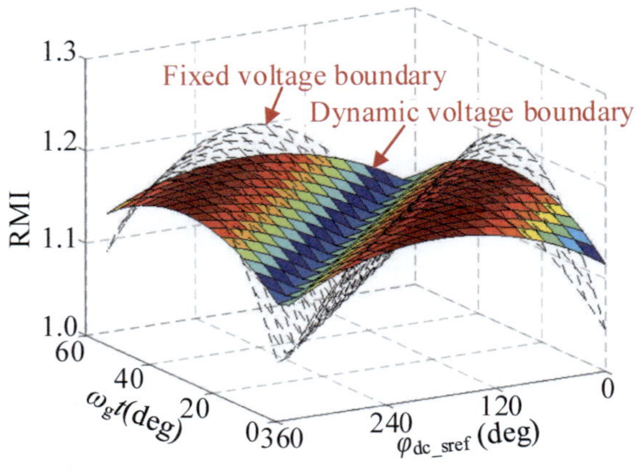

Fig. 6.8 Analysis results of RMI with different phase relationship between the u_{dc} and $\boldsymbol{u}_{\mathbf{sref}}$

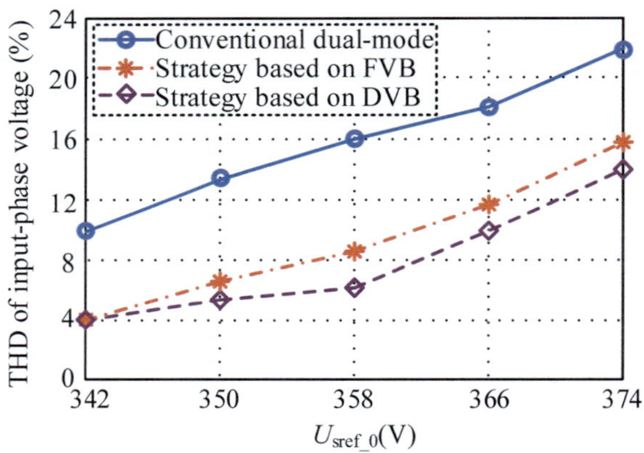

Fig. 6.9 Analysis results of THD in input-phase voltage adopting different strategies

RMI and prevents switching aliasing in the overmodulation region. As a result, the nonlinear harmonics of the stator voltages can be reduced.

Assuming the normal DC-link voltage fluctuation, as shown in Fig. 6.9, the u_{dcmax} is 537 V. Consequently, when the reference voltage amplitude $U_{sref\_0} = 358$ V, the motor operates at the boundary of overmodulation region I. In Fig. 6.9, the THD analysis results of the stator voltages are presented after different overmodulation strategies are adopted. The proposed strategy is capable of further reducing the THD of the input-phase voltage.

6.3.3 Experimental Verification

The experimental results obtained by adopting different strategies within overmodulation region I are presented in Fig. 6.10. In order to indicate the modulation region in which the motor operates, a flag of modulation region (FMR) is defined. It has three values, namely 0, 1, and 2, which respectively signify that the output voltage vector moves along a circular trajectory, a hexagon trajectory, and enters the six-step operation. Figure 6.10a displays the DC-link voltage (u_{dc}), the modulation boundary voltage ($u_{boundary}$), and the RMI. From these results, it can be seen that the conventional dual-mode overmodulation strategy utilizes the DC-link voltage for modulation, and there is a fluctuation of 0.43 in the RMI. After taking into account the fluctuation of the DC-link voltage and selecting the maximum value of the DC-link voltage as the modulation boundary voltage, the fluctuation of the RMI can be decreased to 0.36. The proposed strategy employs a dynamic voltage that is related to the fluctuation of the reference voltage vector for modulation, and as a result, the fluctuation of the RMI is further reduced to 0.15.

The waveforms shown in Fig. 6.10b from top to bottom are the stator current (i_a) and the FMR, respectively. The strategy based on boundary-fixed voltage boundary is able to suppress the switching aliasing that is caused by the fluctuation of the DC-link voltage. However, the reference voltage vector jumps into the overmodulation

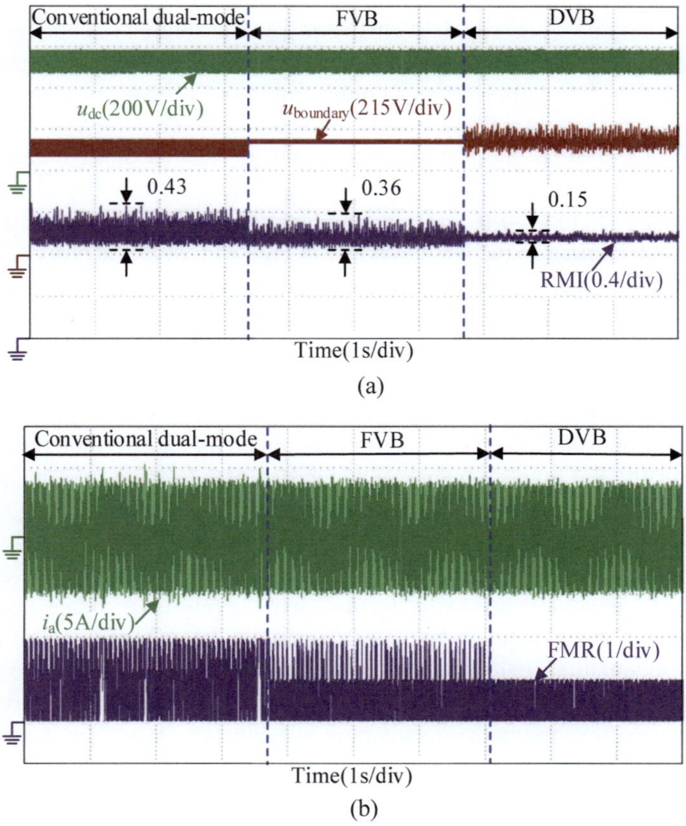

Fig. 6.10 Experimental results adopting different control strategies in OVM region I **a** DC-link voltage, modulation boundary voltage and RMI. **b** Stator current and FMR

region II prematurely. This is because the FMR still shows a value of 2, which indicates that the fluctuation of the reference voltage vector has not been taken into consideration. After adopting the proposed strategy, the problem of switching aliasing can be avoided. At this time, the voltage can be fully utilized within overmodulation region I.

Figure 6.11 illustrates the stator currents and the results of FFT analysis after different strategies are adopted when the motor frequency is 51 Hz. The Reference Modulation Index (RMI) fluctuates around 1.1, indicating that the motor operates close to the boundary of overmodulation region I. After the conventional dual-mode overmodulation strategy is adopted, the motor current becomes distorted because it enters overmodulation region II prematurely. As shown in Fig. 6.11a, the THD of the stator current is 17.1%. When the overmodulation strategy based on boundary-fixed voltage boundary is adopted, the current distortion caused by the fluctuation of the DC-link voltage is eliminated. As depicted in Fig. 6.11b, the THD of the stator current is 14.5%. Nevertheless, the stator current remains distorted after entering overmodulation region II prematurely, which is due to the fluctuation of the reference voltage vector. After the proposed strategy based on dynamic voltage boundary is adopted, the output voltage capability in overmodulation region I can be fully utilized by preventing the motor from jumping into overmodulation region II prematurely. As shown in Fig. 6.11c, this leads to a reduction in the THD of the stator current to 12.4%.

Figure 6.12 presents the stator currents and the results of FFT analysis after different strategies are adopted when the motor frequency is 52 Hz. The RMI fluctuates around 1.15, indicating that the motor operates at the boundary of overmodulation region I. Similar to the results shown in Fig. 6.11, after the proposed strategy based on dynamic voltage boundary is adopted, the output voltage capability in overmodulation region I can be fully utilized by preventing the motor from jumping into overmodulation region II prematurely. As a result, the THD of the stator current is decreased to 14.2%.

Figure 6.13 displays the Lissajous figures of the $\alpha\beta$-axis voltage within one electrical cycle when the motor operates at the boundary of overmodulation region I. Theoretically, the output voltage vector will shift into the vertex-hopping state within the overmodulation region II after traversing a full spatial hexagon. As can be seen from Fig. 6.13a, when fixed voltage boundary is utilized, the phase of the output voltage vector jumps and retreats. Moreover, it even enters the overmodulation region II prematurely due to the fluctuation of the reference voltage vector. After adopting the proposed strategy based on dynamic voltage boundary, the boundary of the overmodulation region changes in line with the fluctuation of the reference voltage vector. From Fig. 6.13b, the trajectory of the output voltage vector presents a better voltage hexagon without switching aliasing, which implies that the output capacity in overmodulation region I is fully exploited.

Figure 6.14 presents the experimental results regarding torque T_e, motor speed ω_e, and d-axis current i_d when the motor operates as it transitions from the linear region to overmodulation region II. By comparing Fig. 6.14a with Fig. 6.14b, it can be observed that the fluctuation of torque T_e is decreased from 7 to 4.2 N m, and the

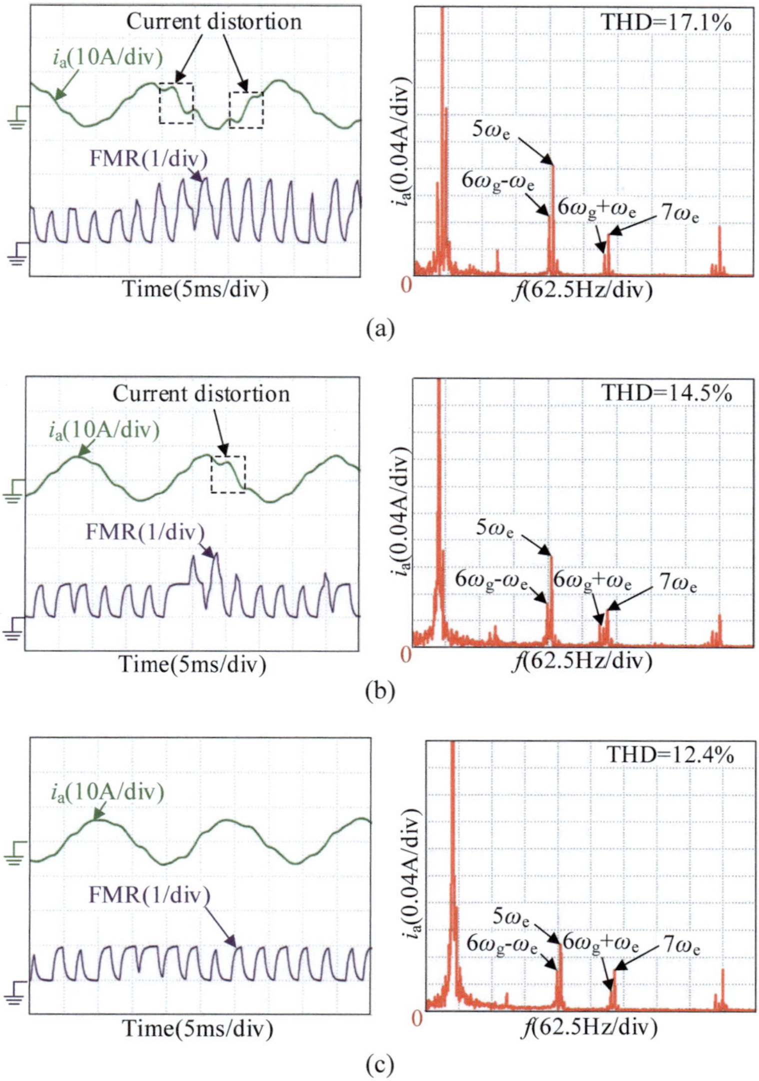

Fig. 6.11 Stator currents and FFT analysis results adopting different strategies at the motor frequency of 51 Hz, reprinted from Ref. [7], copyright 2022, with permission from Nature. **a** Conventional dual-mode overmodulation strategy. **b** Strategy based on FVB. **c** Strategy based on DVB

peak value of the d-axis current i_d is reduced from 3.6 A to 0.9 A. Meanwhile, the ripple in the motor speed does not undergo significant changes.

Figure 6.15 illustrates the dynamic process of the load change, where the load is increased from 15 to 21 N m. As shown in Fig. 6.15a, when the load is 15 N m, the stator current waveform has a fairly satisfactory sinusoidal shape. However, when the

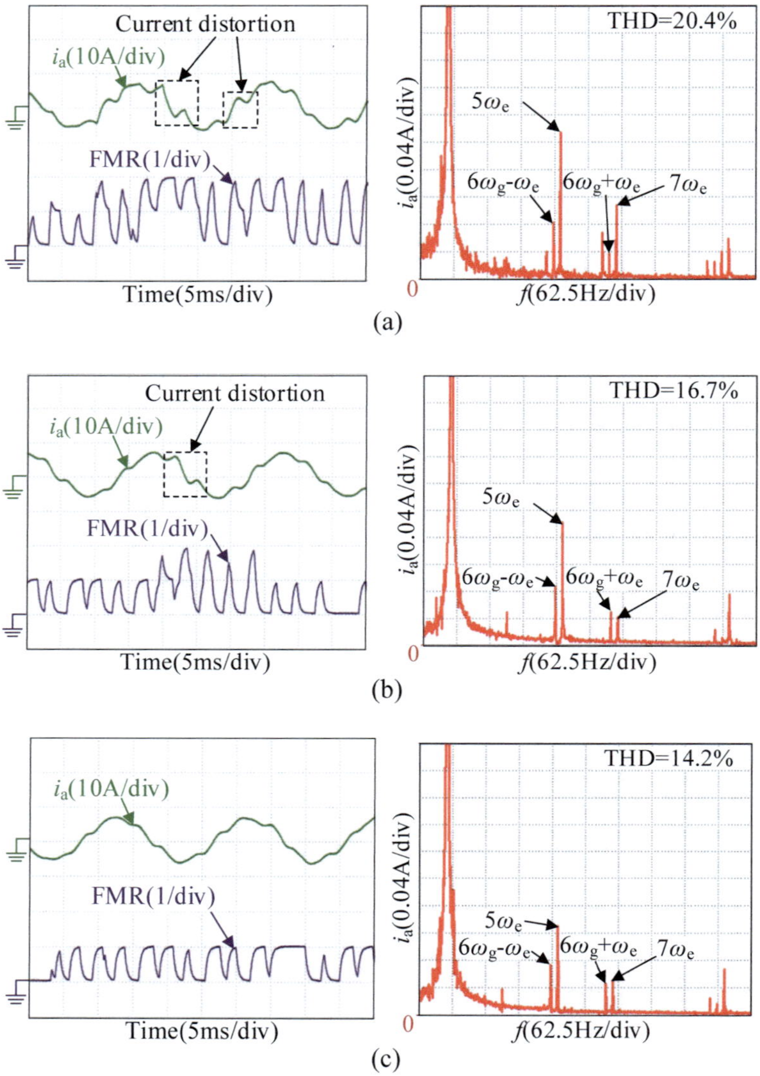

Fig. 6.12 Stator currents and FFT analysis results adopting different strategies at the motor frequency of 52 Hz, reprinted from Ref. [7], copyright 2022, with permission from Nature. **a** Conventional dual-mode overmodulation strategy. **b** Strategy based on FVB. **c** Strategy based on DVB

load becomes 21 N m, the waveform exhibits a significant aberration. After adopting the proposed strategy, as can be seen in Fig. 6.15b, the stator current maintains a satisfactory sinusoidal degree both before and after the load change.

To further confirm the control performance of the motor under other operating conditions, the statistical comparisons of the THD of the stator current when different

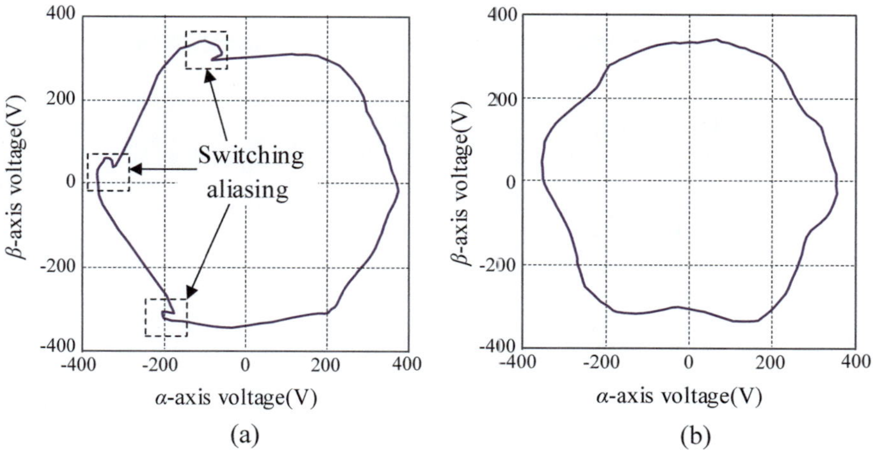

Fig. 6.13 Lissajous figures of αβ-axis voltage in the boundary of OVM region I. **a** Strategy based on FVB. **b** Strategy based on DVB

Fig. 6.14 Experimental results before and after control when the motor operates from linear region to OVM region II. **a** Conventional dual-mode overmodulation strategy. **b** Strategy based on DVB

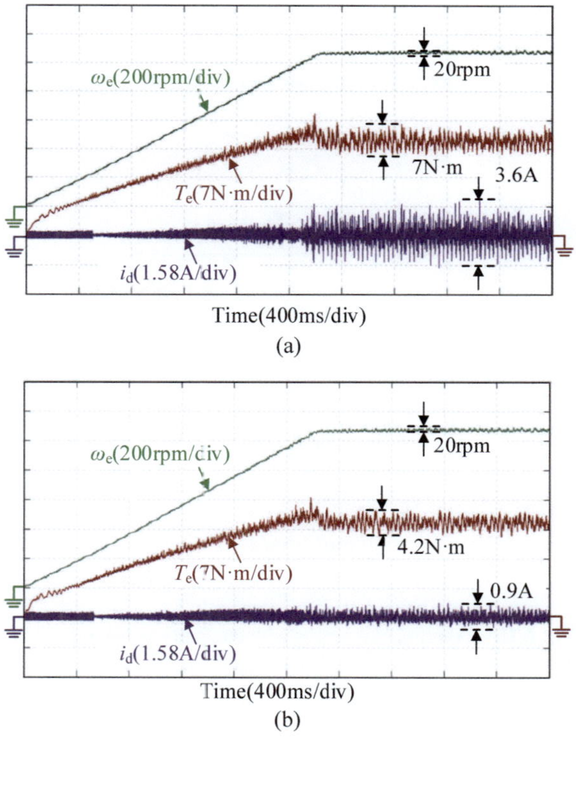

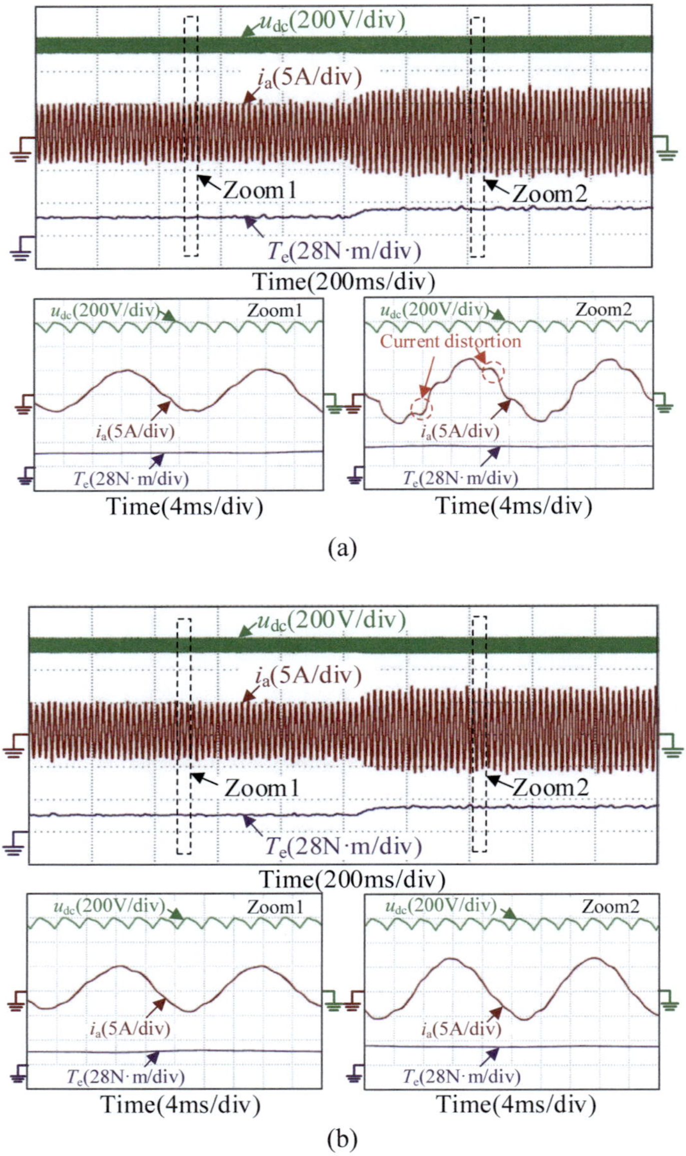

Fig. 6.15 Experimental results with the load change, reprinted from Ref. [7], copyright 2022, with permission from Nature. **a** Conventional dual-mode overmodulation strategy. **b** Strategy based on DVB

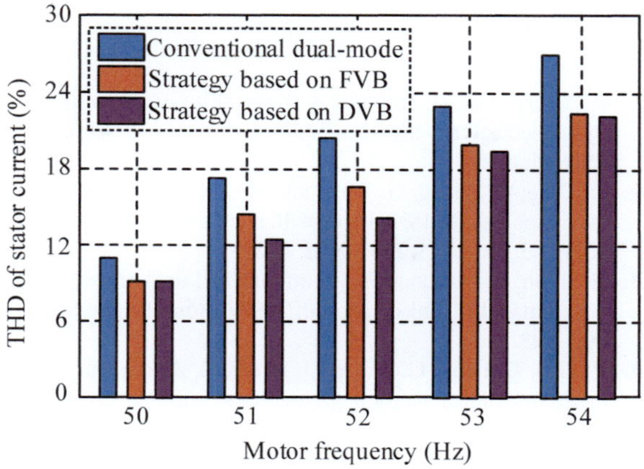

Fig. 6.16 Statistical comparisons of the THD of stator current when adopting different strategies

control strategies are adopted at various speeds are presented in Fig. 6.16. Compared with the conventional two-mode overmodulation control strategy, the strategy based on fixed voltage boundary is capable of reducing the THD of the stator current within the overmodulation region. The proposed strategy further reduces the THD of the stator current. The proposed strategy based on dynamic voltage boundary is primarily designed to address the issue of switching aliasing at the boundary of overmodulation region I, where the reference voltage vector jumps into overmodulation region II prematurely because of its fluctuation. And it can be noticed that the effect of this proposed strategy is more pronounced at motor frequencies of 51 and 52 Hz.

6.4 Summary

In order to suppress the motor voltage and current harmonics of the electrolytic capacitorless drive, the low distortion overmodulation control strategy under the periodic fluctuation of DC-link voltage and reference voltage vector is studied. The THD of the motor voltage and current is reduced while the problem of modulation switching aliasing is suppressed. The main work and conclusions are as follows:

An overmodulation strategy based on dynamic voltage boundary regulation is proposed. By analyzing the motor voltage vector locus, considering the periodic fluctuation of the reference voltage vector and the maximum DC-link voltage, the dynamic voltage fluctuation with the reference voltage vector is constructed. The overmodulation region is modulated by dynamic voltage, and the dynamic boundary is set in real time with the reference voltage vector fluctuation information, so that the direct flow smoothly switches the modulation region with the reference voltage amplitude. By FFT analysis of motor current before and after applying control, experimental results show that this strategy can effectively reduce motor current THD.

References

1. A. Yoo, S. Sul, H. Kim and K. Kim, "Flux-weakening strategy of an induction machine driven by an electrolytic-capacitor-less inverter," *IEEE Trans. Ind. Appl.*, vol. 47, no. 3, pp. 1328–1336, May-June. 2011.
2. D. Ding, G. Zhang, G. Wang and D. Xu, "Dual antiovervoltage control scheme for electrolytic capacitor less IPMSM drives with coefficient autoregulation," *IEEE Trans. Power Electron.*, vol. 35, no. 3, pp. 2895-2907, Mar. 2020.
3. H. Saren, O. Pyrhonen, K. Rauma and O. Laakkonen, "Overmodulation in voltage source inverter with small DC-link capacitor," *IEEE 36th Power Electronics Specialists Conference*, 2005: 892–898.
4. R. Gao, D. Ding, G. Wang, Q. Wang, G. Zhang and D. Xu, "Linear region expansion strategy based on voltage vector angle regulation for PMSM drives with small DC-link capacitor," IEEE Trans. Power Electron., vol. 38, no. 7, pp. 8327-8339, Jul. 2023.
5. G. Wang, H. Hu, D. Ding, N. Zhao, Y. Zou and D. Xu, "Overmodulation strategy for electrolytic capacitorless PMSM drives: voltage distortion analysis and boundary optimization," *IEEE Trans. Power Electron.*, vol. 35, no. 9, pp. 9574-9585, Sept. 2020.
6. R. Jing, G. Zhang, G. Wang, G. Bi, D. Ding and D. Xu, "An overmodulation strategy based on voltage vector space division for high-speed surface-mounted PMSM drives," IEEE Trans. Power Electron., vol. 37, no. 12, pp. 15370-15381, Dec. 2022.
7. W. Yue, D. Ding, B. Li, G. Wang, Y. Yuan, K. Tian, X. Zhang and D. Xu, "Overmodulation Strategy Based on Dynamic Voltage Boundary Regulation for Electrolytic Capacitorless PMSM Drives," IEEE Trans. Transport. Electrific., https://doi.org/10.1109/TTE.2025.3549980.
8. F. Briz, A. Diez, M. W. Degner and R. D. Lorenz, "Current and flux regulation in field-weakening operation [of induction motors]," *IEEE Trans. Ind. Appl.*, vol. 37, no. 1, pp. 42–50, Jan.-Feb. 2001.

Chapter 7
Linear Modulation Region Expansion Strategy

7.1 Introduction

In the electrolytic capacitorless motor drive, reducing the fluctuation of the modulation index can prevent the drive from entering the overmodulation region in advance, and improve the voltage utilization of the drive during the operation of the linear region in the electrolytic capacitorless motor drive. In this chapter, a linear modulation margin enhancement strategy is proposed to improve the voltage utilization of the drive by suppressing the harmonics of modulation index.

7.2 Voltage Vector Angle Adjustment Method

7.2.1 *Modulation Fluctuation Analysis Caused by DC-Link Voltage Pulsation*

In SVPWM control, the given voltage vector $\boldsymbol{u}_{\mathrm{sref}}$ can be obtained by the different combinations in time of the two adjacent basic voltage vectors and the zero vector when rotating to the ith sector ($i = 0$–6). The total dwell time of the voltage vector in a sector can be expressed as

$$T_{\mathrm{v}} = T_{\mathrm{i}} + T_{\mathrm{i+1}} = \frac{\sqrt{3}|\boldsymbol{u}_{\mathrm{sref}}|T_{\mathrm{s}}}{u_{\mathrm{dc\_s}}} \sin\left[\boldsymbol{M}\left(\theta_{\mathrm{u}}, \frac{\pi}{3}\right) + \frac{\pi}{3}\right] \tag{7.1}$$

where T_{i} and $T_{\mathrm{i+1}}$ are the dwell time of two adjacent basic voltage vectors in the ith sector, respectively. T_{v} is the total dwell time. θ_{u} is the angle of the voltage vector reference. $\boldsymbol{M}(x,y)$ is the modulation function.

D. Ding et al., *Electrolytic Capacitorless Permanent Magnet Synchronous Motor Drives*, https://doi.org/10.1007/978-981-95-5601-4_7

The total dwell time T_v should remain consistent in each sector. Therefore, its fundamental frequency is six times the frequency of the voltage vector in (7.1). Assuming the initial phase zero, the FFT of T_v can be expressed as

$$T_{\mathrm{v}} = \frac{3\sqrt{3}|\boldsymbol{u}_{\mathrm{sref}}|T_{\mathrm{s}}}{\pi u_{\mathrm{dc\_s}}} - \sum_{n=1}^{\infty} \frac{6\sqrt{3}|\boldsymbol{u}_{\mathrm{sref}}|T_{\mathrm{s}}}{\pi(36n^2 - 1)u_{\mathrm{dc\_s}}} \cos(6n\omega_{\mathrm{e}}t). \tag{7.2}$$

When there are the large periodical fluctuations in the DC-link voltage, the voltage vector amplitude will contain the harmonics at the same frequency. When there are phase and amplitude differences between the voltage vector amplitude and the fluctuation component of the DC-link voltage, the modulation index m will also contain harmonics at the same frequency, which can be expressed as

$$m = \sqrt{3}|\boldsymbol{u}_{\mathrm{sref}}| \Big/ u_{\mathrm{dc\_s}} = M_0 + \sum_{n=1}^{\infty} M_{\mathrm{n}} \sin(6n\omega_{\mathrm{g}}t + \varphi_{\mathrm{mn}}) \tag{7.3}$$

where M_0 is the average of the modulation index, M_{n} and φ_{mn} are the amplitude and phase of the nth harmonics of the modulation index, respectively.

According to (7.2) and (7.3), the total dwell time will include harmonics at frequencies of $6n\omega_{\mathrm{g}}$, $6(n_2\omega_{\mathrm{g}} - n_1\omega_{\mathrm{e}})$, and $6(n_2\omega_{\mathrm{g}} + n_1\omega_{\mathrm{e}})$ due to the fluctuations of DC-link voltage, as expressed in (7.4). This phenomenon will reduce the voltage utilization of the motor drive, as it can make the motor drive more prone to entering the overmodulation region.

$$\begin{aligned}
\frac{T_{\mathrm{v}}}{T_{\mathrm{s}}} &= \sum_{n=1}^{\infty} \frac{3M_{\mathrm{k}}}{\pi} \sin(6n\omega_{\mathrm{g}}t + \varphi_{\mathrm{mn}}) - \sum_{n=1}^{\infty} \frac{6M_0 \cos(6n\omega_{\mathrm{e}}t)}{\pi(36n^2 - 1)} \\
&\quad - \sum_{n_1=1}^{\infty}\sum_{n_2=1}^{\infty} \frac{3M_{\mathrm{n_2}}/\pi}{(36n_1^2 - 1)} \left\{\sin\left[6\left(n_2\omega_{\mathrm{g}} + n_1\omega_{\mathrm{e}}\right)t + \varphi_{\mathrm{mn_2}}\right]\right. \\
&\quad \left. + \sin\left[6\left(n_2\omega_{\mathrm{g}} - n_1\omega_{\mathrm{e}}\right)t + \varphi_{\mathrm{mn_2}}\right]\right\} + 3M_0/\pi.
\end{aligned} \tag{7.4}$$

The fluctuation of the basic voltage vector can make the size of the voltage boundary hexagon vary with the fluctuation of the DC-link voltage. In order to mitigate the influence of the fluctuation of the DC-link voltage, the modulation vector $\boldsymbol{m}$ can be defined as

$$\boldsymbol{m} = \sqrt{3}\boldsymbol{u}_{\mathrm{sref}} \Big/ u_{\mathrm{dc\_s}} = \sqrt{3}u_{\alpha\mathrm{ref}} \Big/ u_{\mathrm{dc\_s}} + j\sqrt{3}u_{\beta\mathrm{ref}} \Big/ u_{\mathrm{dc\_s}} = m_{\alpha} + jm_{\beta} \tag{7.5}$$

where $u_{\alpha\mathrm{ref}}$, $u_{\beta\mathrm{ref}}$, m_{α}, and m_{β} are the voltage references and modulation indexes in $\alpha\beta$-axis, respectively.

Based on (7.5), the amplitude of the basic modulation vector is constant, indicating that the modulation boundary hexagon remains fixed despite the fluctuations of the

DC-link voltage. Based on (7.1), the maximum modulation index m_{li} in the linear modulation system satisfies

$$m_{li}=1/\sin[\boldsymbol{M}(\theta_u,\pi/3)+\pi/3]. \tag{7.6}$$

The linear modulation margin M_{mar} is defined in (7.7) to represent the modulation region which can be illustrated as the distance between the modulation boundary hexagon and the operating regions of the modulation vectors.

$$M_{mar}=m_{li}-m=\frac{1}{\sin[\boldsymbol{M}(\theta_u,\pi/3)+\pi/3]}-\left[M_0+\sum_{n=1}^{\infty}M_n\sin(6n\omega_g t+\varphi_{mn})\right]. \tag{7.7}$$

According to (7.7), the linear modulation margin M_{mar} under different modulation conditions can be shown in. If M_{mar} is larger than zero, the system will operate in the linear modulation region, as shown in Fig. 7.1a. Otherwise, SVPWM will enter the overmodulation region, as shown in Fig. 7.1b.

The impact of DC-link voltage fluctuation on the modulation index is shown in Fig. 7.2a. Due to the large periodical fluctuation of the modulation index, SVPWM enters the overmodulation region earlier. The linear modulation margin of the DC-link voltage under different voltage vector angles and harmonics is illustrated in Fig. 7.2b.

It can be observed that the linear modulation margin varies in a wide range with different angles of the voltage vectors. For instance, even if U_{dc_1} is relatively small, the system can also enter the overmodulation region when θ_u equals 20°. However, when θ_u equals 35°, the system can always remain in the linear modulation region. This phenomenon indicates that it is essential to study the angle of the voltage vector to improve the linear modulation margin.

The linear modulation margin can be simplified to the minimum value of (7.7), which can be expressed as

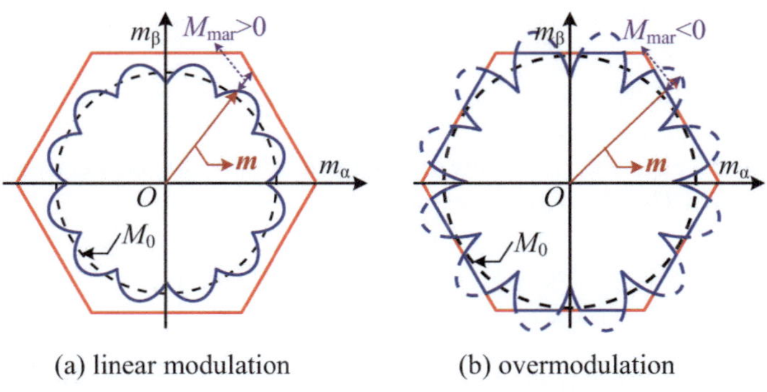

Fig. 7.1 Linear modulation margin under different modulation conditions

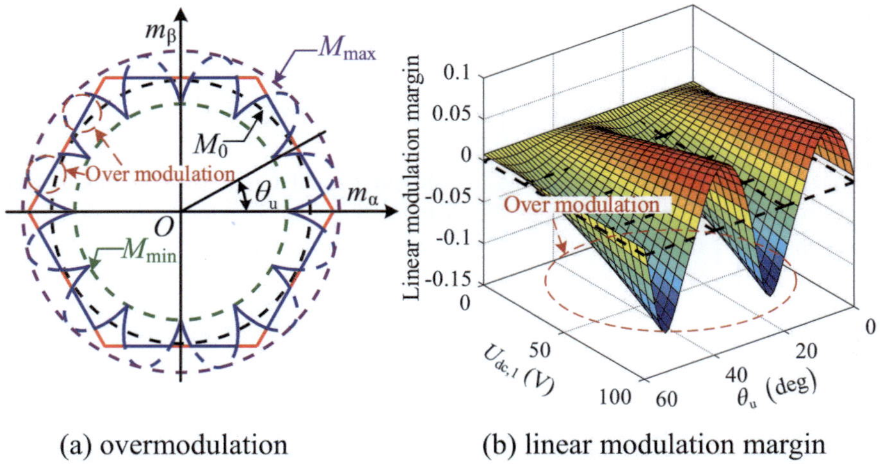

Fig. 7.2 Modulation index under the DC-link voltage fluctuation, reprinted from Ref. [1], copyright 2022, with permission from Nature

$$M_{\mathrm{mar}} = \left\{ \frac{1}{\sin\left[\boldsymbol{M}(\theta_{\mathrm{u}}, \pi/3) + \pi/3\right]} - m \right\}_{\min} = 1 - M_{\max}. \tag{7.8}$$

According to (7.8), when the linear modulation margin becomes lower than zero, the drive system enters the overmodulation region. Besides, the decrease of the linear modulation margin introduces additional fluctuations in the voltage vectors. The distortion of the voltage vectors makes M_0 decrease, resulting in a smaller fundamental voltage in the linear modulation and lower utilization of the DC-link voltage.

7.2.2 Linear Modulation Margin Analysis Based on Impedance Model

In order to quantitatively calculate the modulation harmonics caused by DC-link voltage fluctuations, the linear modulation margin is analyzed with the small signal model of the drive system. According to (2.19), the relationship between the reference voltage and the DC-link voltage is expressed as

$$\begin{bmatrix} \Delta u_{\mathrm{dref}} \\ \Delta u_{\mathrm{qref}} \end{bmatrix} = -\,\mathrm{e}^{-0.5sT_{\mathrm{s}}} \Delta u_{\mathrm{dc}} \begin{bmatrix} \boldsymbol{G}_{\mathrm{d}} \boldsymbol{W}_{\mathrm{d}} \\ \boldsymbol{G}_{\mathrm{q}} \boldsymbol{W}_{\mathrm{q}} \end{bmatrix}. \tag{7.9}$$

The relationship between the amplitude of voltage vector reference and the dq-axis voltage reference is expressed as

$$|\boldsymbol{u}_{\text{sref}}| = \sqrt{u_{\text{dref}}^2 + u_{\text{qref}}^2}. \tag{7.10}$$

By linearizing (7.10), the small-signal component of the amplitude of voltage vector reference $\Delta|\boldsymbol{u}_{\text{sref}}|$ can be expressed as

$$\Delta|\boldsymbol{u}_{\text{sref}}| \approx \left(U_{\text{dref0}}\Delta u_{\text{dref}} + U_{\text{qref0}}\Delta u_{\text{qref}}\right) / U_{\text{sref0}}. \tag{7.11}$$

By combining (7.9) and (7.11), the transfer function $\boldsymbol{G}_{\text{s}}$ between the amplitude of voltage vector reference and the DC-link voltage can be expressed as

$$\boldsymbol{G}_{\text{s}} = \frac{\Delta|\boldsymbol{u}_{\text{sref}}|}{\Delta u_{\text{dc}}} = -\frac{\mathrm{e}^{-0.5sT_{\text{s}}}\left(U_{\text{dref0}}\boldsymbol{G}_{\text{d}}\boldsymbol{W}_{\text{d}} + U_{\text{qref0}}\boldsymbol{G}_{\text{q}}\boldsymbol{W}_{\text{q}}\right)}{U_{\text{sref0}}}. \tag{7.12}$$

In this study, the grid-side voltage is 380 V at 50 Hz. The Bode plots of $\boldsymbol{G}_{\text{s}}$ under different torque conditions at the motor speed of 60 Hz are shown in Fig. 7.3. It can be observed that the frequency characteristics of the impedance model align closely with the simulation results at frequencies of $6\omega_{\text{g}}$ and $12\omega_{\text{g}}$. Besides, the phase difference at the frequency of $6\omega_{\text{g}}$ between the voltage vector harmonics and the DC-link voltage harmonics is 180°, indicating that the maximum of the amplitude of voltage vector reference and the minimum of the DC-link voltage can occur simultaneously.

Figure 7.4 shows the linear modulation margin with different phase differences between the harmonics of u_{dc} and $|\boldsymbol{u}_{\text{sref}}|$. When the phase difference is 180°, the modulation fluctuation can reach maximum, deteriorating the linear modulation margin and decreasing the utilization of the DC-link voltage, as shown in Fig. 7.2. Therefore, it is necessary to restrain modulation fluctuations by adjusting the phase differences.

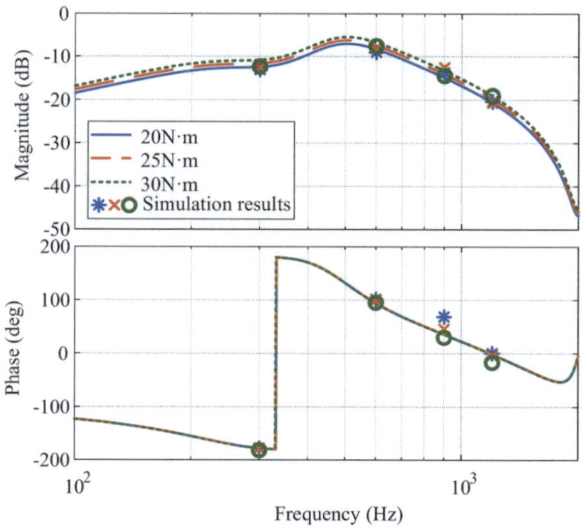

Fig. 7.3 Modulation index under the fluctuation of the DC-link voltage

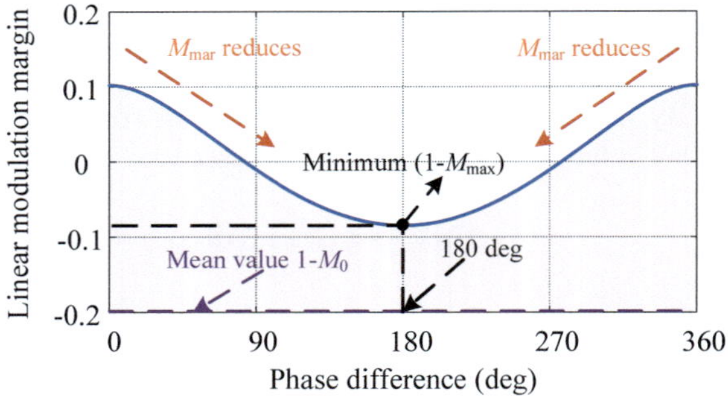

Fig. 7.4 Linear modulation margin with different phase differences

7.2.3 Voltage Vector Angle Adjustment Method Analysis and Performance Verification

Based on the above analysis, the modulation index fluctuations are related to the phase difference between the voltage vector harmonics and the DC-link voltage harmonics at the same frequency. Therefore, the modulation harmonics can be restrained by adjusting the phase difference at the frequencies of $6\omega_g$ and $12\omega_g$. As a result, a linear modulation extension strategy with the voltage reference vector angle adjustment is proposed as shown in Fig. 7.5. The regulated angle $\Delta\theta_u$ can be reconstructed by the DC-link voltage harmonics using the correction components $\boldsymbol{M}_{d\_1}(s)$ and $\boldsymbol{M}_{d\_2}(s)$. The $\alpha\beta$-axis voltage references are generated by the adjusted voltage angle.

Considering that the reduction in linear modulation margin is caused by the modulation harmonics at frequencies of $6\omega_g$ and $12\omega_g$, the voltage vector angle reference θ_u can be superimposed with harmonics at the same frequency, which can be expressed as

$$\begin{aligned}\theta_{un} &= \theta_u + \Delta\theta_u = \theta_u + \Delta\theta_{u1} + \Delta\theta_{u2} \\ &= \theta_u + A_6 \sin\left(6\omega_g t + \varphi_{u6}\right) + A_{12} \sin\left(12\omega_g t + \varphi_{u12}\right).\end{aligned} \tag{7.13}$$

The equivalent voltage references u_{dref_r} and u_{qref_r} are shown in Fig. 7.6, which can be expressed as

$$\begin{cases} u_{dref\_r} = |\boldsymbol{u}_{sref}| \cos(\theta_u + \Delta\theta_u) \\ u_{qref\_r} = |\boldsymbol{u}_{sref}| \sin(\theta_u + \Delta\theta_u) \end{cases}. \tag{7.14}$$

Based on (7.14), the relationship between the dq-axis voltage references before and after applying the control strategy can be expressed as

$$\begin{bmatrix} U_{dref\_r,0} + \Delta u_{dref\_r} \\ U_{qref\_r,0} + \Delta u_{qref\_r} \end{bmatrix} = \begin{bmatrix} \cos\Delta\theta_u & -\sin\Delta\theta_u \\ \sin\Delta\theta_u & \cos\Delta\theta_u \end{bmatrix} \begin{bmatrix} u_{dref} \\ u_{qref} \end{bmatrix}. \tag{7.15}$$

The Bessel function is used to simplify (7.15), which can be expressed as

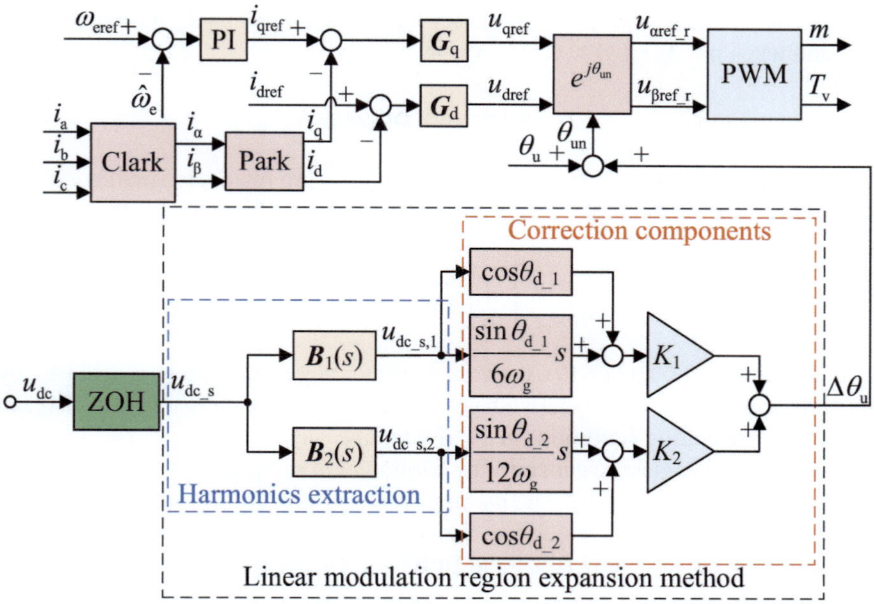

Fig. 7.5 Block diagram of the proposed linear modulation expansion method, reprinted from Ref. [1], copyright 2022, with permission from Nature

Fig. 7.6 Schematic diagram of voltage space vector angle regulation

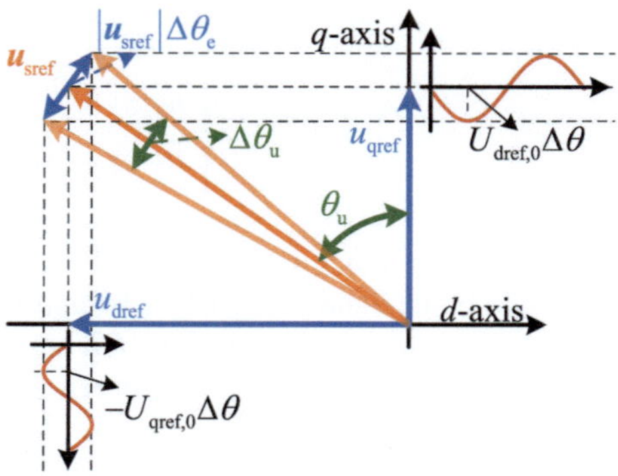

$$
\begin{cases} \cos(A\sin\omega t) = J_0(A) + 2J_2(A)\cos 2\omega t + \cdots \\ \sin(A\sin\omega t) = 2J_1(A)\sin\omega t + 2J_3(A)\sin 3\omega t + \cdots \end{cases}. \tag{7.16}
$$

where,

$$
J_{\mathrm{n}}(x) = \sum_{m=0}^{\infty} \frac{(-1)^m}{m!\Gamma(n+m+1)}\left(\frac{x}{2}\right)^{n+2m}. \tag{7.17}
$$

In the real control system, A is far lower than 1, consequently $J_0(A) = 1$, $J_1(A) = A/2$, and when n > 2, $J_n(A) \approx 0$. As a result,

$$\cos \Delta\theta_e \approx 1, \sin \Delta\theta_e \approx \Delta\theta_e. \tag{7.18}$$

The amplitudes of the regulated angle A_6 and A_{12} are relatively small compared with the voltage vector angle. Based on (7.18), (7.15) can be modified as

$$\begin{bmatrix} U_{\text{dref\_r},0} + \Delta u_{\text{dref\_r}} \\ U_{\text{qref\_r},0} + \Delta u_{\text{qref\_r}} \end{bmatrix} = \begin{bmatrix} U_{\text{dref},0} + \Delta u_{\text{dref}} - U_{\text{qref},0}\Delta\theta_u \\ U_{\text{qref},0} + \Delta u_{\text{qref}} + U_{\text{dref},0}\Delta\theta_u \end{bmatrix}. \tag{7.19}$$

Based on Fig. 7.6 and (7.19), the voltage vector angle adjustment is achieved by injecting additional small signal voltages into dq-axis, the amplitudes of which can change with operating conditions in a self-adaptive way.

The regulated angles are generated by harmonics of sampling DC-link voltages at frequencies of $6\omega_g$ and $12\omega_g$. The harmonics of DC-link voltage $u_{\text{dc\_s},1}$ and $u_{\text{dc\_s},2}$, the central frequencies of which are respectively $6\omega_g$ and $12\omega_g$, can be extracted through two bandpass filters BPF1 and BPF2, which can be expressed as

$$\begin{cases} u_{\text{dc\_s},1} = U_{\text{dc},1}\sin\left(6\omega_g t + \varphi_1 - 3\omega_g T_s\right) \\ u_{\text{dc\_s},2} = U_{\text{dc},2}\sin\left(12\omega_g t + \varphi_2 - 6\omega_g T_s\right) \end{cases}. \tag{7.20}$$

The transfer functions of BPF1 and BPF2 can be expressed as

$$\begin{cases} \boldsymbol{B}_1(s) = \frac{6\xi_1\omega_g s}{s^2 + 6\xi_1\omega_g s + 36\omega_g^2} \\ \boldsymbol{B}_2(s) = \frac{12\xi_2\omega_g s}{s^2 + 12\xi_2\omega_g s + 144\omega_g^2} \end{cases} \tag{7.21}$$

where $\boldsymbol{B}_1(s)$ and $\boldsymbol{B}_2(s)$ are transfer functions of the bandpass filter, ξ_1 and ξ_2 are respectively the bandwidths of BPF1 and BPF2.

In order to calculate the regulated angle of (7.13) by the voltage harmonics of (7.20), the transfer functions of correction components should be designed as

$$\boldsymbol{M}_{\text{d\_1}}(s) = \frac{\boldsymbol{L}(\Delta\theta_{u1})}{\boldsymbol{L}\left(u_{\text{dc\_s},1}\right)}, \boldsymbol{M}_{\text{d\_2}}(s) = \frac{\boldsymbol{L}(\Delta\theta_{u2})}{\boldsymbol{L}\left(u_{\text{dc\_s},2}\right)} \tag{7.22}$$

where $\boldsymbol{M}_{\text{d\_1}}(s)$ and $\boldsymbol{M}_{\text{d\_2}}(s)$ are the transfer functions of correction components, $\boldsymbol{L}(\cdot)$ is the Laplace transformation.

The frequency characteristics of $\boldsymbol{M}_{\text{d\_1}}(s)$ and $\boldsymbol{M}_{\text{d\_2}}(s)$ at frequencies of $6\omega_g$ and $12\omega_g$ can be obtained from (7.16) and (7.20), which can be expressed as

$$\begin{cases} \boldsymbol{M}_{\text{d\_1}}\left(j6\omega_g\right) = K_1\left(\cos\theta_{\text{d\_1}} + j\sin\theta_{\text{d\_1}}\right) \\ \boldsymbol{M}_{\text{d\_2}}\left(j12\omega_g\right) = K_2\left(\cos\theta_{\text{d\_2}} + j\sin\theta_{\text{d\_2}}\right) \end{cases} \tag{7.23}$$

where

$$\begin{cases} K_1 = \dfrac{A_6}{U_{\mathrm{dc\_1}}}, \theta_{\mathrm{d\_1}} = \varphi_{\mathrm{u6}} - \varphi_1 - 3\omega_{\mathrm{g}} T_{\mathrm{s}} \\ K_2 = \dfrac{A_{12}}{U_{\mathrm{dc\_2}}}, \theta_{\mathrm{d\_2}} = \varphi_{\mathrm{u12}} - \varphi_2 - 6\omega_{\mathrm{g}} T_{\mathrm{s}} \end{cases}. \tag{7.24}$$

According to the frequency characteristics of $\boldsymbol{M}_{\mathrm{d\_1}}(s)$ and $\boldsymbol{M}_{\mathrm{d\_2}}(s)$, the transfer functions of correction components can be expressed as

$$\begin{cases} \boldsymbol{M}_{\mathrm{d\_1}}(s) = K_1\left(\cos\theta_{\mathrm{d\_1}} + \frac{\sin\theta_{\mathrm{d\_1}}}{6\omega_{\mathrm{g}}} s\right) \\ \boldsymbol{M}_{\mathrm{d\_2}}(s) = K_2\left(\cos\theta_{\mathrm{d\_2}} + \frac{\sin\theta_{\mathrm{d\_2}}}{12\omega_{\mathrm{g}}} s\right) \end{cases}. \tag{7.25}$$

Consequently, the regulated angle can be expressed as

$$\Delta\theta_{\mathrm{u}}(s) = \boldsymbol{M}_{\mathrm{d\_1}} \boldsymbol{B}_1(s) u_{\mathrm{dc\_s}} + \boldsymbol{M}_{\mathrm{d\_2}} \boldsymbol{B}_2(s) u_{\mathrm{dc\_s}} = \boldsymbol{G}_{\mathrm{mod}}(s) u_{\mathrm{dc}} \tag{7.26}$$

where $\boldsymbol{G}_{\mathrm{mod}}(s)$ is the transfer function between the regulated angle and DC-link voltage.

In order to achieve the best control performance, it is necessary to analyze the parameters of the correction components. The modulation harmonics can serve as the basis for parameter design as the modulation harmonics have been analyzed through the impedance model with the control strategy. According to (2.23), the transfer functions $\boldsymbol{W}_{\mathrm{dmod}}$ and $\boldsymbol{W}_{\mathrm{qmod}}$ between the DC-link voltage and the *dq*-axis currents can be modified as

$$\begin{bmatrix} \boldsymbol{W}_{\mathrm{dmod}} \\ \boldsymbol{W}_{\mathrm{qmod}} \end{bmatrix} = \begin{bmatrix} \boldsymbol{W}_{\mathrm{d}} \\ \boldsymbol{W}_{\mathrm{q}} \end{bmatrix} + \frac{\mathrm{e}^{-1.5sT_{\mathrm{s}}} \boldsymbol{G}_{\mathrm{mod}}}{\boldsymbol{A}_{2\mathrm{q}} \boldsymbol{A}_{1\mathrm{d}} - \boldsymbol{A}_{2\mathrm{d}} \boldsymbol{A}_{1\mathrm{q}}} \begin{bmatrix} -U_{\mathrm{qref0}} \boldsymbol{A}_{2\mathrm{q}} - U_{\mathrm{dref0}} \boldsymbol{A}_{1\mathrm{q}} \\ U_{\mathrm{dref0}} \boldsymbol{A}_{1\mathrm{d}} + U_{\mathrm{qref0}} \boldsymbol{A}_{2\mathrm{d}} \end{bmatrix}. \tag{7.27}$$

Furthermore, the relationship between equivalent values of compensated *dq*-axis voltage references and the small signal of the DC-link voltage can be expressed as

$$\begin{bmatrix} \Delta u_{\mathrm{dref\_r}} / \Delta u_{\mathrm{dc}} \\ \Delta u_{\mathrm{qref\_r}} / \Delta u_{\mathrm{dc}} \end{bmatrix} = \mathrm{e}^{-0.5sT_{\mathrm{s}}} \left\{ -\begin{bmatrix} \boldsymbol{G}_{\mathrm{d}} \boldsymbol{W}_{\mathrm{dmad}} \\ \boldsymbol{G}_{\mathrm{q}} \boldsymbol{W}_{\mathrm{qmad}} \end{bmatrix} + \begin{bmatrix} -U_{\mathrm{qref0}} \\ U_{\mathrm{dref0}} \end{bmatrix} \boldsymbol{G}_{\mathrm{mad}} \right\}. \tag{7.28}$$

By combining (7.12) and (7.28), the transfer function $\boldsymbol{G}_{\mathrm{smad}}$ between amplitudes of compensated *dq*-axis voltage references and the small signal of the DC-link voltage can be expressed as

$$\boldsymbol{G}_{\mathrm{smad}} = \frac{\Delta|\boldsymbol{u}_{\mathrm{sref}}|}{\Delta u_{\mathrm{dc}}} = -\frac{\mathrm{e}^{-0.5sT_{\mathrm{s}}} \begin{bmatrix} U_{\mathrm{dref0}}\left(\boldsymbol{G}_{\mathrm{d}} \boldsymbol{W}_{\mathrm{dmad}} + U_{\mathrm{qref0}} \boldsymbol{G}_{\mathrm{mad}}\right) \\ +U_{\mathrm{qref0}}\left(\boldsymbol{G}_{\mathrm{q}} \boldsymbol{W}_{\mathrm{qmad}} - U_{\mathrm{dref0}} \boldsymbol{G}_{\mathrm{mad}}\right) \end{bmatrix}}{U_{\mathrm{sref0}}}. \tag{7.29}$$

According to (7.29), the transfer function $\boldsymbol{K}_{\text{smad}}$ between the small signal of the modulation index with the linear modulation expansion strategy and the small signal of the DC-link voltage can be expressed as

$$\boldsymbol{K}_{\text{smad}} = \frac{\Delta m}{\Delta u_{\text{dc}}} = \frac{\sqrt{3}\left(U_{\text{dc0}}\boldsymbol{G}_{\text{smad}} - U_{\text{sref0}}\text{e}^{-0.5sT_s}\right)}{U_{\text{dc0}}^2}. \tag{7.30}$$

According to (7.29) and (7.30), the amplitudes and phases of $\boldsymbol{M}_{\text{d\_1}}(s)$ and $\boldsymbol{M}_{\text{d\_2}}(s)$ are designed based on $\boldsymbol{K}_{\text{smad}}(s)$ at frequencies of $6\omega_g$ and $12\omega_g$. The amplitude of $\boldsymbol{K}_{\text{smad}}(j6\omega_g)$ is shown in Fig. 7.7a with different amplitudes and phases of $\boldsymbol{M}_{\text{d\_1}}(s)$. When K_1 and $\theta_{\text{d\_1}}$ are respectively set to 2.2e^{-3} and 72°, $|\boldsymbol{K}_{\text{smad}}(j6\omega_g)|$ decreases from − 57 dB to − 86 dB. Similarly, the amplitude of $\boldsymbol{K}_{\text{smad}}(j12\omega_g)$ is shown in Fig. 7.7b with different amplitudes and phases of $\boldsymbol{M}_{\text{d\_2}}(s)$. When K_2 and $\theta_{\text{d\_2}}$ are respectively set to 1.8e^{-3} and 30°, $|\boldsymbol{K}_{\text{smad}}(j12\omega_g)|$ decreases from − 55 dB to − 79 dB. It can be seen that the proposed control strategy can effectively reduce the modulation harmonics at the frequencies of $6\omega_g$ and $12\omega_g$.

By substituting the designed parameters of correction components into (7.30), the Bode plots of $\boldsymbol{K}_{\text{smad}}(s)$ before and after applying the proposed control strategy are shown in Fig. 7.8. It can be observed that the amplitude of $\boldsymbol{K}_{\text{smad}}(s)$ respectively decreases to −89 dB and −85 dB at frequencies of $6\omega_g$ and $12\omega_g$, improving the linear modulation margin.

The main fluctuations of the linear modulation margin are caused by the harmonics of the DC-link voltage at the frequency of $6\omega_g$. Therefore, the linear modulation margin M_{mar} can be expressed as

$$M_{\text{mar}} = 1 - \frac{\sqrt{3}U_{\text{sref0}}}{U_{\text{dc0}}} - U_{\text{dc\_1}} \cdot \left|\boldsymbol{K}_{\text{smad}}(j6\omega_g)\right| = 1 - M_0 - U_{\text{dc\_1}} \cdot \left|\boldsymbol{K}_{\text{smad}}(j6\omega_g)\right|. \tag{7.31}$$

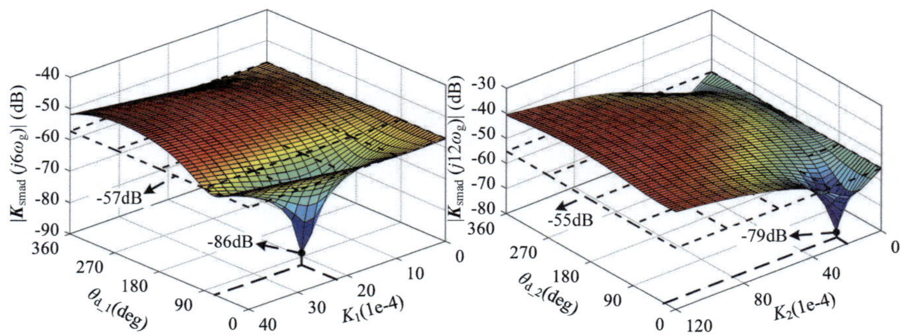

(a) 6ωg (b) 12ωg

Fig. 7.7 Magnitudes of Ksmad(s) with different amplitudes and phases of correction components, reprinted from Ref. [1], copyright 2022, with permission from Nature

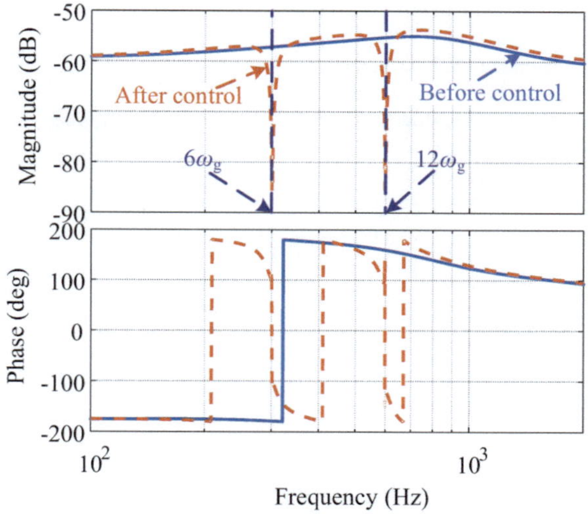

Fig. 7.8 Bode diagrams of $K_{smad}(s)$ before and after applying the proposed control strategy

According to (7.31), the maximum of the average modulation index in the linear modulation system $M_{0,\max}$ can be expressed as

$$M_{0,\max} = 1 - U_{\mathrm{dc\_1}} \cdot \left|\boldsymbol{K}_{\mathrm{smad}}(j6\omega_g)\right|. \tag{7.32}$$

In order to analyze the impact of the proposed control strategy on the utilization of the DC-link voltage, *a*-phase motor voltage u_a can be expressed as

$$u_a = mu_{dc}\sin(\omega_e t + \theta_u + \Delta\theta_u) \approx mu_{dc}[\sin(\omega_e t + \theta_u) + \Delta\theta_u\cos(\omega_e t + \theta_u)]. \tag{7.33}$$

By substituting (7.3) into (7.33), u_a contains the harmonics at the frequencies of ω_e, $(6n\omega_g - \omega_e)$, and $(6n\omega_g + \omega_e)$. the fundamental motor voltage $u_{a,0}$ can be expressed as

$$u_{a,0} = M_0 U_{dc,0}\sin(\omega_e t + \theta_u) + (U_6 + U_{12})\cos(\omega_e t + \theta_u) \tag{7.34}$$

where

$$\begin{cases} U_6 = 0.5A_6\left[M_0 U_{dc,1}\cos(\varphi_1 - \varphi_{u6}) + M_1 U_{dc,0}\cos(\varphi_{g1} - \varphi_{u6})\right] \\ U_{12} = 0.5A_{12}\left[M_0 U_{dc,2}\cos(\varphi_2 - \varphi_{u12}) + M_2 U_{dc,0}\cos(\varphi_{g2} - \varphi_{u12})\right] \end{cases}. \tag{7.35}$$

According to (7.32) and (7.34), the maximum of the fundamental motor voltage during the linear modulation can be expressed as

$$U_{\mathrm{amax},0} = \sqrt{U_{dc,0}^2\left[1 - U_{dc,1}\left|\boldsymbol{K}_{\mathrm{smad}}(j6\omega_g)\right|\right]^2 + (U_6 + U_{12})^2}. \tag{7.36}$$

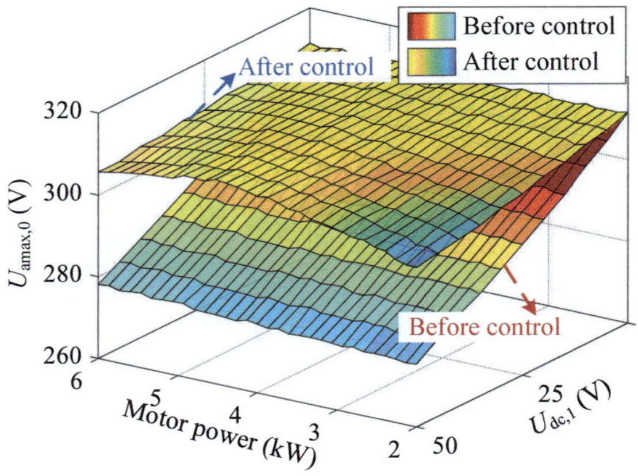

Fig. 7.9 Maximum amplitude of the fundamental motor voltage under different powers and harmonic of DC-link voltage, reprinted from Ref. [1], copyright 2022, with permission from Nature

According to (7.36), the maximum amplitude of the fundamental voltage at different motor frequencies and DC-link voltage harmonics is shown in Fig. 7.9. Due to the fluctuation of DC-link voltage, the maximum amplitude of the fundamental motor voltage without control can decrease to 270 V when the amplitude of the DC-link voltage harmonic increases to 50 V. The linear modulation expansion method can make the maximum amplitude increase to over 300 V, maintaining control performance despite variations in the motor power.

7.2.4 Experimental Verification

Figure 7.10a shows the experimental results at an operating frequency of 74 Hz. The waveforms from top to bottom are the DC-link voltage, the total dwell time, and the amplitude of the voltage vector, respectively. The total dwell time before control reaches $1.25e^{-4}$ s, indicating that the system is operating in the overmodulation region. The control strategy makes the total time of action less than $1.25e^{-4}$ s, indicating that the SVPWM is operating in the linear modulation region. Meanwhile, the peak-to-peak value of the voltage vector amplitude is reduced from 64.2 V to 40.9 V, and the voltage distortion is suppressed.

Figure 7.10b and c show enlarged views of the waveforms before and after control. The fluctuation frequency of the voltage vector amplitude is the same as that of the DC-link voltage. The maximum value of the voltage vector amplitude corresponds to the minimum value of the DC-link voltage, indicating that the phase difference between the two fluctuations is 180°, leading to the deterioration of the waveform of the regulation system. The fundamental frequency of the total dwell time is six times the motor frequency. After applying the control, the maximum value of the voltage vector amplitude corresponds to the maximum value of the DC-link voltage, and the fluctuation amplitude of the modulation index is suppressed from $0.50e^{-4}$ s to $0.30e^{-4}$ s.

Fig. 7.10 Experimental results at the operation frequency of 74 Hz, reprinted from Ref. [1], copyright 2022, with permission from Nature

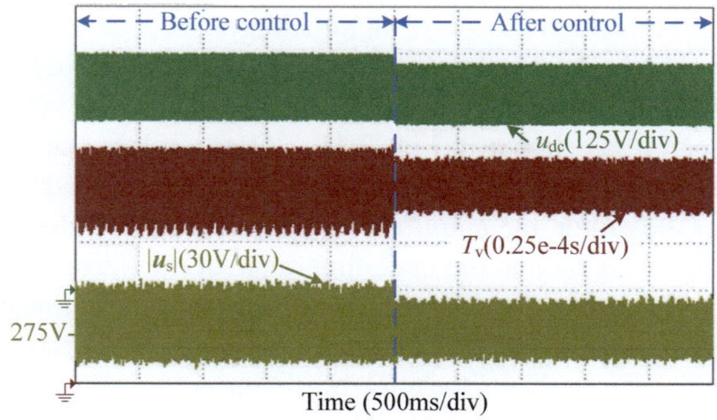

(a) Enable the proposed strategy

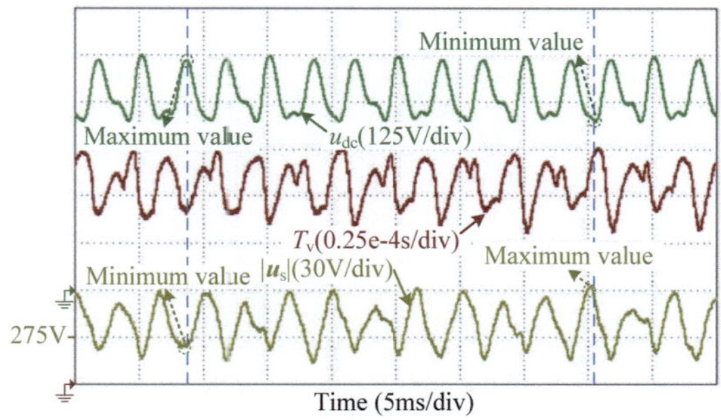

(b) Zoomed view with the proposed strategy

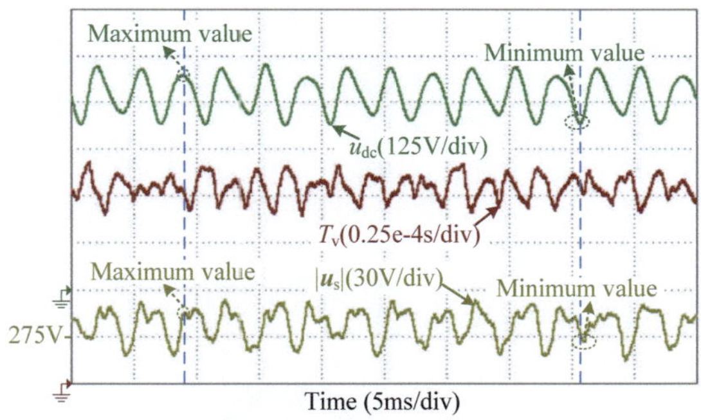

(c) Zoomed view without the proposed strategy

Figure 7.11a and b show the FFT of the total dwell time before and after applying the control strategy at a motor frequency of 74 Hz. As shown in Fig. 7.11a, the total dwell time consists of the fundamental component at a frequency of 444 Hz as well as the other harmonics at frequencies of $6\omega_g$ and $12\omega_g$. The frequencies of the main harmonics are the same as those of the DC-link voltage harmonics. The linear modulation expansion strategy reduces the harmonics at $6\omega_g$ and $12\omega_g$ frequencies from $1.08e^{-5}$ s and $0.24e^{-5}$ s to $0.18e^{-5}$ s and $0.14e^{-5}$ s, respectively. Figure 7.11c and d show the αβ-axis modulation indexes before and after the control strategy is applied. The amplitude of the modulation index before applying control will lie on the boundary of the hexagon, indicating that the system enters the overmodulation region, as shown in Fig. 7.11c. After applying the control, the fluctuation of the modulation index decreases from 0.22 to 0.10. Under the same operating conditions of the motor, the amplitude of the modulation index will not reach the boundary of the hexagon, and the operating range of the linear modulation region increases.

To demonstrate the control performance in the linear modulation region, Fig. 7.12 shows the experimental results for the motor frequency condition of 62 Hz. The maximum value of the total dwell time before applying control is less than 1.25e−4 s, indicating that the system is linearly modulated. The control strategy can reduce the maximum value of the total dwell time from 1.14e−4 s to 1.02e−4 s, increasing the linear modulation margin. As can be seen in Fig. 7.12a, the peak-to-peak value of

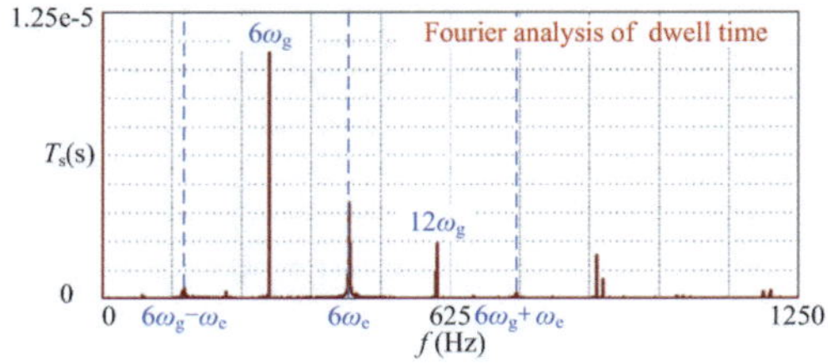

(a) Fourier analysis of dwell time without the method

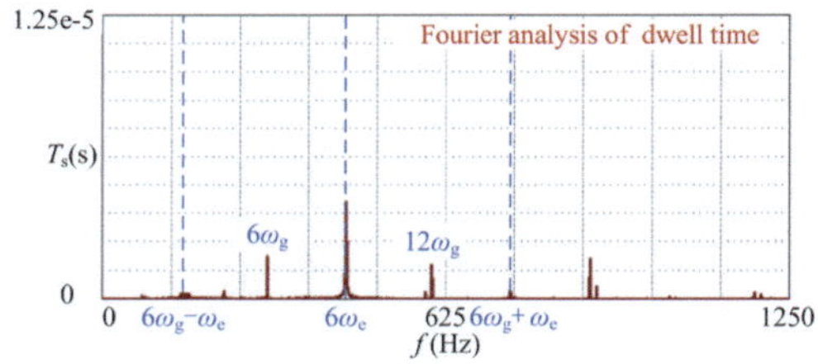

(b) Fourier analysis of dwell time with the method

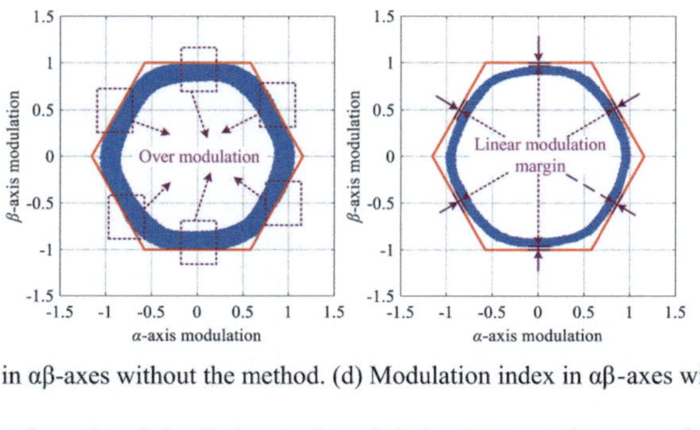

(c) Modulation index in αβ-axes without the method. (d) Modulation index in αβ-axes with the method

Fig. 7.11 Experimental results of dwell time and modulation index at the motor frequency of 74 Hz, reprinted from Ref. [1], copyright 2022, with permission from Nature the voltage vector amplitude is limited to between 44.8 and 27 V. Figure 7.12b and c show the α-β axis components of the modulation index. The method suppresses the fluctuation of the modulation index from 0.18 to 0.07 and the distance between the boundary hexagon and the modulation index amplitude is increased after applying the method, thus improving the linear modulation margin.

To evaluate the performance at the point in time when the control strategy is enabled, the waveforms of DC-link voltage, voltage vector amplitude, motor position and motor speed are shown in Fig. 7.13. It can be seen that the peak value of the motor speed ripple decreases from 56 r/min to 24 r/min and there is no overshoot during the enabling process. As a result, the DC link voltage and speed fluctuations can be effectively reduced due to the extension of the linear modulation region.

Figure 7.14 shows the experimental results for the maximum operating frequency that can be achieved with linear modulation. From Fig. 7.14a, it can be seen that the maximum motor frequency that can be achieved without applying the control strategy is 71 Hz, when the linear modulation margin is close to 0 and the voltage vector action time is severely distorted. Figure 7.14b shows the results after applying the proposed method, where the maximum motor frequency that can be achieved is 80 Hz when the linear modulation margin is close to 0 and the modulation index degree is suppressed. Therefore, by applying this method, the limiting speed of the motor can be increased by 12.7%. A Phase voltage amplitude is increased from 265

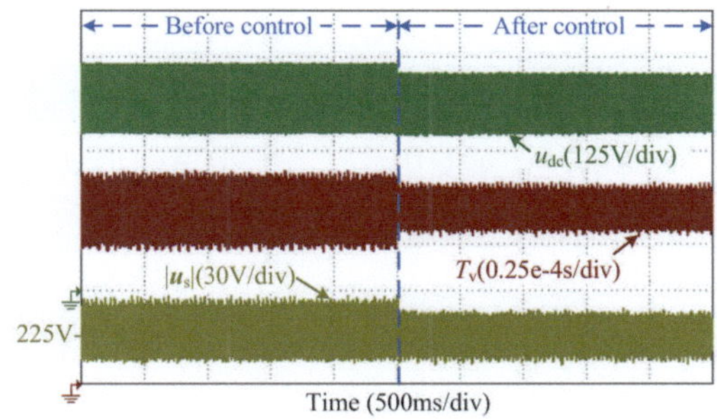

(a) Enable the proposed strategy

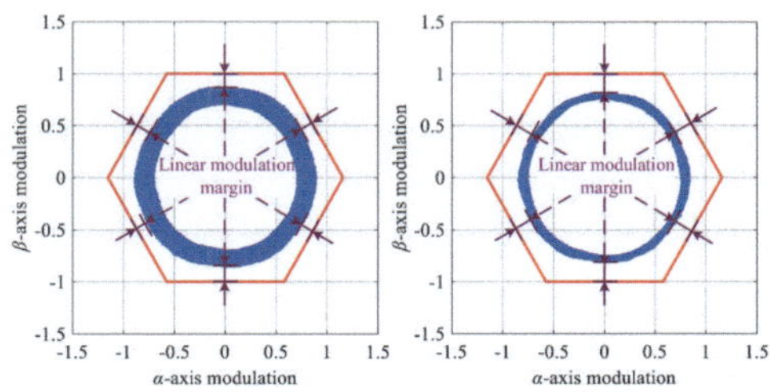

(b) Modulation index in αβ-axes without the method. (c) Modulation index in αβ-axes with the method

Fig. 7.12 Experimental results at the operation frequency of 62 Hz, reprinted from Ref. [1], copyright 2022, with permission from Nature

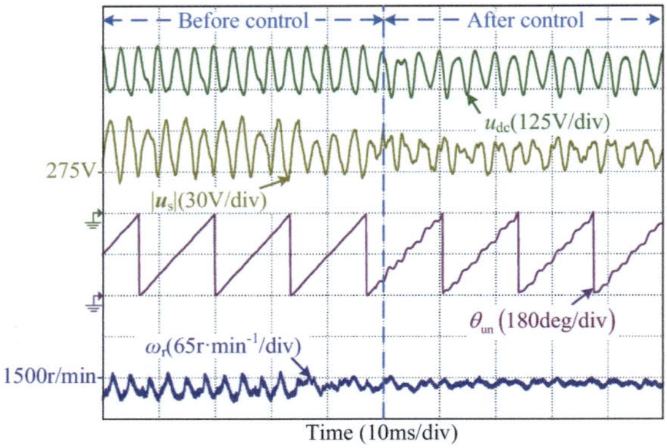

Fig. 7.13 Experimental results at the time point of enabling the proposed method

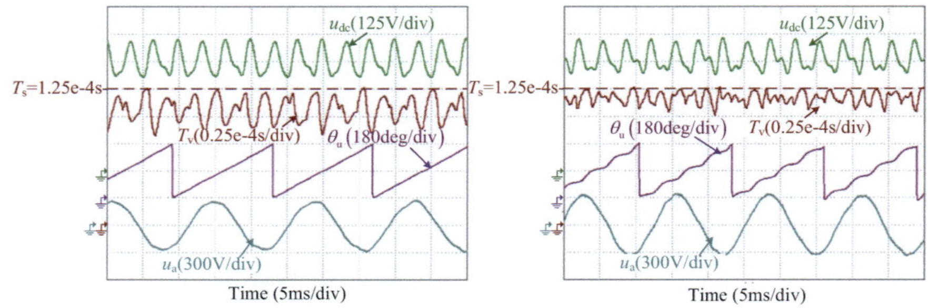

(a) Without the control method (71Hz) (b) With the control method (80Hz)

Fig. 7.14 Experimental results of attenable motor frequency, reprinted from Ref. [1], copyright 2022, with permission from Nature

to 302 V, which indicates that the utilization of the DC-link voltage can be increased by 13.2%.

The experimental waveform of the speed change process is shown in Fig. 7.15a when the motor speed changes from 1240 to 1480 r/min and back to 1240 r/min. From the rising region, it can be seen that the amplitude of DC-link voltage fluctuation increases from 74 to 81 V, while the opposite is true in the falling region. Figure 7.15b shows the experimental waveforms when the load torque varies between 25 and 32 N m. There is no deterioration in the stability of the motor speed and there is no drop in the DC-link voltage during the torque and speed variations. Therefore, the proposed method maintains stable control performance in applications and is adaptable to torque and speed variations.

Figure 7.16 shows the maximum value of the modulation index under different operating frequency conditions. It can be seen that without applying the control strategy, the SVPWM will enter the over-modulation region when the motor frequency is 65 Hz. The linear region extension strategy can reduce the maximum

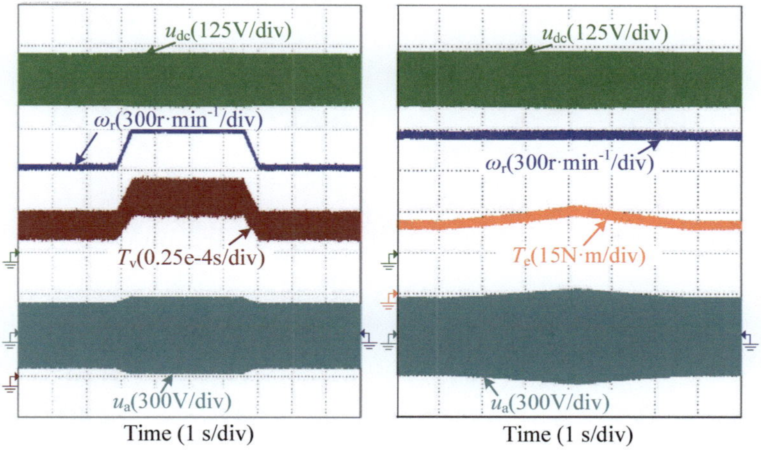

(a) Speed dynamic process (b) Load torque dynamic process

Fig. 7.15 Dynamic experimental waveforms of the proposed method

Fig. 7.16 Maximum value of modulation index under different motor frequencies

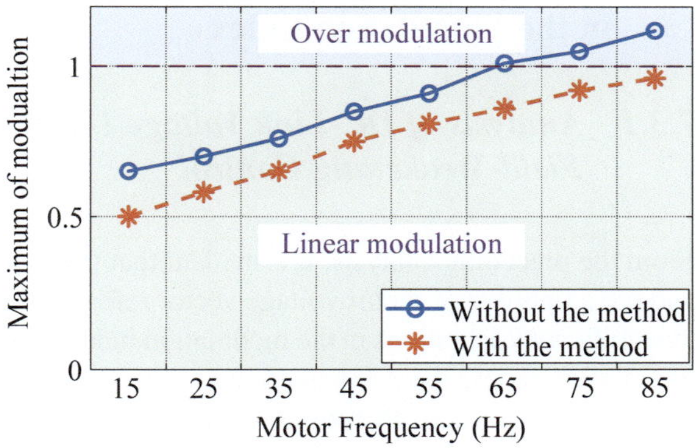

value of the modulation index by more than 0.1 over the entire frequency range, and the frequency boundary of linear modulation is increased from 65 to 85 Hz.

Figure 7.17a shows the maximum value of the attainable operating frequency for different load torques. It can be seen that without applying the proposed method, the maximum frequency of the motor will be reduced to 65 Hz when the load torque is increased to 36 N m. The method increases the maximum value of the motor frequency by more than 5 Hz in the torque range from 22 to 36 N m. Figure 7.17b shows the copper losses at different load torques. Before applying the method, the power loss in the stator resistor will be more than 200 W when the load torque is increased to 32 N m. The use of the linear zone expansion strategy reduces the copper loss by more than 13 W. The stator resistor will be more than 200 W when the load torque is increased to 32 N m.

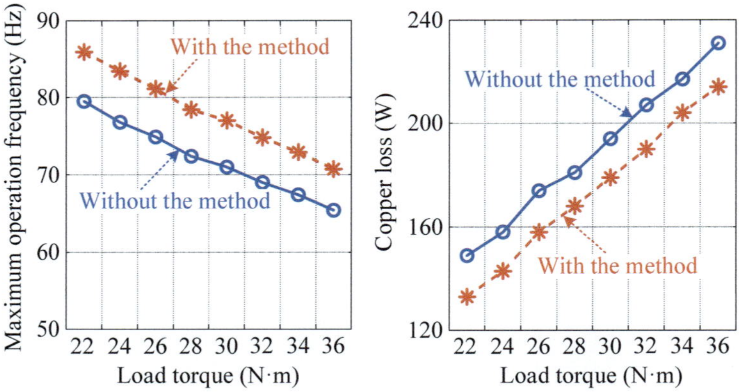

(a) Maximum value of attainable operation frequency (b) Copper loss of the stator resistance

Fig. 7.17 Experimental results of maximum operation frequency and copper loss under different load torques

7.3 Adaptive Suppression of Fluctuations in the Modulation Index

7.3.1 Analysis of DC-Link Voltage Utilization Under Field-Weakening Control

From the preceding analysis, it is evident that the harmonics of the DC-link voltage and the amplitude of motor voltage vector reference are nearly out of phase, leading to significant fluctuations in the modulation index. Under the constraint of maximum modulation index, these fluctuations result in a reduction of the DC component of the modulation index, thereby lowering the utilization rate of the DC-link voltage. To quantitatively analyze DC-link voltage utilization, it is necessary to establish the transfer function between the output voltage of the uncontrolled rectifier bridge and the modulation index.

The voltage-closed-loop field-weakening control method is less dependent on parameter tuning and widely applicable. When adopting this control strategy, the reference d-axis current can be expressed as

$$i_{\text{dref}} = \boldsymbol{G}_{\text{f}}\left(K_{\text{dc}} u_{\text{dc\_s}} - \sqrt{3}|u_{\text{sref}}|\right). \tag{7.37}$$

Considering sampling delays, the small-signal expression for the d-axis current is given by

$$\Delta i_{\text{dref}} = \boldsymbol{G}_{\text{f}}\left(K_{\text{dc}} T_{\text{d}} \Delta u_{\text{dc}} - \sqrt{3}\frac{U_{\text{d\_0}}\Delta u_{\text{dref}} + U_{\text{q\_0}}\Delta u_{\text{qref}}}{U_{\text{s\_0}}}\right). \tag{7.38}$$

Under field-weakening control, the small-signal expressions for the dq-axis voltage references are

$$\begin{bmatrix} \Delta u_{\mathrm{dref}} \\ \Delta u_{\mathrm{qref}} \end{bmatrix} = \begin{bmatrix} -\boldsymbol{G}_{\mathrm{d}} & -\omega_{\mathrm{e}}L_{\mathrm{q}} \\ \omega_{\mathrm{e}}L_{\mathrm{d}} & -\boldsymbol{G}_{\mathrm{q}} \end{bmatrix} \begin{bmatrix} \Delta i_{\mathrm{d}} \\ \Delta i_{\mathrm{q}} \end{bmatrix} T_{\mathrm{d}} + \begin{bmatrix} \boldsymbol{G}_{\mathrm{d}}\Delta i_{\mathrm{dref}} \\ 0 \end{bmatrix}. \quad (7.39)$$

Combining (7.38) and (7.39) with the motor voltage equations, the transfer function between the DC-link voltage and the d–q-axis currents can be derived as

$$\begin{bmatrix} \boldsymbol{W}_{\mathrm{d}} \\ \boldsymbol{W}_{\mathrm{q}} \end{bmatrix} = \frac{1}{\Delta u_{\mathrm{dc}}} \begin{bmatrix} \Delta i_{\mathrm{d}} \\ \Delta i_{\mathrm{q}} \end{bmatrix} = \left\{ k_{\mathrm{f}} U_{\mathrm{s\_0}} T_{\mathrm{d}} \boldsymbol{G}_{\mathrm{dc}} \begin{bmatrix} \boldsymbol{G}_{\mathrm{d}}\boldsymbol{A}_{22} \\ -\boldsymbol{G}_{\mathrm{d}}\boldsymbol{A}_{21} \end{bmatrix} + \frac{1-\mathrm{e}^{-1.5sT_{\mathrm{s}}}}{U_{\mathrm{dc\_0}}} \begin{bmatrix} U_{\mathrm{d\_0}}\boldsymbol{A}_{22} - U_{\mathrm{q\_0}}\boldsymbol{A}_{12} \\ -U_{\mathrm{d\_0}}\boldsymbol{A}_{21} + U_{\mathrm{q\_0}}\boldsymbol{A}_{11} \end{bmatrix} \right\} \frac{1}{\boldsymbol{A}_{22}\boldsymbol{A}_{11} - \boldsymbol{A}_{12}\boldsymbol{A}_{21}}. \quad (7.40)$$

where $\boldsymbol{G}_{\mathrm{dc}}$ can be expressed as

$$\boldsymbol{G}_{\mathrm{dc}} = \boldsymbol{G}_{\mathrm{f}} \Big/ \left(U_{\mathrm{s\_0}} + \sqrt{3} U_{\mathrm{d\_0}} \boldsymbol{G}_{\mathrm{d}} \boldsymbol{G}_{\mathrm{f}} \right) \quad (7.41)$$

where the elements of matrix $\boldsymbol{A}$ are

$$\begin{cases} \boldsymbol{A}_{11} = R_{\mathrm{s}} + L_{\mathrm{d}}s + \left[\boldsymbol{G}_{\mathrm{d}} - \sqrt{3}\boldsymbol{G}_{\mathrm{dc}}\boldsymbol{G}_{\mathrm{d}}\left(U_{\mathrm{d\_0}}\boldsymbol{G}_{\mathrm{d}} - U_{\mathrm{q\_0}}\omega_{\mathrm{e}}L_{\mathrm{d}}\right) \right] T_{\mathrm{d}} \\ \boldsymbol{A}_{12} = T_{\mathrm{d}}\left[\omega_{\mathrm{e}}L_{\mathrm{q}} - \sqrt{3}\boldsymbol{G}_{\mathrm{dc}}\boldsymbol{G}_{\mathrm{d}}\left(U_{\mathrm{q\_0}}\boldsymbol{G}_{\mathrm{q}} + U_{\mathrm{d\_0}}\omega_{\mathrm{e}}L_{\mathrm{q}}\right) \right] - \omega_{\mathrm{e}}L_{\mathrm{q}} \\ \boldsymbol{A}_{21} = (1 - T_{\mathrm{d}})\omega_{\mathrm{e}}L_{\mathrm{d}} \\ \boldsymbol{A}_{22} = R_{\mathrm{s}} + L_{\mathrm{q}}s + \boldsymbol{G}_{\mathrm{q}}T_{\mathrm{d}} \end{cases}. \quad (7.42)$$

By further combining (7.38), (7.39) and (7.40), the transfer function between the d-q-axis voltage commands and the bus voltage under field-weakening control can be obtained

$$\begin{bmatrix} \boldsymbol{B}_{11} & \boldsymbol{B}_{12} \\ \boldsymbol{B}_{21} & \boldsymbol{B}_{22} \end{bmatrix} \begin{bmatrix} \Delta u_{\mathrm{dref}} \\ \Delta u_{\mathrm{qref}} \end{bmatrix} = \begin{bmatrix} \boldsymbol{C}_{11} \\ \boldsymbol{C}_{21} \end{bmatrix} \Delta u_{\mathrm{dc}} \quad (7.43)$$

where the elements of matrices $\boldsymbol{B}$ and $\boldsymbol{C}$ are

$$\begin{cases} \boldsymbol{B}_{11} = 1 + \sqrt{3}\boldsymbol{G}_{\mathrm{d}}\boldsymbol{G}_{\mathrm{f}}U_{\mathrm{d\_0}}/U_{\mathrm{s\_0}} \\ \boldsymbol{B}_{12} = \sqrt{3}\boldsymbol{G}_{\mathrm{d}}\boldsymbol{G}_{\mathrm{f}}U_{\mathrm{q\_0}}/U_{\mathrm{s\_0}} \\ \boldsymbol{B}_{21} = 0 \\ \boldsymbol{B}_{22} = 1 \end{cases}, \quad (7.44)$$

$$\begin{cases} \boldsymbol{C}_{11} = \left(-\boldsymbol{G}_{\mathrm{d}}\boldsymbol{W}_{\mathrm{d}} - \omega_{\mathrm{e}}L_{\mathrm{q}}\boldsymbol{W}_{\mathrm{q}} + \boldsymbol{G}_{\mathrm{d}}\boldsymbol{G}_{\mathrm{f}}k_{\mathrm{f}}\right)T_{\mathrm{d}} \\ \boldsymbol{C}_{21} = \left(\omega_{\mathrm{e}}L_{\mathrm{d}}\boldsymbol{W}_{\mathrm{d}} - \boldsymbol{G}_{\mathrm{q}}\boldsymbol{W}_{\mathrm{q}}\right)T_{\mathrm{d}} \end{cases}. \quad (7.45)$$

From (7.43), the transfer function of small-signals between the motor voltage vector command amplitude and the DC-link voltage, $\boldsymbol{G}_{1\mathrm{f}}$, can be expressed as

$$\begin{aligned}\boldsymbol{G}_{1\mathrm{f}} &= \frac{\Delta|u_{\mathrm{sref}}|}{\Delta u_{\mathrm{dc}}} \\ &= [(-\boldsymbol{G}_{\mathrm{d}}W_{\mathrm{d}} + \boldsymbol{G}_{\mathrm{dc}}K_{\mathrm{dc}}U_{\mathrm{sref0}} + \boldsymbol{G}_{\mathrm{dc}}U_{\mathrm{dref0}}\boldsymbol{G}_{\mathrm{d}}W_{\mathrm{d}} \\ &\quad + \boldsymbol{G}_{\mathrm{dc}}U_{\mathrm{qref0}}\boldsymbol{G}_{\mathrm{q}}W_{\mathrm{q}})U_{\mathrm{dref0}} - \boldsymbol{G}_{\mathrm{q}}W_{\mathrm{q}}U_{\mathrm{qref0}}]\mathrm{e}^{-0.5sT_{\mathrm{s}}}/U_{\mathrm{sref0}}. \end{aligned} \tag{7.46}$$

Subsequently, the transfer function between the small-signal modulation index Δm and the small-signal output voltage of the rectifier bridge Δu_{grec} is given as

$$\boldsymbol{G}_{\mathrm{m}}(s) = \frac{\Delta m}{\Delta u_{\mathrm{g}}} = \frac{\sqrt{3}(U_{\mathrm{dc0}}\boldsymbol{G}_{1\mathrm{f}} - U_{\mathrm{sref0}})}{\left[(L_{\mathrm{g}}s + R_{\mathrm{g}})(C_{\mathrm{dc}}s + \boldsymbol{Y}_{\mathrm{m}}) + 1\right]U_{\mathrm{dc0}}^2}. \tag{7.47}$$

The modulation index fluctuations are primarily caused by the $6\omega_{\mathrm{g}}$-order harmonic in the DC-link voltage. The maximum average value of the motor voltage vector command, U_{smax}, can be expressed as

$$U_{\mathrm{smax}} = \frac{(M_{\mathrm{b}} - M_1)U_{\mathrm{dc0}}}{\sqrt{3}} = \frac{\left(M_{\mathrm{b}} - U_{\mathrm{grec1}}\boldsymbol{G}_{\mathrm{m}}(j6\omega_{\mathrm{g}})\right)U_{\mathrm{dc0}}}{\sqrt{3}} \tag{7.48}$$

where M_{b} is the upper limit of the modulation index, M_1 is the amplitude of the $6\omega_{\mathrm{g}}$-order harmonic in the modulation index, and $U_{\mathrm{grec\_1}}$ is the amplitude of the $6\omega_{\mathrm{g}}$-order harmonic in the rectifier bridge output voltage.

Excessive modulation index fluctuations force the system to frequently enter the overmodulation region, resulting in current harmonics related to the operating frequency, which further degrades motor torque and speed performance. Therefore, in this study, the maximum modulation index is restricted within the linear region for voltage utilization analysis. This analysis method also applies to the overmodulation region. From (7.48), under a supply voltage of 380 V and a frequency of 50 Hz, the relationship between the maximum mean value of the motor voltage vector command and the load torque and operating frequency is depicted in Fig. 7.18. Large fluctuations in the modulation index cause a significant gap between the mean motor voltage vector and the voltage boundary. Hence, suppressing modulation index fluctuations can improve voltage utilization during field-weakening control.

7.3.2 Analysis and Performance Verification of Modulation Index Fluctuation Suppression Method

To suppress the modulation index fluctuations and improve voltage utilization, a motor voltage vector angle compensation method is proposed, as illustrated in the

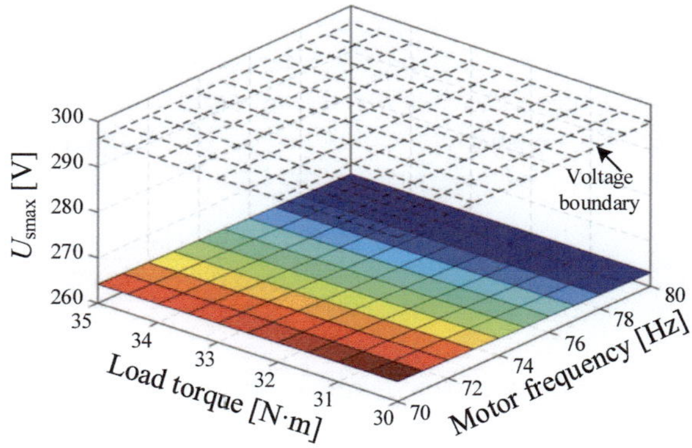

Fig. 7.18 Maximum average of the motor voltage vector reference

control block diagram in Fig. 7.19. The input to the angle compensation controller is the harmonic component at frequencies $6\omega_g$ and $12\omega_g$ of the product of the DC-link voltage and the DC component of the modulation index, while the feedback is the harmonic component at the same frequencies of amplitude of the motor voltage vector reference. The compensated angle is superimposed on the motor voltage vector angle. The adaptive field-weakening coefficient controller uses the peak modulation index as feedback, with the modulation index upper boundary M_b as the reference.

The compensation angle should adjust the amplitude and phase difference between the DC-link voltage and the motor voltage vector command amplitude at frequencies

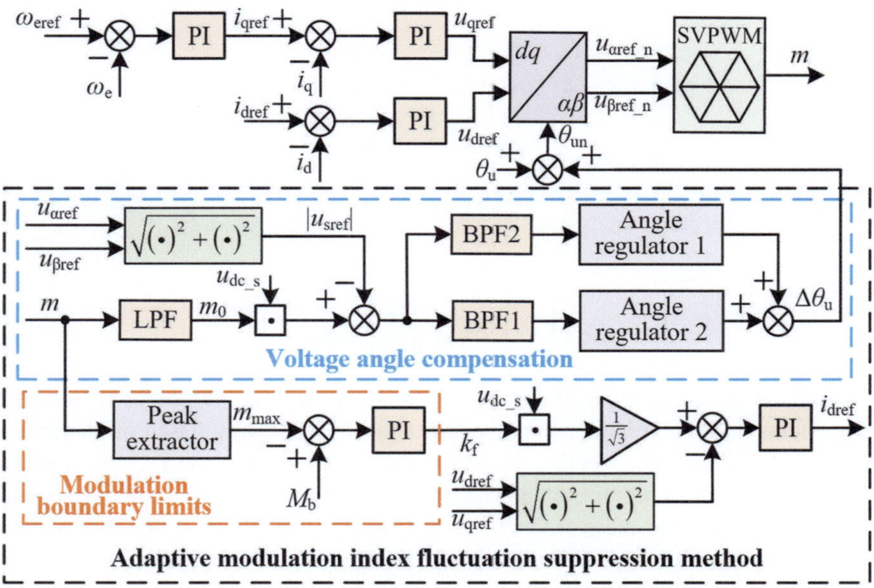

Fig. 7.19 Block diagram of the proposed adaptive MI fluctuation suppression strategy, reprinted from Ref. [2], copyright 2022, with permission from Nature

$6\omega_g$ and $12\omega_g$. Therefore, the compensation angle $\Delta\theta_e$ should include harmonics at these frequencies. Bandpass filters BPF1 and BPF2 are centered at $6\omega_g$ and $12\omega_g$, respectively. Consequently, the input and output of the angle compensation controller share the same harmonic frequencies. The transfer function of the angle compensation controller can be derived via the Laplace transform

$$\begin{aligned} G_{\theta 1}(s) &= \frac{\theta_{u1}}{U_c 1}\frac{s\cos\varphi_{\theta u1}-6\omega_g\sin\varphi_{\theta u1}}{s\cos\varphi_{c1}-6\omega_g\sin\varphi_{c1}} \\ G_{\theta 2}(s) &= \frac{\theta_{u2}}{U_{c2}}\frac{s\cos\varphi_{\theta u12}-12\omega_g\sin\varphi_{\theta u2}}{s\cos\varphi_{c2}-12\omega_g\sin\varphi_{c2}} \end{aligned} \tag{7.49}$$

Given the dominant harmonic frequencies $6\omega_g$ and $12\omega_g$, the transfer function at these points can be expressed as

$$\begin{cases} G_{\theta 1}(j6\omega_g) = \frac{\theta_{u1}}{U_{c1}}\frac{j6\omega_g\cos\varphi_{\theta u1}-6\omega_g\sin\varphi_{\theta u1}}{j6\omega_g\cos\varphi_{c1}-6\omega_g\sin\varphi_{c1}} \\ G_{\theta 2}(j12\omega_g) = \frac{\theta_{u2}}{U_{c2}}\frac{j12\omega_g\cos\varphi_{\theta u2}-12\omega_g\sin\varphi_{\theta u2}}{j12\omega_g\cos\varphi_{c2}-12\omega_g\sin\varphi_{c2}} \end{cases}. \tag{7.50}$$

Using Euler's formula, the controller transfer function is expressed as

$$\begin{cases} G_{\theta 1}(s) = \frac{\theta_{u1}}{U_{c1}}\left(\cos(\varphi_{\theta u1}-\varphi_{c1})+\frac{s}{6\omega_g}\sin(\varphi_{\theta u1}-\varphi_{c1})\right) \\ G_{\theta 2}(s) = \frac{\theta_{u2}}{U_{c2}}\left(\cos(\varphi_{\theta u2}-\varphi_{c2})+\frac{s}{12\omega_g}\sin(\varphi_{\theta u2}-\varphi_{c2})\right) \end{cases}. \tag{7.51}$$

The angle compensation regulator has a proportional-derivative form, where $k_1 = \theta_{e1}/c_1$, $k_2 = \theta_{e2}/c_2$, $\theta_1 = \varphi_{e1}-\varphi_{c1}$, $\theta_2 = \varphi_{e2}-\varphi_{c2}$. Thus, the parameters of the angle compensation controller are

$$\begin{cases} k_{P1} = k_1\cos\theta_1 \\ k_{D1} = \dfrac{k_1}{6\omega_g}\sin\theta_1 \end{cases}, \begin{cases} k_{P2} = k_2\cos\theta_2 \\ k_{D2} = \dfrac{k_2}{12\omega_g}\sin\theta_2 \end{cases}. \tag{7.52}$$

From the preceding analysis

$$\begin{bmatrix} \Delta u_{dref\_n} \\ \Delta u_{qref\_n} \end{bmatrix} = \begin{bmatrix} \Delta u_{dref} \\ \Delta u_{qref} \end{bmatrix} + \begin{bmatrix} -U_{qref0} \\ U_{dref0} \end{bmatrix}\Delta\theta_u. \tag{7.53}$$

From (7.53), the compensation angle effectively injects harmonics of relevant frequencies into the voltage command, automatically adapting based on the motor voltage's DC component. When the phase difference between the DC-link voltage and the amplitude of motor voltage vector reference at $6\omega_g$ and $12\omega_g$ decreases and the amplitude ratio at these frequencies approaches the DC ratio, the modulation index fluctuations are suppressed. Through the adaptive field-weakening coefficient controller, the motor operates within the linear modulation region. Once the field-weakening loop converges, the field-weakening coefficient k_f can be expressed as

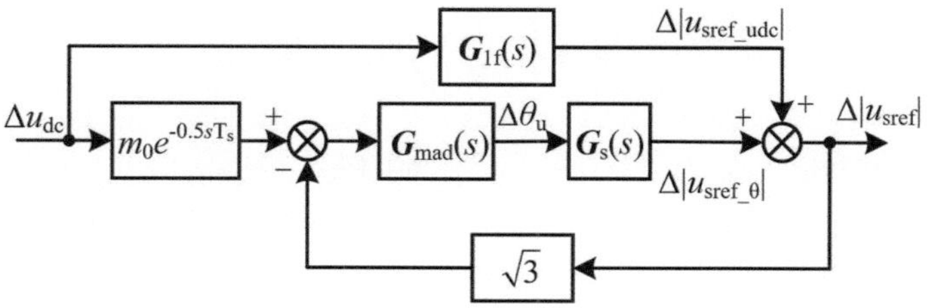

Fig. 7.20 Small signal block diagram of DC-link voltage and motor voltage vector reference amplitude

$$k_{\mathrm{f}} = \boldsymbol{G}_{\mathrm{fw}}\left(m_{\mathrm{boundary}} - m_{\mathrm{max}}\right) = \frac{\sqrt{3}U_{\mathrm{sref0}}}{U_{\mathrm{dc0}}} = m_0. \tag{7.54}$$

Indicating that k_{f} is positively correlated with the DC component of the modulation index m_0. Voltage utilization is negatively correlated with modulation index fluctuations. After compensating for the angle in the inverse Park transformation, the small-signal block diagram for the DC-link voltage and the motor voltage vector reference amplitude is shown in Fig. 7.20. The motor voltage vector reference amplitude is influenced by both the DC-link voltage and the compensation angle. The portion influenced by the bus voltage can be expressed by (7.46). The influence of the compensation angle's small-signal on the motor voltage vector reference amplitude is analyzed as follows. In Fig. 7.20, the transfer function $G_{\mathrm{mad}}(\mathrm{s})$ represents the cascade of the bandpass filter and the angle compensation controller

$$\boldsymbol{G}_{\mathrm{mad}}(s) = \boldsymbol{G}_{\mathrm{BPF1}}(s)(k_{\mathrm{P1}} + k_{\mathrm{D1}}s) + \boldsymbol{G}_{\mathrm{BPF2}}(s)(k_{\mathrm{P2}} + k_{\mathrm{D2}}s). \tag{7.55}$$

After compensating for the angle, combined with (7.53), (7.40) can be rewritten as:

$$\begin{bmatrix} \boldsymbol{A}_{1\mathrm{d}} & \boldsymbol{A}_{1\mathrm{q}} \\ \boldsymbol{A}_{2\mathrm{d}} & \boldsymbol{A}_{2\mathrm{q}} \end{bmatrix} \begin{bmatrix} \Delta i_{\mathrm{d}} \\ \Delta i_{\mathrm{q}} \end{bmatrix} = \left\{ \frac{1 - \mathrm{e}^{-1.5sT_{\mathrm{s}}}}{U_{\mathrm{dc0}}} \begin{bmatrix} U_{\mathrm{dref0}} \\ U_{\mathrm{qref0}} \end{bmatrix} + \begin{bmatrix} K_{\mathrm{dc}} U_{\mathrm{sref0}} \mathrm{e}^{-1.5sT_{\mathrm{s}}} \boldsymbol{G}_{\mathrm{dc}} \\ 0 \end{bmatrix} \right\} \Delta u_{\mathrm{dc}} + \begin{bmatrix} -U_{\mathrm{qref0}} \\ U_{\mathrm{dref0}} \end{bmatrix} \Delta\theta_{\mathrm{u}} \tag{7.56}$$

Then the transfer function between the angular small signal $\Delta\theta_{\mathrm{e}}$ and its induced additional small signal $\Delta i_{\mathrm{d,q\theta}}$ of the d–q axis current can be expressed as follows.

$$\begin{bmatrix} \boldsymbol{W}_{\mathrm{d}\theta} \\ \boldsymbol{W}_{\mathrm{q}\theta} \end{bmatrix} = \frac{1}{\Delta\theta_{\mathrm{u}}} \begin{bmatrix} \Delta i_{\mathrm{d}\theta} \\ \Delta i_{\mathrm{q}\theta} \end{bmatrix} = \begin{bmatrix} \frac{-\boldsymbol{A}_{2\mathrm{q}} U_{\mathrm{qref0}} - \boldsymbol{A}_{1\mathrm{q}} U_{\mathrm{dref0}}}{\boldsymbol{A}_{2\mathrm{q}} \boldsymbol{A}_{1\mathrm{d}}} \\ \frac{\boldsymbol{A}_{1\mathrm{d}} U_{\mathrm{dref0}}}{\boldsymbol{A}_{2\mathrm{q}} \boldsymbol{A}_{1\mathrm{d}}} \end{bmatrix}. \tag{7.57}$$

Bringing (7.57) into (2.19) yields a transfer function between the angular small signal $\Delta\theta_{\mathrm{u}}$ and its induced additional small signal of the dq axis voltage reference $\Delta u_{\mathrm{d,qref\theta}}$ can be expressed as follows

$$\begin{bmatrix} \Delta u_{\text{dref}\theta} \\ \Delta u_{\text{qref}\theta} \end{bmatrix} = \begin{bmatrix} -\boldsymbol{W}_{\text{d}\theta}\boldsymbol{G}_{\text{d}} \\ -\boldsymbol{W}_{\text{q}\theta}\boldsymbol{G}_{\text{q}} \end{bmatrix} \text{e}^{-0.5sT_{\text{s}}} \Delta\theta_{\text{u}}. \tag{7.58}$$

The transfer function $\boldsymbol{G}_{\text{s}}$ between the angular small signal $\Delta\theta_{\text{u}}$ and its induced additional small signal $\Delta|u_{\text{sref\_}\theta}|$ of the commanded amplitude of the motor voltage vector can be expressed as follows

$$\begin{aligned} \boldsymbol{G}_{\text{s}}(s) &= \frac{\Delta\left|u_{\text{sref\_}\theta}\right|}{\Delta\theta_{\text{u}}} \\ &= \frac{1}{U_{\text{sref0}}}\left[U_{\text{dref0}}\left(-\boldsymbol{W}_{\text{d}\theta}\boldsymbol{G}_{\text{d}} - \boldsymbol{W}_{\text{q}\theta}\omega_{\text{e}}L_{\text{q}}\right)\right. \\ &\quad \left.+U_{\text{qref0}}\left(\boldsymbol{W}_{\text{d}\theta}\omega_{\text{e}}L_{\text{d}} - \boldsymbol{W}_{\text{q}\theta}\boldsymbol{G}_{\text{q}}\right)\right] \end{aligned} \tag{7.59}$$

Combined with the control block diagram, the closed-loop transfer function between the angle-compensated motor voltage vector reference amplitude of the small signal $\Delta|u_{\text{sref}}|$ and the DC-link voltage small signal can be obtained as follows.

$$\boldsymbol{G}_{\text{b}} = \frac{\Delta|u_{\text{sref}}|}{\Delta u_{\text{dc}}} = \frac{m_0\text{e}^{-0.5sT_{\text{s}}}\boldsymbol{G}_{\text{mad}}\boldsymbol{G}_{\text{s}} + \boldsymbol{G}_{\text{1f}}}{1 + \sqrt{3}\boldsymbol{G}_{\text{mad}}\boldsymbol{G}_{\text{s}}}. \tag{7.60}$$

Analyzing the Bode plot of $\boldsymbol{G}_{\text{b}}(s)$, as shown in Fig. 7.21, the phase difference at $6\omega_{\text{g}}$ between the DC-link voltage and the motor voltage vector reference amplitude decreases from 162° (uncontrolled) to 8° (controlled). At $12\omega_{\text{g}}$, the phase difference reduces from 177° to 12°. The DC-link voltage and motor voltage vector reference amplitude become nearly in-phase, suppressing modulation index fluctuations.

To evaluate the impact of modulation index fluctuation suppression on voltage utilization, the transfer function between the modulation index and the rectifier output

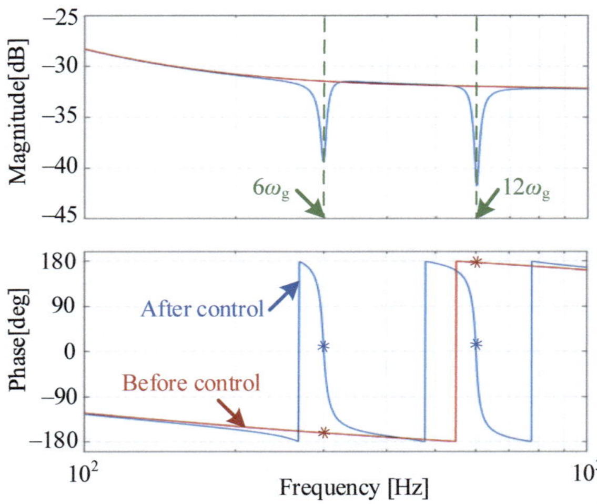

Fig. 7.21 Bode diagram of the transfer function between the amplitude of motor voltage vector reference and the DC-link voltage

voltage after control is analyzed. The small signal $\Delta\theta_u$ of the compensation angle can be expressed as a function of the small signal Δu_{dc} of the DC-link voltage

$$\Delta\theta_u = \Delta u_{dc}\boldsymbol{G}_{mad}\left(m_0 - \sqrt{3}\boldsymbol{G}_b\right). \tag{7.61}$$

Combining Eqs. (2.36), (7.57), and (7.61), the transfer function between the d–q-axis current small signal $\Delta i_{d,q}$ and the bus voltage small signal Δu_{dc} after compensation is

$$\begin{bmatrix} \boldsymbol{W}_{dmad} \\ \boldsymbol{W}_{qmad} \end{bmatrix} = \frac{1}{\Delta u_{dc}}\begin{bmatrix} \Delta i_d \\ \Delta i_q \end{bmatrix} = \begin{bmatrix} \boldsymbol{W}_d \\ \boldsymbol{W}_q \end{bmatrix} + \boldsymbol{G}_{mad}\left(m_0 - \sqrt{3}\boldsymbol{G}_b\right)\begin{bmatrix} \boldsymbol{W}_{d\theta} \\ \boldsymbol{W}_{q\theta} \end{bmatrix}. \tag{7.62}$$

The inverter-side admittance $\boldsymbol{Y}_m$ after control is modified as

$$\begin{aligned} \boldsymbol{Y}_{mmad} = \frac{\Delta i_{inv}}{\Delta u_{dc}} &= \frac{3\boldsymbol{W}_{dmad}\left(U_{d0} + \omega_e L_d I_{q0} + L_d I_{d0} s + I_{d0} R_s\right)}{2U_{dc0}} \\ &+ \frac{3\boldsymbol{W}_{qmad}\left(U_{q0} - \omega_e L_q I_{d0} + L_q I_{q0} s + I_{q0} R_s\right) - 2I_{inv0}}{2U_{dc0}}. \end{aligned} \tag{7.63}$$

The transfer function between the modulation index small signal Δm and the rectifier output voltage small signal Δu_g is

$$\boldsymbol{G}_{ms}(s) = \frac{\Delta m}{\Delta u_g} = \frac{\sqrt{3}(U_{dc0}\boldsymbol{G}_b - U_{sref0})}{\left[(L_g s + R_g)(C_{dc}s + \boldsymbol{Y}_{mmad}) + 1\right]U_{dc0}^2}. \tag{7.64}$$

From (7.48), Fig. 7.22a shows the reduction in the $6\omega_g$-order harmonic amplitude of the modulation index under various operating conditions. Under a load torque of 35 N m and an operating frequency of 80 Hz, the $6\omega_g$-order harmonic amplitude decreases by 80% after control. Compared to Fig. 7.18, Fig. 7.22b illustrates that the maximum achievable mean value of the motor voltage vector reference increases by 10% due to the significant reduction in modulation index fluctuations.

From (7.47) and (7.65), Fig. 7.23 shows the amplitude-frequency characteristics of the transfer function between the modulation index and rectifier voltage under different DC-link capacitor values. As the DC-link capacitor value increases, the harmonic amplitude of the modulation index decreases, but the harmonic amplitude near the resonance frequency rises. Meanwhile, after control, the harmonic amplitudes of $6\omega_g$ and $12\omega_g$ are reduced.

According to the topology of the electroless system, the transfer function between the DC-link voltage and rectifier voltage is

$$\boldsymbol{K}_m(s) = \frac{\Delta u_{dc}}{\Delta u_g} = \frac{1}{1 + (L_g s + R_g)[\boldsymbol{Y}_m(s) + C_{dc}s]}. \tag{7.65}$$

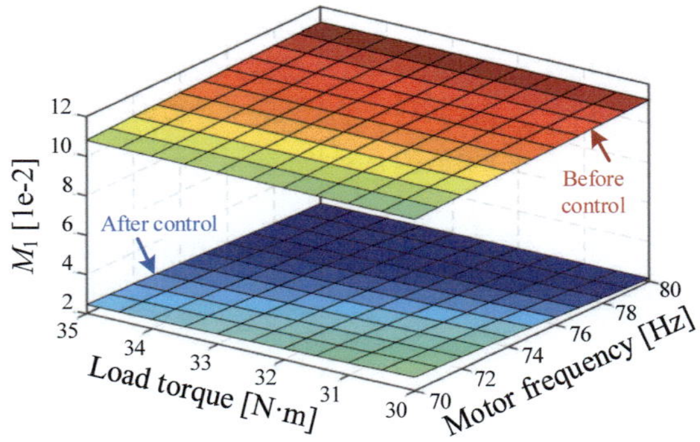

(a) $6\omega_g$-order harmonic amplitude of the modulation index

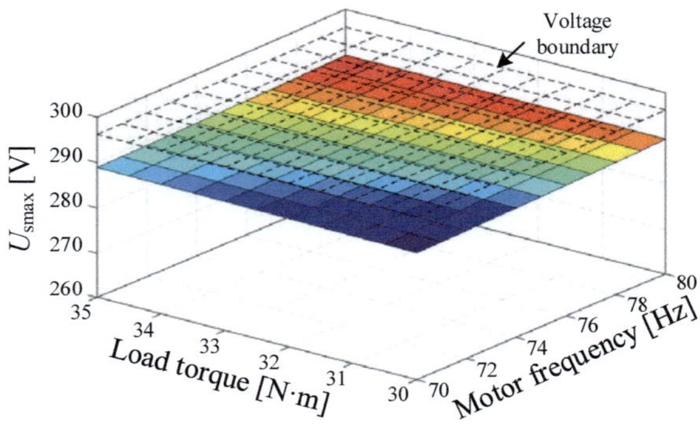

(b) Maximum average of the motor voltage vector reference

Fig. 7.22 Modulation of the system after control, reprinted from Ref. [2], copyright 2022, with permission from Nature

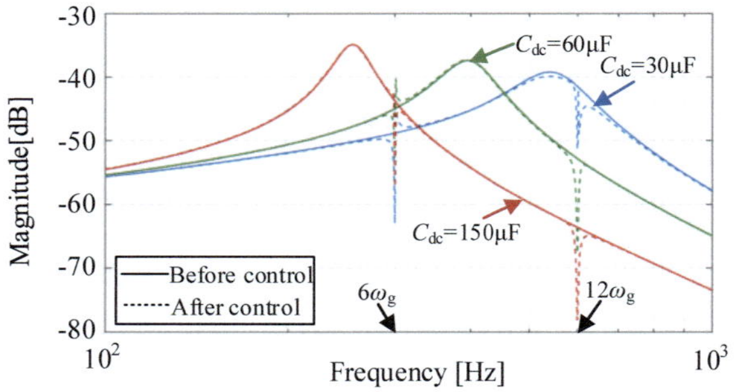

Fig. 7.23 Magnitude-frequency characteristics of $\boldsymbol{G}_m(s)$ and $\boldsymbol{G}_{ms}(s)$ under different DC-link capacitance

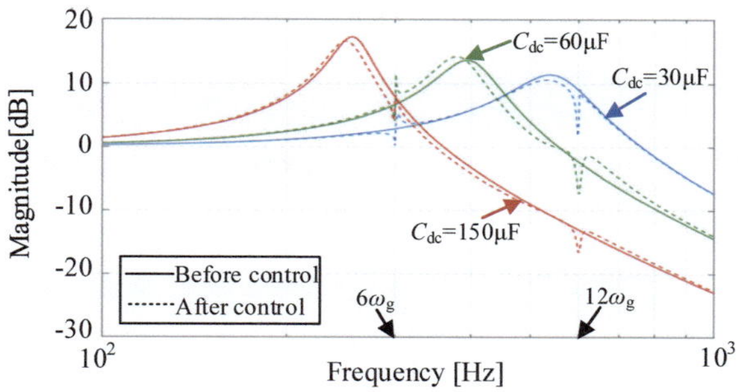

Fig. 7.24 Magnitude-frequency characteristics of Km(s) under different DC-link capacitance

According to (7.65), Fig. 7.24 depicts the amplitude-frequency characteristics of this transfer function under different DC-link capacitor values. As the DC-link capacitor value increases, the high-frequency components of the DC-link voltage harmonics decrease, reducing DC-link voltage fluctuations. After compensating the $6\omega_g$ and $12\omega_g$-order harmonics in the voltage angle, the harmonic amplitudes of these frequencies in the DC-link voltage decrease, demonstrating that the proposed control method also suppresses bus voltage fluctuations.

7.3.3 Experimental Results

Figure 7.25a shows the experimental waveforms before and after control at an operating frequency of 70 Hz and a power of 4.4 kW, and the waveforms of the DC-link voltage, the motor voltage vector reference amplitude and the modulation index are shown in order from the top to the bottom, and the system is just in the weak magnetic region before control, and out of the weak magnetic region after control. After control, the DC-link voltage fluctuation decreases from 118 to 104 V, the motor voltage vector reference amplitude fluctuation increases from 35.4 V to 41.9 V, and the modulation index fluctuation decreases from 0.261 to 0.111. Figure 7.25b and c show the enlarged graphs of the experimental waveforms before and after control, and the phase difference between the DC-link voltage and the motor voltage vector reference amplitude is 168° before control, and after control is 13°, and the fluctuation of the modulation index is reduced. Figure 7.25d shows the electric angle and speed waveforms before and after control, the main harmonic of 6ωg order is superimposed on the electric angle after control, and the speed pulsation does not increase after control.

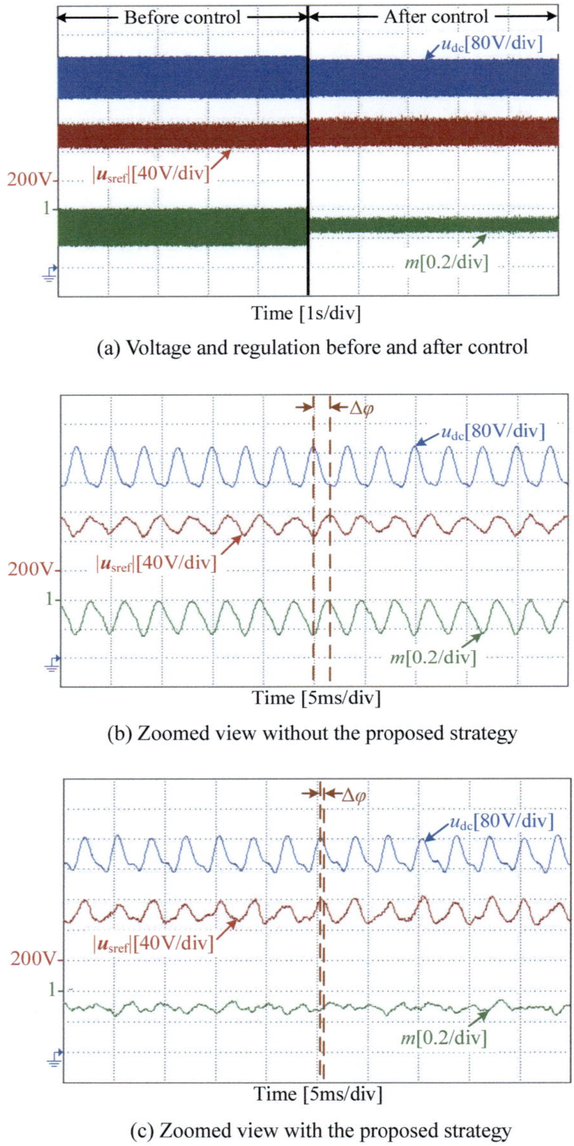

Fig. 7.25 Experimental results at the motor frequency of 70 Hz, reprinted from Ref. [2], copyright 2022, with permission from Nature

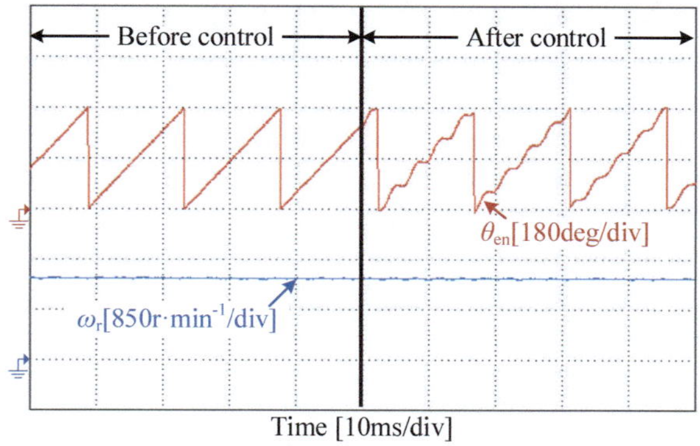

(d) Waveforms of electrical angle and motor frequency.

Fig. 7.25 (continued)

Figure 7.26a and b show the fluctuation of the driving system modulation before and after control at an operating frequency of 70 Hz and a power of 4.4 kW, respectively. The pre-control tuning regime reaches its upper boundary, and the distance from the maximum value of the post-control tuning regime to its upper boundary is 0.07.

Figure 7.27 shows the experimental results for the operation frequency of 80 Hz, and the waveforms of the DC-link voltage, the motor voltage vector reference amplitude and the modulation index are shown in order from the top to the bottom, at which time the motor enters the weak magnetic region. As shown in Fig. 7.27a, the fluctuation of the DC-link voltage decreases from 160 to 120 V, the fluctuation of the motor voltage vector reference amplitude decreases from 70.2 V to 43.8 V, and the direct flow of the motor voltage vector reference amplitude improves from 256.5 V to 286.7 V. The fluctuation of the modulation index decreases from 0.367 to 0.124.

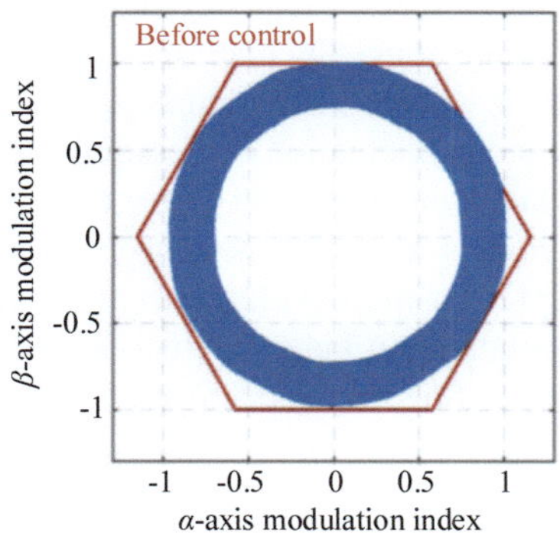

(a) Before control.

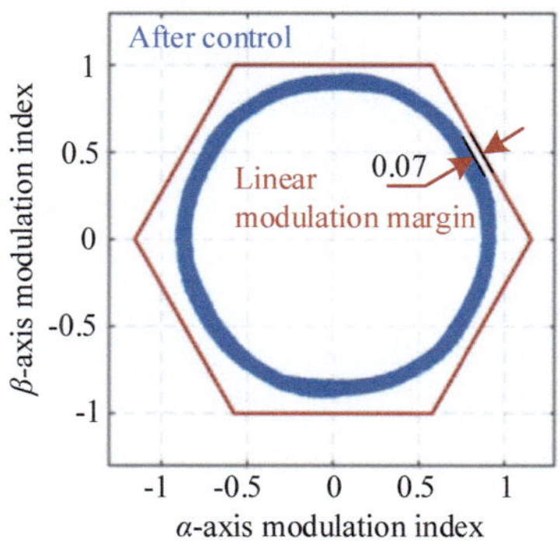

(b) After control.

Fig. 7.26 Lissajous figure of MI at the motor frequency of 70 Hz

Figure 7.27b and c show the enlargement of the experimental waveforms before and after control graphs, the phase difference between the DC-link voltage and the motor voltage vector reference amplitude is reduced from 151° before control to 17°, which significantly reduces the fluctuation of the modulation index. Figure 7.27d shows the electric angle and speed waveforms before and after control, and there is no obvious speed fluctuation before and after control.

Figure 7.28 shows the waveforms of d-axis current reference, q-axis current reference and motor current at the operating frequency of 80 Hz. Before and after the control, the amplitude of the motor current decreases from 28.9 A to 24.4 A, the direct flow of q-axis current reference decreases from 26.4 A to 21.7 A, and the direct flow of d-axis current reference decreases from 6.5 A to 2.4 A. The increase in voltage utilization reduces the amplitude of the motor current, and at the same time reduces the copper loss of the motor, and the input power decreases by 0.2 kW, and the operating efficiency of the motor increases by 3.3%.

Figure 7.29a and b show the fluctuation of the modulation index of the drive system at the operating frequency of 80 Hz. The fluctuation of the modulation index before control is large, and the direct flow rate is 0.866, and the fluctuation of the modulation index after control is reduced, and the direct flow rate of the modulation index can be increased to 0.968, and the voltage utilization is increased by 11.8%, and the voltage utilization of the system in the linear modulation region is maximized.

Figure 7.30 summarizes the changes in the direct flow versus the $6\omega_g$-order harmonic amplitude of the modulation index before and after control for different operating conditions. Figure 7.30a shows the $6\omega_g$-order harmonic amplitude of the modulation index before and after control, and the proposed method can reduce the $6\omega_g$ subharmonic of the regulation regime to less than 30% of the original. Figure 7.30b shows the average of modulation index before and after control. At the motor frequency of 70–72 Hz, after applying the proposed method, the system is out of the weak magnetization region, and at this time, the direct flow of the modulation index is not significantly improved. In weak magnetic operation, the proposed method can enhance the average of modulation index to 0.968.

Figure 7.31a shows the experiment results for a motor frequency of 70 Hz and an input voltage of 342 V. After control, the DC-link voltage fluctuation decreases from 120.7 V to 96.3 V, the direct flow of the motor voltage vector amplitude increases from 234.2 V to 255.5 V, and the modulation index fluctuation decreases from 0.321 to 0.122. Figure 7.31b shows the experimental results when the motor frequency is 80 Hz and the input voltage is 418 V. After control, the DC-link voltage fluctuation decreases from 142.3 V to 120.7 V, the average of the motor voltage vector amplitude increases from 285.1 V to 307.2 V, and the modulation index fluctuation decreases from 0.301 to 0.096. The experimental results show the versatility of the control strategy for different input voltages.

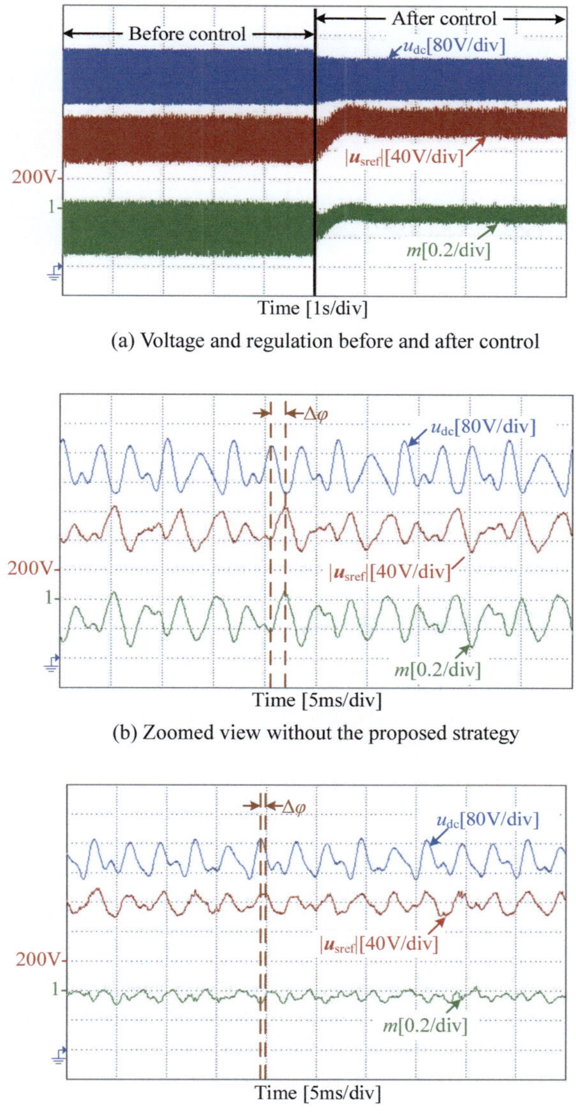

(a) Voltage and regulation before and after control

(b) Zoomed view without the proposed strategy

(c) Zoomed view with the proposed strategy

Fig. 7.27 Experimental results at the motor frequency of 80 Hz, reprinted from Ref. [2], copyright 2022, with permission from Nature

Figure 7.32 shows the experimental results when the motor frequency is 60 Hz and the grid voltage dips to 80% of the rated value. As shown in Fig. 7.32a, when the grid voltage decreases, the modulation depth deepens and the fluctuation of the modulation index increases from 0.217 to 0.289. The direct current of the motor voltage vector amplitude decreases from 230.3 V to 206.4 V, which leads to an

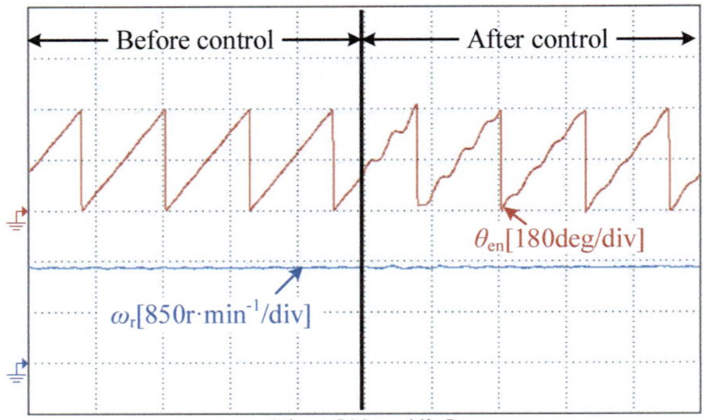

(d) Waveforms of electrical angle and motor frequency.

Fig. 7.27 (continued)

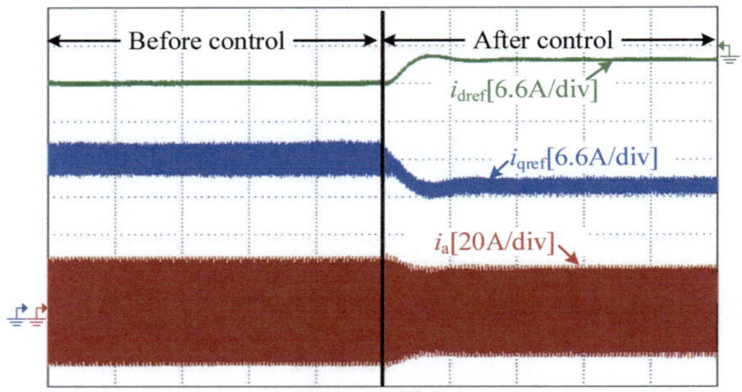

Fig. 7.28 Experimental results of motor currents

increase in the motor current. The experimental results after control are shown in Fig. 7.32b. When the grid voltage decreases, the modulation depth deepens and the fluctuation of the modulation index increases from 0.072 to 0.122. The direct current of the motor voltage vector amplitude decreases from 229.3 V to 217.4 V. The higher utilization of the DC-link voltage and the smaller fluctuation of the modulation index after the control imply a lower output current and a more stable system.

7.4 Summary

The DC-link voltage fluctuations in electrolytic capacitorless motor drives introduce additional fluctuations to modulation index and motor-side currents, resulting in reduced linear modulation margins and beat phenomenon. Combined with the principle of SVPWM extension, the impedance model of the system is established,

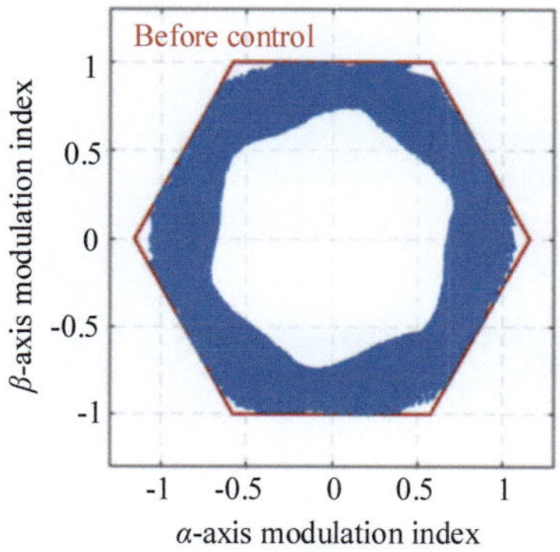

(a) Before control.

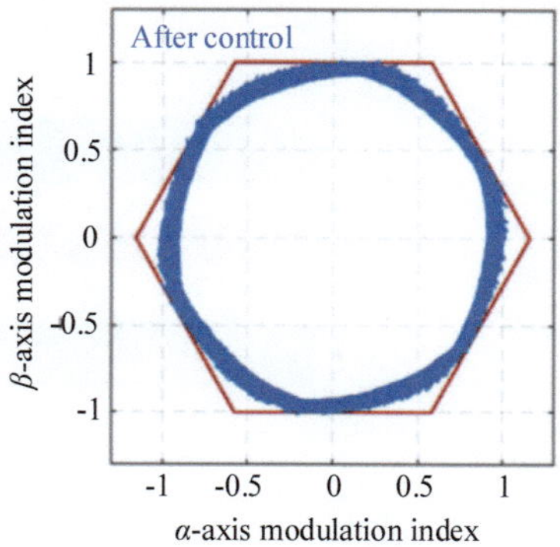

(b) After control.

Fig. 7.29 Lissajous figure of MI at the motor frequency of 80 Hz

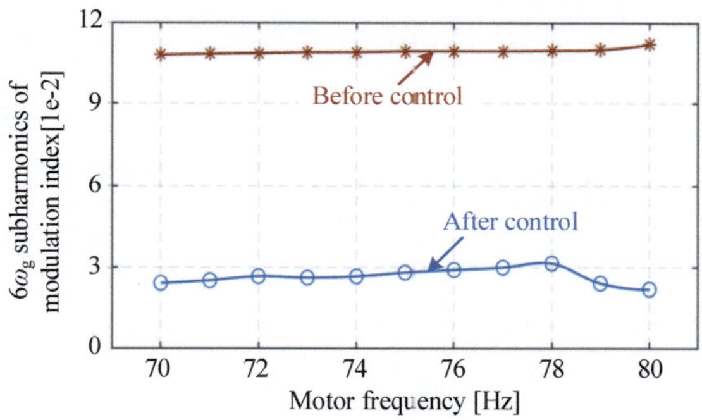

(a) 6ωg-order harmonic amplitude of the modulation index

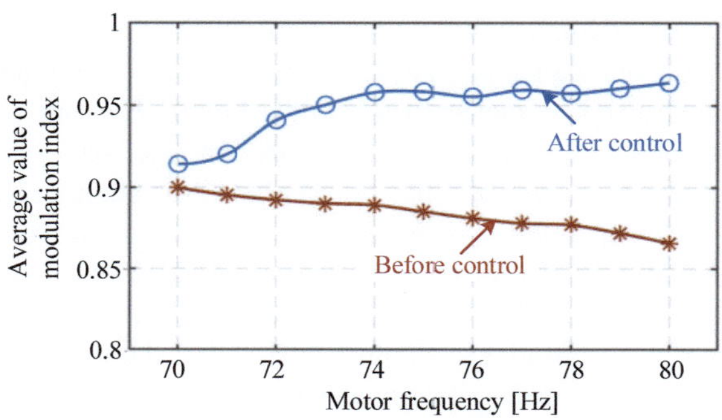

(b) Average of modulation index.

Fig. 7.30 Statistical results of modulation index under different working conditions

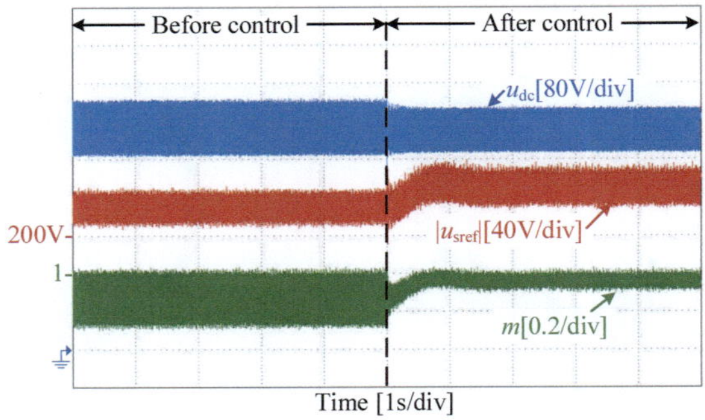

(a) The input voltage is 342V at the motor frequency of 70 Hz.

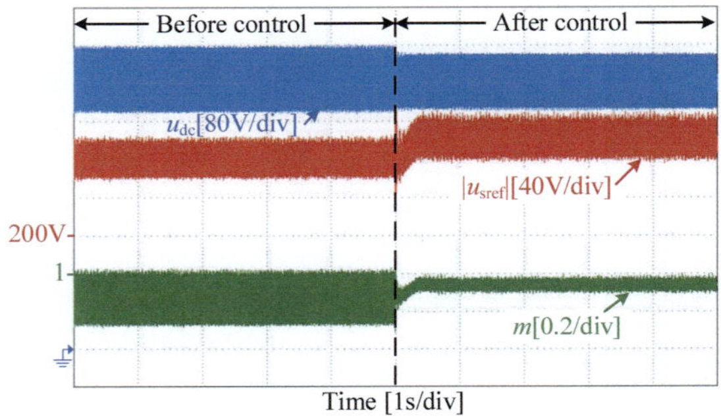

(b) The input voltage is 418V at the motor frequency of 80 Hz.

Fig. 7.31 Experimental results under different input voltage, reprinted from Ref. [2], copyright 2022, with permission from Nature

and the influence of DC-link voltage fluctuation on the amplitude of modulation index harmonic is mathematically analyzed. The harmonic components of the DC-link voltage are extracted by means of a bandpass filter. The amplitude and phase of the voltage vector are changed by the frequency characteristic correction link. By analyzing the impedance characteristics after control, the suppression effect of the control strategy on the modulation index and the current harmonics on the motor side is verified. Through the closed-loop method of voltage harmonics, the harmonics at the inherent frequency of the DC-link voltage are extracted, and the electrical angle of the iPark link is compensated to adaptively suppress the fluctuation of the field weakening region. And according to the peak value of the modulation index, the coefficient of field weakening is adaptively corrected to maximize the voltage utilization. Experimental results show that the proposed strategy can effectively suppress modulation index fluctuations and improve linear modulation margin, voltage utilization and motor side performance.

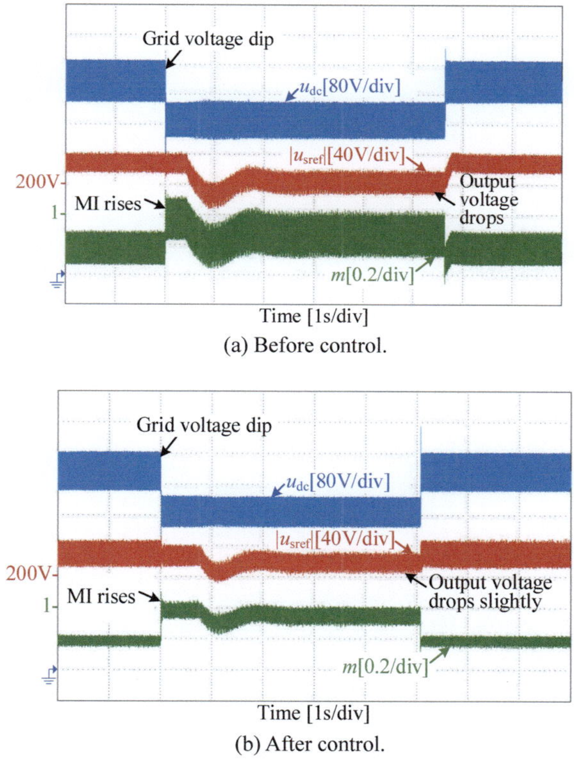

Fig. 7.32 Experimental results when the grid voltage dips, reprinted from Ref. [2], copyright 2022, with permission from Nature

References

1. R. Gao, D. Ding, G. Wang, Q. Wang, G. Zhang and D. Xu, "Linear Modulation Region Expansion Strategy Based on Voltage Vector Angle Regulation for PMSM Drives with Small DC-Link Capacitor," in IEEE Transactions on Power Electronics, vol. 38, no. 7, pp. 8327-8339, July 2023.
2. D. Ding, T. Zhang, B. Li, G. Wang, Q. Wang, G. Zhang and D. Xu, "Adaptive Modulation Index Fluctuation Suppression Strategy for Electrolytic Capacitorless PMSM Drives," in IEEE Transactions on Transportation Electrification, vol. 11, no. 2, pp. 6406-6417, April 2025.

MIX
Papier aus verantwortungsvollen Quellen
Paper from responsible sources
FSC® C105338
FSC
www.fsc.org

If you have any concerns about our products,
you can contact us on
ProductSafety@springernature.com

In case Publisher is established outside the EU,
the EU authorized representative is:
Springer Nature Customer Service Center GmbH
Europaplatz 3, 69115 Heidelberg, Germany

Printed by Libri Plureos GmbH
in Hamburg, Germany